Realism, rationalism and scientific method

Realism, rationalism and scientific method

Philosophical papers

Volume 1

PAUL K. FEYERABEND

The right of the
University of Cambridge
to print and sell
all manner of books
was granted by
Henry VIII in 1534.
The University has printed
and published continuously
since 1584.

CAMBRIDGE UNIVERSITY PRESS

CAMBRIDGE
NEW YORK PORT CHESTER MELBOURNE
SYDNEY

Published by the Press Syndicate of the University of Cambridge
The Pitt Building, Trumpington Street, Cambridge CB2 1RP
40 West 20th Street, New York, NY 10011, USA
10 Stamford Road, Oakleigh, Melbourne 3166, Australia

First published 1981
First paperback edition 1985
Reprinted 1986, 1987, 1990

Printed in the United States of America

British Library Cataloguing in Publication Data
Feyerabend, Paul Karl
Philosophical papers.
Vol. 1: Realism, rationalism and scientific method
1. Science – Methodology
I. Title
501'.8   Q175   80-41931

ISBN 0-521-22897-2 hardback
ISBN 0-521-31642-1 paperback

Contents

Introduction to volumes 1 and 2

The present volume and its companion discuss three ideas that have played an important role in the history of science, philosophy and civilization: criticism, proliferation and reality. The ideas are presented, explained and made the starting points of argumentative chains.

The first idea, that of *criticism*, is found in almost all civilizations. It plays an important role in philosophies such as Buddhism and Mysticism, it is the cornerstone of late nineteenth-century science and philosophy of science, and it has been applied to the theatre by Diderot and Brecht.[1] Criticism means that we do not simply accept the phenomena, processes, institutions that surround us but we examine them and try to change them. Criticism is facilitated by *proliferation* (vol. 1, ch. 8): we do not work with a single theory, system of thought, institutional framework until circumstances force us to modify it or to give it up; we use a plurality of theories (systems of thought, institutional frameworks) from the very beginning. The theories (systems of thought, forms of life, frameworks) are used in their strongest form, not as schemes for the processing of events whose nature is determined by other considerations, but as accounts or determinants of this very nature (*realism*, see vol. 1, chs. 11.15f[2]). One chain of argument is therefore

$$\text{criticism} \Longrightarrow \text{proliferation} \Longrightarrow \text{realism} \qquad \text{(i)}$$

In the first volume this chain is applied to a rather narrow and technical problem, viz. the interpretation of *scientific theories*.

None of the ideas is defined in a precise fashion. This is quite intentional. For although some papers, especially the early ones, are fairly abstract and 'philosophical', they still try to stay close to scientific practice which means that their concepts try to preserve the fruitful imprecision of this practice (cf. vol. 2, ch. 5 on the ways of the scientist and the ways of the philosopher; cf. also vol. 2, ch. 6, nn. 47ff and text).

Nor does the arrow in (i) express a well-defined connection such as

---

[1] This wider function of criticism is explained in my essay 'On the Improvement of the Sciences and the Arts and the Possible Identity of the Two' in *Boston Studies in the Philosophy of Science* (New York, 1965), III.

[2] ch. 11.15 means section 15 of ch. 11. This method of reference is used throughout both volumes.

logical implication. It rather suggests that starting with the left hand side and adding physical principles, psychological assumptions, plausible cosmological conjectures, absurd guesses and plain commonsense views, a dialectical debate will eventually arrive at the right hand side. Examples are the arguments for proliferation in vol. 1, ch. 6.1, ch. 8, n.14 and text, ch. 4.6 as well as the arguments for realism in vol. 1, chs. 11, 14 and 15. The meaning of the arrow emerges from these examples and not from example-independent attempts at 'clarification'.

Chapters 2–7 of vol. 1, which are some of the oldest papers, deal mainly with the interpretation of theories (for the notion of 'theory' used cf. the remarks in the preceding paragraph and in vol. 1, ch. 6, n.5). Chapter 1 of the first volume shows how the realism that is asserted in thesis 1 of vol. 1, ch. 2.6 and again in ch. 11, is related to other types of realism that have been discussed by scientists. The thesis can be read as a philosophical thesis about the influence of theories on our observations. It then asserts that observations (observation terms) are not merely theory-*laden* (the position of Hanson, Hesse and others) but *fully theoretical* (observation statements have no 'observational core'[3]). But the thesis can also be read as a historical thesis concerning the use of theoretical terms by scientists. In this case it asserts that scientists often use theories to restructure abstract matters *as well as* phenomena, and that no part of the phenomena is exempt from the possibility of being restructured in this way. My discussion of the relation between impetus and momentum in vol. 1, ch. 4.5 is entirely of the second kind. It is not an attempt to draw consequences from a contextual theory of meaning – theories of meaning play no role in this discussion – it simply shows that both facts and the laws of Newtonian mechanics prevent us from using the concept of impetus as part of Newton's theory of motion. Nor is the result generalized to all competing theories. It is merely argued that certain popular views on explanation and the relation between theories in the same domain that claim to be universally valid fail for important scientific developments. *General* assertions about incommensurability are more characteristic for Kuhn whose ideas differ from mine and were developed independently (cf. my *Science in a Free Society*,[4] 65ff for a comparison and a

---

[3] Or, to express it differently: there are only theoretical terms (for his version of the thesis see my 'Das Problem der Existenz theoretischer Entitäten' in *Probleme der Wissenschaftstheorie*, ed. E. Topitsch (Vienna, 1960), 35ff). There is of course a distinction between theoretical terms and observation terms, but it is a psychological distinction, dealing with the psychological processes that accompany their use, but having nothing to do with their content (for details see vol. 1, ch. 6, section 6). This feature of the thesis has been overlooked by some more recent critics who ascribed to me the 'triviality that theoretical terms are theoretical'. The best and most concise expression of the thesis can be found in Goethe: 'Das Hoechste zu begreifen waere, dass alles Faktische schon Theorie ist' ('Aus den Wanderjahren', *Insel Werkausgabe* (Frankfurt, 1970), vi, 468).   [4] (London, 1978), hereafter referred to as *SFS*.

brief history. Chapter 17 of my *Against Method*[5] discusses a special case that shows what elements must be considered in any detailed discussion of incommensurability). There do exist cases where not only do *some* older concepts break the framework of a new theory, but where *an entire theory*, all its observation statements included, is incommensurable with the theory that succeeds it, but such cases are rare and need special analysis. Using the terms of vol. 1, ch. 2.2 one can tentatively say that a theory is incommensurable with another theory if its ontological consequences are incompatible with the ontological consequences of the latter (cf. also the considerations in vol. 1, ch. 4.7 as well as the more concrete definition in *A M*, 269 and the appendix to ch. 8 of vol. 2). But even in this case incommensurability does not lead to complete disjointness, as the phenomenon depends on a rather subtle connection between the more subterranean machinery of the two theories (cf. again *A M*, 269). Besides there are many ways of comparing incommensurable frameworks, and scientists make full use of them (vol. 1, ch. 1, n.39; cf. also vol. 1, ch. 2.6, n.21 and ch. 4.8). Incommensurability is a difficulty for some rather simpleminded philosophical views (on explanation, verisimilitude, progress in terms of content increase); it shows that these views fail when applied to scientific practice; it does not create any difficulty for scientific practice itself (see vol. 2, ch. 11.2, comments on incommensurability).

Chapters 8–15 of vol. 1 apply chain (1) to the mind–body problem, commonsense, the problem of induction, far-reaching changes in outlook such as the Copernican revolution and the quantum theory. The procedure is always the same: attempts to retain well-entrenched conceptions are criticized by pointing out that the excellence of a view can be asserted only *after* alternatives have been given a chance, that the process of knowledge acquisition and knowledge improvement must be kept in motion and that even the most familiar practices and the most evident forms of thought are not strong enough to deflect it from its path. The cosmologies and forms of life that are used as alternatives need not be newly invented; they may be parts of older traditions that were pushed aside by overly eager inventors of New Things. The whole history is mobilized in probing what is plausible, well established and generally accepted (vol. 1, ch. 4, n.67 and ch. 6.1).

There is much to be said in favour of a pluralistic realism of this kind. John Stuart Mill has explained the arguments in his immortal essay *On Liberty* which is still the best modern exposition and defence of a critical philosophy (see vol. 1, ch. 8 and vol. 2, ch. 4 and ch. 9.13). *But the drawbacks are considerable.* To start with, modern philosophers of science, 'critical' rationalists included, base their arguments on only a tiny part of Mill's scheme; they uncritically adopt some standards, which they use for weeding

[5] (London, 1975), hereafter referred to as *A M*.

out conflicting ideas, but they hardly ever examine the standards themselves. Secondly, and much more importantly, there may be excellent reasons for resisting the universal application of the realism of thesis 1 (vol. 1, ch. 2.6). Take the case of the quantum theory. Interpreted in accordance with vol. 1, ch. 2.6 wave mechanics does not permit the existence of well-defined objects (see vol. 1, ch. 16.3 and especially nn.26, 27 and text). Commonsense (as refined by classical physics) tells us that there are such objects. The realism of vol. 1, ch. 2.6, chs. 4.6f, chs. 9 and 11 invites us to reject commonsense and to announce the discovery: objective reality has been found to be a metaphysical mistake.

Physicists did not go that way, however. They demanded that some fundamental properties of commonsense be preserved and so they either added a further postulate (reduction of the wave packet) leading to the desired result, or constructed a 'generalized quantum theory' whose propositions no longer form an irreducible atomic lattice. The proponents of hidden variables, too, want to retain some features of the classical (commonsense) level and they propose to change the theory accordingly. In all these cases (excepting, perhaps, the last) a realistic interpretation of the quantum theory is replaced by a partial instrumentalism.

Two elements are contained in this procedure and they are not always clearly separated. The first element which affects the actions of the physicists is *factual*: *there are* relatively isolated objects in the world and physics must be capable of describing them. (Commonsense arguments, though more complex, often boil down to the same assertion.) But Buddhist exercises create an experience that no longer contains the customary distinctions between subject and object on the one hand and distinct objects on the other. Indeed many philosophies deny separate existence and regard it as illusion only: the existence of separate objects and the experiences confirming it are not tradition-independent 'facts'; they are parts of special traditions. Physicists *choose* one of these traditions (without realizing that a choice is being made) and turn it into a boundary condition of research. This is the second element. The transition to a partial instrumentalism therefore consists of a *choice* and the utilization of the *facts* that belong to the tradition chosen.

The history of philosophy offers many cases for the study of both elements. The debates about the quantum theory have much in common with the ancient issue between Parmenides (and his followers) on the one hand and Aristotle on the other and the more recent issue between Reason and the (Roman) Church. Parmenides showed (with additional arguments provided by Melissus and Zeno) that there is no change and that Being has no parts. But we deal with change and with objects and processes that differ in many respects. Our lives as human beings are directed towards taking change and division into account. Are we to admit that we live an illusion, that the truth is hidden from us and that it must be discovered by special means? Or should we not rather assert the reality of our common views over the reality of some specialist conceptions? Must we adapt our lives to the ideas and rules devised by small groups of intellectuals (physicians, medical researchers, socio-biologists, 'rationalists' of all sorts) or should we not rather demand that intellectuals be mindful of circumstances that matter to their fellow human beings? Or, to consider a more important dichotomy: can we regard our lives on this earth and the ideas we have developed to cope with the accidents we encounter as measures of reality, or are they of only secondary importance when compared with the conditions of the soul as described in religious beliefs? These are the questions which arise when we compare commonsense with religious notions or with the abstract ideas that intellectuals have tried to put over on us ever since the so-called rise of rationalism in the West (see vol. 2, ch. 1, sections 1f and 7). They involve both a choice between forms of life and an adaptation of our ideas and habits to the ideas (perceptions, intuitions) of the tradition chosen: *we decide to regard those things as real which play an important role in the kind of life we prefer.*

Making the decision we start a reverse argumentative chain of the form

$$L \implies \overline{\text{criticism}} \implies \overline{\text{realism}_I} \tag{ii}$$

or, in words: accepting a form of life L we reject a universal criticism and the realistic interpretation of theories not in agreement with L. Proceeding in this way we notice that instrumentalism is not a philosophy of defeat; it is often the result of far-reaching ethical and political decisions. Realism, on the other hand, only reflects the wish of certain groups to have their ideas accepted as the foundations of an entire civilization and even of life itself.

Chapters 16 and 17 of vol. 1 as well as the essays in vol. 2 contain first steps towards undermining this intellectual arrogance. They contain instances of the use of the reverse chain (ii). It is argued that science never obeys, and cannot be made to obey, stable and research independent standards (vol. 2, ch. 1.5, chs. 8, 10, 11): scientific standards are subjected to the process of research just as scientific theories are subjected to that process (vol. 1, chs. 1.3f; cf. also part 1 of *SFS*); they do not guide the process from the outside (cf. vol. 2, ch. 7 on rules and vol. 2, ch. 5 on the difference between the scientists' way and the philosophers' way of solving problems). It is also shown that philosophers of science who tried to understand and to tame science with the help of standards and methodologies that transcend research, *have failed* (vol. 2, chs. 9, 10, 11; cf. vol. 2, ch. 1.5f and part 1 of *SFS*): *one of the most important and influential institutions of our times is beyond the reach of reason as interpreted by most contemporary rationalists.* The failure does not put an end to our attempts to adapt science to our favourite forms of life. Quite the contrary: it frees the attempt from irrelevant restrictions. This is in perfect agreement with the Aristotelian philosophy which also limits science by reference to commonsense except that conceptions of an individual philosopher (Aristotle) are now replaced by the *political decisions* emerging from the institutions of a free society (vol. 2, ch. 1.7).

Most articles were written with support from the National Science Foundation; some articles were written while I held a Humanities Research Fellowship and a Humanities Research Professorship at the University of California at Berkeley. The reader will notice that some articles defend ideas which are attacked in others. This reflects my belief (which seems to have been held by Protagoras) that good arguments can be found for the opposite sides of any issue. It is also connected with my 'development' (details in *SFS*, 107ff). I have occasionally made extensive changes both in the text and in the footnotes but I have not always given the place and nature of such changes. Chapters 1 and 8 of vol. 1 and ch. 1 of vol. 2 are new and prepare a longer case study of the rise of rationalism in the West and its drawbacks. An account of, and arguments for, my present position on the structure and authority of science can be found in vol. 2 of *Versuchungen*, ed. H. P. Dürr (Frankfurt, 1981) which contains essays by various authors commenting on and criticizing my earlier views on these matters.

Part I

On the interpretation of scientific theories

I

Introduction: scientific realism and philosophical realism

1. HISTORICAL BACKGROUND

Scientific realism is a *general* theory of (scientific) knowledge. In one of its forms it assumes that the world is independent of our knowledge-gathering activities and that science is the best way to explore it. Science not only produces predictions, it is also about the nature of things; it is metaphysics and engineering theory in one.

As will be shown in vol. 2, ch. 1.1 scientific realism owes its existence and its concepts to an ancient antagonism between commonsense and comprehensive theories. It arose when Greek intellectuals, guided by a love for abstractions, new kinds of stories (now called 'arguments') and new values for life,[1] denied the traditional views and tried to replace them by their own accounts. It was the fight between tradition and these accounts, 'the ancient battle between philosophy and poetry',[2] that led to a consideration of traditions *as a whole* and introduced *general* notions of existence and reality.[3]

Scientific realism has had a considerable influence on the development of science. It was not only a way of describing results after they had been obtained by other means, it also provided strategies for research and suggestions for the solution of special problems. Thus *Copernicus'* claim that his new astronomy reflected the true arrangement of the spheres raised dynamical, methodological as well as exegetic problems (*SFS*, 40ff). His ideas were in conflict with physics, epistemology and theological doctrine, all of which were important boundary conditions of research. Copernicus created these problems but he also gave hints for their solution and thereby

---

[1] The conflict between city life and heroic virtues is one of the main subjects of Greek tragedy. Cf. the analysis of the *Oresteia* and of Euripides' *Medea* and *Alkestis* in Kurt von Fritz's *Antike und Moderne Tragoedie* (Berlin, 1962) as well as George Thomson, *Aeschylus and Athens* (London, 1966). Gerald Else, *The Origin and Early Form of Tragedy* (Cambridge, 1965) traces the history back to Solon.

[2] Plato, *Republic*, 607B6.

[3] The earlier investigations of the Ionian historians led in the same direction but without any *explicit* discussion of the new and more general concepts used. There existed therefore two different movements towards abstraction, a 'natural' development, and the artificial and explicit considerations of the Eleatics which imposed entirely new ideas (cf. also vol. 2, ch. 1.1).

initiated new research traditions. In the nineteenth century, the *atomic theory* raised philosophical, physical, chemical and metaphysical problems and there were many scientists who wanted either to abandon it as false, or to use it as a convenient scheme for the ordering of facts.[4] Realists developed it further and could finally demonstrate the limitations of a purely phenomenological view. *Einstein's* criticism of the quantum theory initiated interesting theoretical developments and delicate experiments and clarified the basic concepts of the theory (cf. ch. 2.8). In all these cases scientific realism produced discoveries and contributed to the development of science.

Only a few philosophers have examined this fruitful interaction between scientific realism and scientific practice. The reason is that scientists and philosophers are interested in different things and approach their problems in different ways. A scientist deals with concrete difficulties and he judges assumptions, theories, world views, rules of procedure by the way in which they affect his problem situation. His judgement may change from one case to the next for he may find that while an idea such as scientific realism is useful on some occasions it only complicates matters on others (cf. the quotations in vol. 2, ch. 6.9).

A philosopher also wants to solve problems, but they are problems of an entirely different kind. They concern abstract ideas such as 'rationality', 'determinism', 'reality' and so forth. The philosopher examines the ideas with great vigour and, occasionally, in a critical spirit, but he also believes that the very generality of his inquiry gives him the right to impose the achieved results on all subjects without regard for their particular problems, methods, assumptions. He simply assumes that a general discussion of general ideas covers all particular applications.

While this assumption may be correct for *abstract traditions* which are developed from principles and can therefore be expected to agree with them, it is not correct for *historical traditions* where particular cases, including the use of laws and theories, are treated in accordance with the particular circumstances in which they occur and where principles are modified, or provided with exceptions in order to agree with the requirements of these circumstances. More recent research (vol. 2, chs. 4, 5, 6, 8, 9, 11—remarks on Kuhn; cf. vol. 2, ch. 1.2 for general considerations) has made us realize that scientific practice, even the practice of the natural sciences, is a tightly woven net of historical traditions (in mathematics this was first pointed out by the intuitionists; Kuhn has popularized the results for the natural sciences while Wittgenstein has developed the philosophical background). This means that general statements *about* science, statements of logic included, cannot without further ado be taken to agree with scientific practice (the attempt to apply them to this practice and at the same time to give a historically correct account of it has led to the decline of rationalism

[4] An excellent survey is Mary Jo Nye, *Molecular Reality* (London, 1972).

described in vol. 2, chs. 1.6, 10 and 11). For example, we cannot be satisfied with arguments of type (i) (ch. 1). We must inquire how scientists actually think about 'reality' and what notions of realism they employ. We must study the various versions of scientific realism.

2. TYPES OF REALISM

For the Copernicans the issue is about the *truth of theories*. While the followers of Aristotle looked to physics and basic philosophy for information about the structure of the world, Copernicus and Kepler claimed truth for a point of view that did not belong to the basic theories of the time. As in antiquity the clash was not between a realist position and an absolute instrumentalism, it was 'between two realist positions',[5] i.e. between two different claims to truth.

Claims to truth can be raised only with regard to particular theories. The first version of scientific realism therefore does not lead to a realistic interpretation for *all* theories, but only for those which have been chosen as a basis for research. It may be asserted (a) that the chosen theory *has been shown* to be true or (b) that it is possible to *assume* its truth, even though (ba) the theory has not been established or (bb) is in conflict with facts and established views.

As far as I can see, (a) is adopted by Kepler:[6] Copernicus' views are true not simply because they fit the facts – any false theory can be made to fit the facts – but because they have led to novel predictions *and* because they do not fail when applied to topics similar to those where success was achieved. *They remain true in whatever direction one decides to pass through them.*[7] While the rivals can assert the truth of some parts of their theories (e.g. longitudes and latitudes of the planets) but not of others (mutual penetration of the paths of Venus and Mercury), the Copernican view is found to be true in all its parts and therefore true *simpliciter*.[8]

---

[5] P. Duhem, *To Save the Phenomena* (Chicago, 1969), 106.

[6] *Mysterium Cosmographicum*, ch. 1 and Kepler's footnotes to that chapter.

[7] 'Nam jube quidlibet eorum, quae revera in coelo apparent, ex semel posita hypothesi demonstrare, regredi, progredi, unum ex alio colligere, et quidvis agere, quae veritas rerum patitur; neque ille hesitabat in ullo, si genuinum sit, et vel ex intricatissimis demonstrationum anfractibus in se unum constatissime revertetur.' *Ibid.*

[8] According to (b), the Copernican hypothesis has been found to be true in more of its parts than any alternative, it is stronger than the alternatives, its strength is not due to 'an arbitrary addition of many false statements designed to repair whatever faults might turn up' (Kepler) but to the nature of the basic postulates, and these postulates can therefore be *assumed* to be true. It is Popper's merit to have stated in the philosophy of science what is an ancient triviality in mathematics and even in certain forms of scepticism (Carneades): that one *may* (tentatively) assert the truth of a statement not all of whose parts have yet been examined. Popper adds that this is also *required* because of the way in which scientific hypotheses are used (*Conjectures and Refutations* (New York, 1962), 112f): they are not tested like instruments (which we want to retain after some modification) but by selecting crucial

A second version of scientific realism assumes that *scientific theories introduce new entities with new properties and new causal effects*. This version is often identified with the first, but mistakenly so: false theories can introduce new entities (almost all ingredients of our physical universe were introduced by theories now believed to be false), theories containing theoretical terms as syncategorematic terms can be true, not every theory introduces entities and, most importantly, theories can be formulated in different ways, using different theoretical entities and it is not at all clear which entities are supposed to be the 'real' ones (the first known example was the use of an excentre or of an epicycle for the path of the sun). Kepler's interpretation of Copernicus establishes a relation between version one and version two in this special case: the theory is true in all its parts which means that, in the formulation given by Copernicus, all its theoretical entities can be assumed to represent real entities.

The situation is not always that simple, however. A theoretical entity may represent a real entity – but not in the theory in which it was first proposed. An example is the (vector) potential in electrodynamics. Using Stokes' theorem together with div $B = 0$ (non-existence of magnetic charges) we can present every magnetic field as the curl of a vector field, just as any electrostatic field can be presented as the gradient of a scalar. Many physicists have interpreted the potentials as auxiliary magnitudes, i.e. as theoretical entities only indirectly linked to real entities such as charges, currents, fields. Faraday, who introduced the 'electrotonic state'[9] that was later represented by the vector potential,[10] assumed it to be a real state of matter and looked for effects. The *change* of the state has clearly indentifiable effects (induction currents) – but Faraday also looked for effects of the state 'while it continued', and he regarded such effects as necessary conditions of its existence. The criterion behind the search (which I shall call *Faraday's criterion*) is that a theoretical entity represents a real entity only if it can be shown to have effects *by itself* and not merely while changing, or acting in concert with other entities. The criterion considerably complicates the application of the second version of scientific realism.

cases in which the thesis is expected to fail if not true. This alternative is hardly convincing: some artifacts are withdrawn from circulation after a single decisive test (example: drugs), while hypotheses are modified and improved after crucial experiments (e.g. Lorentz's content-increasing modification of the theory of electrons after the Michelson–Morley experiment). A much better argument is (bb), that ascribing truth to an unsupported hypothesis that conflicts with facts and well-supported alternatives increases the number of possible tests and thereby the empirical content of the latter. This argument is prepared in ch. 2, described in greater detail in ch. 3 and applied to Copernicus and the quantum theory in ch. 11.

[9] *Experimental Researches in Electricity* series 1, sections 60ff. The brief quotation further below is from section 61, first sentence.

[10] A. M. Bork, 'Maxwell and the Vector Potential', *Isis*, 58 (1967), 210ff.

It also makes us understand why so many scientists rejected the atomic theory as an account of the constitution of matter despite its ability to explain familiar facts and to predict unfamiliar ones (independence, over a wide range of values, of the density and the viscosity of a gas): the predictions involved mass phenomena and did not depend on the peculiarities of individual molecular (atomic) processes.[11] These enter only in Brownian motion – which therefore became a crucial phenomenon for the kinetic theory. Furthermore, we realize that it may be reasonable to retain theoretical entities not satisfying Faraday's criterion: new theories might introduce new connections and provide means for finding the needed effects. The potentials are a good example for the developments I have in mind.

The electric potential 'became real' when the theory of relativity turned differences of potential energy into measurable mass differences (mass-defect of nuclei). The vector potential 'became real' when Bohm and Aharonov[12] showed the existence of quantum effects, as follows: in quantum theory the phase change along a trajectory passing a magnetic field is:

Fig. 1

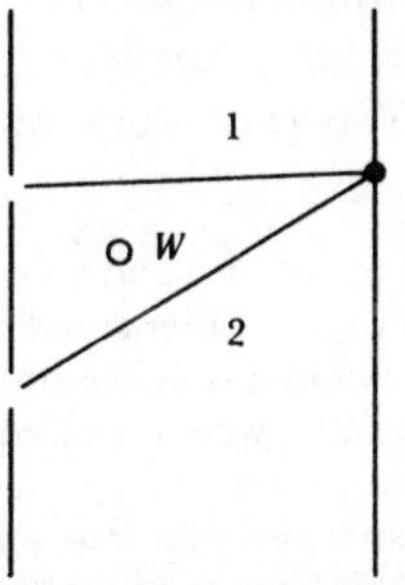

$$\delta = \frac{q}{\hbar} \int_{\text{trajectory}} A\,ds$$

and the total phase change in an interference pattern:

$$\delta = \frac{q}{\hbar} \oint A\,ds = \frac{q}{\hbar} \text{ (flux of } B \text{ between } 1 \text{ and } 2)$$

Now assume that B is the field of a solenoid W situated between path 1 and path 2 (fig. 1). The $B = 0$ along the paths and we can either assume that B acts at a distance or that the observed phase changes are due to the potential A, for $A \neq 0$ along 1 and 2.

Examples such as these show that a direct application of the second version of scientific realism ('theories always introduce new entities') and a corresponding abstract criticism of 'positivistic' tendencies are too crude to fit scientific practice. What one needs are not philosophical slogans but a more detailed examination of historical phenomena.

[11] Berthelot, Mach and others pointed out that nobody had ever 'seen' an atom – a somewhat crude but sensible application of Faraday's criterion.
[12] 'Significance of Electromagnetic Potentials in the Quantum Theory', *Phys. Rev.*, 115 (1959), 485ff.

The crudity of a purely philosophical approach becomes even clearer when we turn to a *third version of scientific realism* which is found in Maxwell, Helmholtz, Hertz, Boltzmann and Einstein.[13]

Naive realists – and many scientists and philosophers supporting the second version belong to this group – assume that there are certain objects in the world and that some theories have managed to represent them correctly. These theories speak about reality. The task of science is to discover laws and phenomena and to reduce them to those theories. Newton's theory was for a long time regarded as a basic theory in the sense just described. Today many scientists, especially in chemistry and molecular biology, have the same attitude towards the quantum theory. Seen from such a point of view, the nineteenth-century quarrels about atomism were quarrels about the *nature of things*, carried out with the help of experiment and basic theory.

Naive realism occurs in commonsense as well as in the sciences and it has been criticized in both. In the nineteenth century, the scientific criticism consisted in pointing out that theoretical entities and especially the theoretical entities of mathematical physics have a life of their own which may conceal the matter under examination. 'Whoever does mathematics', writes Ernst Mach on this point,[14]

> will occasionally have the uncanny feeling that his science and even his pencil are more clever than he, a feeling which even the great Euler could not always overcome. The feeling is justified to a certain extent if only we consider how many of the ideas we use in the most familiar manner were invented centuries ago. It is indeed a partly alien intelligence that confronts us in science. But recognizing this state of affairs removes all mysticism and all the magic of the first impression[15] especially as we are able to rethink the alien thought as often as we wish.

Rethinking the alien thought means trying to view reality in a different way; it means trying to separate concepts and things conceptualized.

A well-known example of this attempt at a separation are Hertz's remarks in the introduction to his version of classical mechanics. According to Hertz, 'we make ourselves inner phantom pictures [*Scheinbilder*] or symbols of the outer objects of such a kind that the logically necessary [*denknotwendigen*] consequences of the picture are always pictures of the physically necessary [*naturnotwendigen*] consequences of the objects pictured . . . Experience shows that the demand can be satisfied and that

---

[13] My attention was drawn to this version by C. M. Curd's excellent thesis *Ludwig Boltzmann's Philosophy of Science* (Pittsburgh, 1978).

[14] 'Die oekonomische Natur der physikalischen Forschung', lecture before the Vienna Academy of May 25, 1882, quoted from *Populaerwissenschaftliche Vorlesungen* (Leipzig, 1896), 213.

[15] This unanalysed magic and mysticism is the starting point of Popper's world three: cf. vol. 2, ch. 9.10.

such correspondences do in fact exist.'[16] Pictures are judged by their logical properties; they must be consistent, correct and distinct. 'Considering two pictures of the same object . . . we shall call the one more distinct that reflects more relations of the object than the other. Considering two pictures which are equally distinct we shall call the picture that . . . contains fewer superfluous or empty relations the more appropriate.' Using these terms we can say, on the basis of what is believed to be the case today, that a picture of quantum-mechanical processes that does not contain any 'hidden variables' is more appropriate than a picture that does, while a picture of gases that contains atoms such as the kinetic picture is more distinct than a phenomenological picture that does not. Note that the theoretical entities of a distinct and appropriate picture are still separated from the objects represented and that their nature as 'phantom pictures' or fictions is never forgotten.

According to Boltzmann, who accepted Hertz's account of scientific theories,

> the lack of clarity in the principles of mechanics may be explained by the fact that one did not at once introduce hypothetical mental pictures but tried to start from experience. One then tried to conceal the transition to hypotheses or even to find some sham proof to the effect that. . . no hypotheses had been used, creating unclarity by this very step.[17]

Boltzmann adds[18] that the use of partial differential equations (in the phenomenological approach to thermodynamics) instead of mechanical models does not eliminate pictures but simply introduces pictures of a different kind, and he sums up Hertz's position:

> Hertz made it quite clear to physicists (though philosophers most likely anticipated him long ago) that a theory cannot be an objective thing that really agrees with nature [etwas mit der Natur sich wirklich Deckendes] but must rather be regarded as merely a mental picture of phenomena that is related to them in the same way in which a symbol is related to the thing symbolised. It follows that it cannot be our task to find an absolutely correct theory – all we can do is to find a picture that represents phenomena in as simple a way as possible.[19]

Note the similarity between this point of view and that of Duhem. 'Theoretical Physics', writes Duhem,[20] 'does not have the power to grasp the real properties of bodies underneath the observable appearances; it cannot, therefore, without going beyond the legitimate scope of its

---

[16] Die Prinzipien der Mechanik (Leipzig, 1894), 1ff.

[17] L. Boltzmann, Vorlesungen ueber die Principe der Mechanik (Leipzig, 1897), I, 2.

[18] Ibid., 3. Cf. Populaere Vorlesungen (Leipzig, 1905), 142f, 144, 225f.

[19] Populaere Vorlesungen, 215f. Note that the distribution between the picture and the things pictured remains even if one denies, as Boltzmann did, that theories can ever be 'absolutely correct'.

[20] The Aim and Structure of Physical Theory (New York, 1962), 115.

methods, decide whether these properties are qualitative or quantitative. . . . Theoretical physics is limited to representing observable appearances by signs and symbols.'

The accounts just given assume two different domains, or layers. On the
one side we have phenomena, facts, things, qualities as well as concepts for
the direct expression of their properties and relations. On the other side we
have an abstract (quantitative) language in which the 'phantom pictures,'
i.e. scientific theories, are formulated. The pictures are correlated to the
phenomena, facts, things, qualities of the first domain. Attention is paid to
the language of the pictures or the 'theoretical language', as one might call
it, and one considers ways of modifying and improving it. Little attention is
paid to the 'observation language'. Vol. 2, ch. 2 describes Newton's
version of this *two layer model* of scientific knowledge (which *does* pay
attention to the observational level, or the 'phenomena'), vol. 2 ch. 3
describes Nagel's more technical presentation of the model, chs. 2, 4 and 6
criticize the technical presentation. I shall presently return to this point.

I am now ready to state the *third version of scientific realism* which one might
call, somewhat paradoxically, the *positivistic version of scientific realism*. It was
this version which was most frequently used in connection with the debates
about atomic reality and the reality of hidden parameters in the quantum
theory. Making judgements of reality here amounts to asserting that a
particular 'phantom picture' (e.g. the phantom picture containing the
locations of numerous mass points) is preferable to another phantom
picture. 'The differential equations of the phenomenological approach',
writes Boltzmann on this point,[21] 'are obviously nothing but rules for the
forming of numbers and for connecting them with other numbers and
geometrical concepts which in turn are nothing but thought pictures
[*Gedankenbilder*] for the presentation of phenomena. *Exactly the same applies to
the atomic conceptions* [*Vorstellungen der Atomistik*] so that I cannot see any
difference in this respect.' According to Boltzmann even the general idea of
the reality of the external world is but a (very abstract) picture,[22] and the
philosophical doctrine of the reality of the external world asserts no more
than that this picture, this *Scheinbild*, is preferable to other pictures such as
solipsism.

The clearest and most concise account of the positivistic version is found
in Einstein (cf. vol. 2, ch. 6.4). In his essay 'Physics and Reality',[23] Einstein
criticizes the quantum theory for its '*incomplete* representation of real
things'.[24] but explains at once what is meant by 'real existence':

Out of the multitude of our sense experiences we take, mentally and arbi-

---

[21] *Populaere Vorlesungen*, 142; my italics.   [22] *Ibid.*, ch. 12.

[23] *J. Frankl. Inst.*, 221 (1936), reprinted in *Ideas and Opinions* (New York, 1954), 290ff. I am
quoting from the latter source. Einstein was thoroughly familiar with the writings of
Boltzmann and Mach.   [24] *Ibid.*, 325f.

trarily, certain repeatedly occurring complexes of sense impressions . . . and correlate to them a concept – the concept of a bodily object. Considered logically this concept is not identical with the totality of sense impressions referred to; but it is a free creation of the human (animal) mind. On the other hand, this concept owes its meaning and its justification exclusively to the totality of the sense impressions we associate with it. The second step is to be found in the fact that, in our thinking (which determines our expectations), we attribute to this concept of a bodily object a significance which is to a high degree independent of the sense impressions which originally gave rise to it. This is what we mean when we attribute to the bodily object a 'real existence'.[25]

We see that according to Einstein the quantum theoretical issue is not an 'ontological' issue; it is an issue over the choice of systems for the correlation of 'impressions'.

3. MAXWELL AND MACH

The ideas of Maxwell and Mach differ from all the versions I have explained so far. They are also more subtle. They were developed in close connection with research and it is therefore somewhat difficult to isolate their philosophical components. But one feels a sense of relief when transferred from the fruitless technicalities and ontological primitivisms of modern 'philosophers' to the brief, simple, but profound remarks of these scientists.

Maxwell introduced his philosophy before and not after he had made his discoveries, as a guide for finding a new theory of electromagnetic phenomena. He distinguishes between 'mathematical formulae', 'physical hypotheses' and 'analogies'.[26] Mathematical formulae may help us to 'trace out the consequences of given laws' but at the expense of 'los[ing] sight of the phenomena to be explained'. Also 'we can never obtain more extended views of the connections of the subject'. What Maxwell means is that mathematical formulae fail to keep the subject matter before the eye of the scientist, and they also lack in heuristic potential. This is a brief and powerful criticism of theories such as the one proposed (much later) by Hertz and of more recent formalistic tendencies.

A physical hypothesis does provide a guide and it also keeps the subject matter before our eyes. However, it makes us see the phenomena 'only through a medium'. Maxwell seems to fear that physical hypotheses may be imposed upon the phenomena without the possibility of checking them independently. As a result we cannot decide whether the phenomena are correctly represented by these hypotheses.

---

[25] *Ibid.*, 291.
[26] 'On Faraday's Lines of Force', *Trans. Camb. Phil. Soc.*, 10, part 1, read on Dec. 10, 1855 and Feb. 11, 1856 and quoted from *The Scientific Papers of James Clerk Maxwell* (Dover, 1965), 155f.

Analogies avoid the drawbacks of mathematical formulae and of physical hypotheses. They are hypotheses in Mill's sense of the word (ch. 8, nn.12ff and text), i.e. assumptions about the nature of things which have been examined and have passed tests. They have heuristic potential, but they don't blind us. 'The changes of direction', writes Maxwell on this point,

> which light undergoes in passing from one medium to another, are identical with the deviations of the path of a particle in moving through a narrow space in which intense forces act. This analogy, which extends only to the direction and not to the velocity of the motion, was long believed to be the true explanation of the refraction of light; and we still find it useful in the solution of certain problems, in which we employ it without danger, as an artificial method. The other analogy, between light and the vibrations of an elastic medium, extends much farther out, though its importance and fruitfulness cannot be overestimated, we must recollect that it is founded only on a resemblance *in form* between the laws of light and those of vibrations. By stripping it of its physical dress and reducing it to a theory of 'transverse alternations' we might obtain a system of truth strictly founded on observation, but probably deficient both in the vividness of its conceptions and the fertility of its method . . .

These remarks which Boltzmann regarded as 'path breaking for epistemology as well as for theoretical physics.[27] and which, according to him, 'clearly adumbrated the development of epistemology during the next 40 years'[28] show that Maxwell wants a conception that *guides* the researcher without *forcing* him into a definite path; that makes suggestions without eliminating the means of controlling them. The research instruments recommended by Maxwell differ from physical hypotheses not in *content* but in *use*: one follows the suggestions made by an analogy but checks them at every step by a comparison with independently described phenomena. Analogies are physical hypotheses restricted by such a process of checking and used with the thought of possible further restrictions firmly in mind. Their theoretical entities do not represent any real entities unless it turns out that the phenomena follow the hypothesis in every detail. It seems that Maxwell hoped some day to find such a real physical theory.[29]

[27] In the footnotes to his translation of Maxwell's essay: 'Ueber Faradays Kraftlinien', *Ostwalds Klassiker der Exakten Wissenschaften* (Leipzig, 1895), 100.

[28] *Ibid.* Boltzmann adds: 'The later epistemologists treated all that in much greater detail but also mostly in a more one sided way and they introduced their rules for the development of theories only after that development had taken place and not, as here, before.' That is certainly true of later-nineteenth-century philosophy of science which Boltzmann had in mind. It is also true of Popper and the positivists.

[29] Boltzmann, on the other, emphasized the difference between analogies and 'hypotheses in the older sense of the world' ('Ueber die Methoden der Theoretischen Physik' in *Populaere Vorlesungen*, 8). He pointed out that 'Maxwell's gas molecules which repel each other with a force inversely proportional to the fifth power of their distance' are analogies and not real things, and he merged Hertz's idea of phantom pictures with Maxwell's very different idea of an analogy, thus staying firmly within the positivistic version of scientific realism. Bohr's 'pictures' (the wave picture; the particle picture; etc.) are exact modern repetitions of Maxwell's analogies.

Mach differs from the positivists (third version, explained above) in two ways: he does not assume a two layer model of knowledge (except locally: cf. vol. 2, ch. 5) and he examines the historical (physiological, psychological) determinants of scientific change. He gives an account of this change and of the difficulties facing the individual scientist that is much more realistic than the accounts of the philosophers (vol. 2, chs. 5 and 6). According to Mach it is our task not only to classify, correlate and predict phenomena, but also to examine and to analyse them. And this task is not a matter for philosophy, but for science. For example, Mach points out that the 'mental field', i.e. the domain where thoughts, emotions, sensations appear, 'can never be fully explored by introspection. But introspection combined with physiological research which examines the physical connections can put this field clearly before us and only thereby makes us acquainted with our inner being.'[30] In other words, *science explores all aspects of knowledge, 'phenomena' as well as theories, 'foundations' as well as standards; it is an autonomous enterprise not dependent on principles taken from other fields.* This idea according to which all concepts are theoretical concepts, at least in principle, is definitely in conflict with the positivistic version of scientific realism and it is very close to the point of view of ch. 2.6, thesis I. It was this idea of Mach's which led to the conceptual revolutions of the twentieth century. And it is this idea which also explains Mach's opposition to nineteenth-century atomism and the more dogmatic versions of Einstein's theory of relativity (cf. vol. 2, chs. 5 and 6).

4. THE DOUBLE LANGUAGE MODEL

The two layer model of scientific knowledge assumes a domain of phenomena and 'phantom pictures' for their prediction. Not many scientists regarded this distinction as absolute. 'In my opinion' writes Boltzmann,[31] 'we cannot utter a single statement that would be a pure fact of experience.' For Duhem, primary qualities are only 'provisional'[32] and can be subdivided by further research. For Mach all concepts are theoretical, as we have seen; even sensation talk involves a 'one sided theory'.[33] A distinction is recognized – but it is regarded as temporary and as being subjected to further research.

The double language model 'clarifies' the distinction by cutting it off from scientific research and reformulating it in epistemological, i.e. non-scientific, terms. A 'clarification' is certainly achieved – simpleminded notions are always more easy to understand than complex ones – but the

[30] *Populaerwissenschaftliche Vorlesungen*, 228.
[31] *Populaere Vorlesungen*, 286. 'According to Goethe all experience is only half experience.'
[32] *Aim and Structure*, 128.
[33] *Analyse der Empfindungen* (Jena, 1922), 18. Cf. vol. 2, chs. 5 and 6.

result has little to do with scientific practice. This is my main criticism of
the double language model.

To elaborate: there is no doubt that the double layer model which
scientists discussed in the last century captured certain features of scientific
knowledge. The *concepts* used on the observational level are often quite
different from the 'theoretical entities' of a newly introduced abstract
theory – after all, they belong to an earlier stage of knowledge, they are
familiar, their application may be connected with perceptual processes
while the application of theoretical terms, especially of newly introduced
theoretical terms, is mostly perception free (cf. the explanations in ch. 2.1).
But a closer looks reveals that the situation is much more complex. Thus in
thermodynamics we have an observational level (reading of dials, thermometers, pressure gauges, etc. plus operations such as building the necessary instruments), followed by a phenomenological layer where changes of
the observable entities of the first layer (temperature, pressure, free energy
and so on) are connected by partial differential equations, the kinetic
theory introduces theoretical entities of an entirely different kind obeying
the familiar equations of mechanics, there is quantum mechanics, quantum
statistics, there are gravitational effects in the large, and so on. The example also shows that the idea of neatly separated layers often breaks down
and gives way to much more complex arrangements: kinetic effects and
quantum effects occasionally bypass the phenomenological layer (Brownian motion, specific heats), quantum effects turn up on the observational
level (superconductivity), while the structure of the world at large may
reach down and shape even elementary particles (Eddington). Scientific
knowledge is not arranged in layers and is not conceptually unified.

On the other hand, scientists often attempt to overcome this variety and
incoherence and to unify disparate domains by a single point of view.
Examples are the unification of physics and astronomy by Galileo's investigations and Newton's mechanics, the unification of electrostatics, mag-
netostatics, electrodynamics and optics by Maxwell's theory, the unified
treatment of mechanical and electromagnetic phenomena by the special
theory of relativity, and the more recent attempt to find a unified account of
elementary particles and fields. What happens here is not the absorption of
an *unchanged* conceptual system into a wider context (e.g. one did not
continue to use Aristotelian concepts together with the new astronomy of
Newton; nor did all mechanical notions survive relativity), but an *entire
reorganization* both of observations and of theoretical ideas (as an example,
cf. the conceptual changes that were necessary to adapt the old impetus
mechanics to Newton's theory as described in ch. 4.5). This suggests that
*the layer model, while giving a correct account of passing stages of science, states a
problem rather than an intrinsic feature of knowledge and that the problem has often
been solved by developments within the sciences themselves.*

There are *two requirements* which must be satisfied if such developments are to occur.

First, there must be a theory whose concepts are sufficiently rich and flexible to help us reorganize, represent and combine the domains to be unified. There is no need to restate every fact and every problem in the new terms. For example, there is no need for a relational theory of spacetime to be able to answer questions about absolute motion. But important experiments and problems must be accounted for.[34]

Secondly, the theory satisfying the first requirement must be used to the limits of its capacity. The organizing and explanatory power of Newton's theory is wasted if we employ it only to calculate the paths of the planets and leave the behaviour of cranes, cannonballs, skeletons, gases to the Aristotelians. And the notions of the special theory of relativity are wasted if we refuse to formulate observations and experimental results in its terms. There may be theoretical reasons for restricting a new theory – Aristotle, for example, regarded mathematical considerations as instruments because he was convinced that perception, which was his criterion of reality, gave rise to qualities only – but even they may be overruled in an attempt to find the limits (cf. vol. 2, ch. 1). This is the main point of thesis I of ch. 2.6 as well as of the arguments in chs. 4 and 6.[35] The extension of a theory into new domains may of course take considerable time – but the difficulties that such an extension meets are scientific difficulties and not proof of philosophical impossibilities.[36]

5. INCOMMENSURABILITY

To repeat: thesis I is not merely a philosophical thesis; it is also a summary of a rather widespread scientific procedure that has often been successful. A *general* attack on thesis I is therefore not merely an attack against a philosophical position (for example against a philosophical 'realism'), it is also an attack on science; it amounts to no less than a criticism of procedures that have brought us a great number of superb scientific achievements. *Conversely, it is quite possible to reject thesis I on special grounds* (examples of such special arguments are found in ch. 16). Chs. 2 and 11 are therefore somewhat misleading. Producing philosophical arguments for a point of view whose applicability has to be decided by concrete scientific research, they suggest that scientific realism is the only reasonable position

---

[34] Ideally the theory should be a complete theory in the sense of vol. 2, ch. 8 (appendix).

[35] Note that such a determined application of a theory means using a pragmatic theory of observation instead of a semantic theory (for explanation and arguments cf. ch. 4.1 and ch. 6.7). Note also that the difficulties mentioned in the next sentence of the text may make it advisable to return to a semantic theory.

[36] Cf. the difficulties created for theories of measurement in the traditional sense by the laws of the microlevel (ch. 13.3) which have led to a restriction of thesis I.

to take, come what may, and inject a dogmatic element into scientific discussions (this dogmatism is responsible for the less than satisfactory nature of discussions about the foundations of the quantum theory). Of course, philosophical arguments should not be avoided; *but they have to pass the test of scientific practice.* They are welcome if they help the practice; they must be withdrawn if they hinder it, or deflect it in undesirable directions.[37]

These remarks apply especially to philosophical views about explanation, reduction and theory comparison. Many such views assume that a comparison of rival theories involves logical relations between their statements. But adopting the second requirement of section 4 above may make theories incommensurable in the sense that such relations cease to exist. To many philosophers this is the end of the world.[38] But the fact that an abstract philosophical theory has been found to conflict with scientific practice does not mean that the practice is without a guide. Incommensurability only shows that scientific discourse *which contains detailed and highly sophisticated discussions concerning the comparative advantages of paradigms* obeys laws and standards that have little in common with the naive models that philosophers of science have constructed for that purpose.[39]

[37] Cf. vol. 2, ch. 7 and part 1 of *SFS* as well as vol. 2, ch. 1.

[38] All the existing attempts to overcome incommensurability in the sense just described assume that the concepts of the systems to be connected *can be used at the same time*, and that the only problem is how to establish relations between them. But the example I explained in ch. 17 of *A M* and the relation between relativity and classical physics which shows similar features (cf. the appendix to ch. 8 of vol. 2) show that there are cases which do not agree with this assumption. There exist pairs of theories (world views; forms of life) such that using terms in accordance with the rules of the one theory (world view; form of life) makes it impossible to construct and even to think of the concepts that arise when terms are used in accordance with the rules of the other theory (world view; form of life). Logicians have not yet found any remedy for dealing with this situation – and there is no need to, for the practice of science is not hindered by it. The only difficulty that arises is for certain abstract semantical views.

[39] For details on 'crucial experiments' cf. vol. 2, ch. 8.9, 8.10 as well as *A M*, 282ff. Moreover, there are *formal criteria*: a linear theory (theory with linear differential equations as basic equations) is preferable to non-linear theories because solutions can be obtained much more easily. This was one of the main arguments against the non-linear electrodynamics of Mie, Born and Infeld. The argument was also used against the general theory of relativity until the development of high speed computers simplified numerical calculations. Or, a 'coherent' account is preferable to a non-coherent one (this was one of Einstein's main criteria in favour of his approach). A theory using many and daring approximations to reach its 'facts' is, to some, much less likeable than a theory that uses only a few safe approximations. Number of facts predicted may be another criterion. *Non-formal criteria* usually demand conformity with basic theory (relativistic invariance; agreement with the quantum laws) or with metaphysical principles (such as Einstein's 'principle of reality' or his principle that physical entities such as space which *have effects* must also be capable of *being affected*). It is interesting to see that the criteria often give conflicting results so that a *choice* becomes necessary.

2

An attempt at a realistic interpretation of experience*

1. INTRODUCTION

'The task of science', writes Niels Bohr,[1] 'is both to extend the range of our experience and to reduce it to order.' 'Science', echoes a modern philosopher,[2] 'is ultimately intended to systematize the data of our experience.' In the following paper I shall try to show that these two statements, in spite of their simplicity and their apparently innocuous character, have consequences which are at variance with scientific method and reasonable philosophy.

For convenience I shall call any interpretation of science (and of theoretical knowledge in general), which implies an assumption equivalent to the two statements quoted above, a *positivistic interpretation*. Examples of positivistic interpretations in this sense are (1) instrumentalism, i.e. the view that scientific theories are instruments of prediction which do not possess any descriptive meaning; and (2) the more sophisticated view that scientific theories do possess meaning, but that their meaning is due to the connection with experience only.[3]

I shall proceed in the following way. After a few preliminary remarks on the notion of observability I shall develop some consequences of positivism. These consequences will be expressed in the form of a thesis (stability thesis, section 3). It will be shown that there exist serious objections against the stability thesis as well as against the customary attempts to defend it (sections 4 and 5). An alternative thesis will be considered and its consequences developed (section 6). This latter thesis may be said to be an attempt at a realistic interpretation of experience. I shall conclude with a discussion of the logical status of the arguments against the stability thesis, and of the issue between positivism and realism in general.

2. OBSERVATION LANGUAGES

Within science a rough distinction is drawn between theory and observation. This distinction can best be explained by formulating the conditions

* This is a very much abbreviated version of my thesis, *Zur Theorie der Basissätze* (Vienna, 1951).
[1] Niels Bohr, *Atomic Theory and the Description of Nature* (Cambridge, 1934), 1.
[2] C. G. Hempel in *International Encyclopedia of Unified Science* (Chicago, 1952), II 7, 21.
[3] This is Carnap's view. Cf. the discussion in n.7.

which a language must satisfy in order to be acceptable as a means of describing the results of observation and experiment. Any language satisfying those conditions will be called an observation language.

We may distinguish two sets of conditions for observation languages. The conditions of the first set are *pragmatic* (psychological, sociological) conditions. They stipulate what is to be the relation between the (verbal or sensory) behaviour of human beings of a class C (the observers) and a set of physical situations S (the situations observed). It is demanded that for every atomic sentence a (of a class A) of the language considered there exists a situation s (a so-called appropriate situation) such that every C, when presented with a in s will run through a series of states and operations which terminates either in the acceptance of a or in its rejection by the C chosen.[4] This we call the condition of decidability. Any series of the kind mentioned will be called a C-series associated with a or simply an associated series. The function correlating atomic sentences with associated series will be called the associating function of the language concerned and it will be designated by the letter F. Secondly, it is demanded that in the appropriate situation the associated series should be passed through fairly quickly. This we call the condition of quick decidability.[5] Thirdly, we shall have to stipulate that if (in an appropriate situation) an atomic sentence is accepted (or rejected) by some C, it will be accepted (or rejected) by (nearly) every C. This we call the condition of unanimous decidability. Finally, we must stipulate that the decision made be (causally) dependent upon the situation and not only upon the atomic sentence presented or the internal state of the C chosen. This we call the condition of relevance. Any function correlating situations with either acceptance or rejection of a given sentence will be called a relevance-function and it will be designated by the letter R.

Summarizing the four pragmatic conditions just stated we may say that, given three classes, A, C and S, the class A will be called a class of *observable sentences* (used by observers C in situations S) only if, given some S, every C is able to come to a quick, unanimous and relevant decision with respect to those A for which the chosen S is appropriate. The pragmatic properties of a given observation language will then be fully characterized by the set $\{C, A, S, F, R\}$. Any such set will be called a *characteristic*. The characteristic of an observation language completely determines the 'use' of each of its atomic sentences.

As stated above, the pragmatic conditions concern the relation between observation *sentences* (*not* statements) and human beings with-

---

[4] The terms 'acceptance' and 'rejection' are pragmatic terms and they refer to two specific and clearly distinguishable types of reaction.

[5] It should be noted that this condition does not contain any restriction as to the *complexity* of the associated series.

out making any stipulation as to what those sentences are supposed to assert. Further conditions will have to be added if we want to obtain a fully fledged language. Any complete class of such further conditions will be called an *interpretation*. A particular observation language is completely specified by its characteristic together with its interpretation.

The distinction between the pragmatic properties of a language and its interpretation is clear and unambiguous. Yet in view of the fact that some influential doctrines to be discussed later in this paper owe their existence to the neglect of this very distinction, a few more words of explanation seem to be required.

Observability is a pragmatic concept. Whether or not a situation s is observable for an organism O can be ascertained by investigating the behaviour of O, mental (sensations) or otherwise; more especially, it can be ascertained by investigating O's ability to distinguish between s and other situations. And we shall say that O is able to distinguish between s and situations different from s if it can be conditioned such that it (conditionally or unconditionally) produces a specific reaction r whenever s is present, and does not produce r when s is absent.

Exactly the same considerations apply if O happens to be a human observer and r one of the atomic sentences of his observation language. It is of course true that in this case r, apart from satisfying the pragmatic criterion of observability outlined above, will also be interpreted. But from this neither can we derive, as has frequently been done, that its interpretation is *logically* determined by the observational situation, nor is the assumption correct that man is capable of reactions of a very sublime kind (sensations, abstract ideas) which by their very nature allow us to confer meaning upon those expressions which are their verbal manifestations. What the observational situation determines (causally) is the acceptance or the rejection of a sentence, i.e. a physical event. In so far as this causal chain involves our own organism we are on a par with physical instruments. But we also interpret the indications of these instruments (i.e. either the sensations which occur during observation, or the observational sentence uttered) and this interpretation is an additional act, whether now the instrument used is some apparatus or our own sensory organization (our own body).[6]

---

[6] Two attempts to overcome this dualism (which is only another form of the dualism between nature and convention (cf. K. R. Popper, *The Open Society and Its Enemies* (London, 1946), ch. 5) are (1) the attempt to 'naturalize' the conventional element – this is done by the behaviourists – or (2) the converse attempt to 'spiritualize' parts of nature (example: doctrine of abstract ideas). Both attempts suffer from fundamental difficulties some of which will be discussed later (cf. also S. Körner, *Conceptual Thinking* (London, 1955), especially chs. 7 and 17, for a similar distinction between (a) descriptive and (b) interpretative and non-ostensive, concepts and propositions.

3. THE STABILITY THESIS

Any philosopher who holds that scientific theories and other general assumptions are nothing but convenient means for the systematization of the data of our experience is thereby committed to the view (which I shall call the stability thesis) that *interpretations* (in the sense explained above) *do not depend upon the status of our theoretical knowledge.*[7] Our first attack against positivism will consist in showing that the stability thesis has undesirable consequences.

For this purpose it is sufficient to point out that we make assertions not only by *formulating* (with the help of a certain language) *a sentence* (or a theory) and *asserting* that it is true, but also by *using a language* as a means of communication. Thus when using natural numbers for counting objects and reporting the result we assume (*inter alia*) (without explicitly saying so and perhaps without even being able to state this assumption within the language used) (1) that those objects are discrete entities which can always be arranged in a series, and (2) that the result of our counting is independent of the order in which we proceed as well as of the particular method of counting (the particular method of 'observing' the number of a certain class) used. However plausible these two assumptions may be, there is no *a priori* reason why they should be true. Conversely, the discovery that, for example assumption (1) is incorrect for every set of objects amounts to the discovery that no observational language containing natural numbers for the purpose of counting can be applied to reality.

We shall call any statement which is implied by the statement that a certain language *L* is applicable (either universally, or in a certain domain) an *ontological consequence* of *L*. The existence of ontological consequences

---

[7] That the stability thesis is a consequence of positivism may be seen from a closer look at two positivistic philosophies. Take first *instrumentalism*. According to instrumentalism theories are tools for the prediction of events of a certain kind. Hence, a language is required for the description of those events whose sentences are (a) observable, and (b) interpreted. On the other hand it is denied that theories have descriptive meaning, i.e. it is denied that they possess an interpretation (in the sense in which the word has been introduced in the text above). If this is correct, then they cannot provide an interpretation for any other language either. Consequently, whatever interpretation an observation language may possess, it will not depend upon the theoretical 'superstructure'.

As a second example we take *Carnap's method* of reconstructing the language of science by a dual scheme, consisting of an interpreted observation language and of a theoretical language *T*. In this method it is assumed that the interpretation of the primitive descriptive terms of *T* can be completely accounted for by pointing to the fact 'that some of these are connected . . . with observational terms' (cf. Carnap's essay in the *Minnesota Studies in the Philosophy of Science* (Minneapolis, 1956), I, 47). No independent interpretation is given for the theoretical terms (*ibid.*). This implies that the interpretation of a theory depends upon the interpretation of the observation language used, but not the other way round. And as it is stipulated that the observation language be completely interpreted (*ibid.*, 40) it follows also that Carnap's more sophisticated account is based upon an observation language whose interpretation has been introduced independently of the state of the theoretical 'superstructure'.

which are not logically true leads to the first difficulty of the stability thesis.

For let us assume (a) that the observation language has ontological consequences; (b) that it satisfies the stability thesis (that it is a positivistic observation language, as we shall put it); and (c) that it is applicable, was applicable and will always be applicable.[8] Then it follows, (1) that those ontological consequences cannot have emerged as the result of empirical research (for if this were the case, the stability-thesis would have been violated at some time in the past); (2) nor will it ever be possible to show by empirical research that they are incorrect (for if this were the case the stability thesis would be violated at some time in the future). Hence, if the ontological consequences of a given language are not all logically true statements (in which latter case the language would be applicable for purely logical reasons which seems implausible) we arrive at the result that *every positivistic observation language is based upon a metaphysical ontology*. This is the first undesirable consequence of the stability thesis (undesirable, that is, for the positivists who hold the thesis).[9]

This consequence leads at once to the question: how does a positivist justify the particular interpretation which he has chosen for his observation language? In the next two sections I shall attempt to give a tentative answer to this question.

4. PRAGMATIC MEANING; COMPLEMENTARITY

The most primitive ways of introducing an interpretation consists in the uncritical acceptance of a certain ontology, with or without the comment that it would be 'unnatural' to use a different one. Many forms of phenomenalism ('experiences exist and nothing else exists') are of this kind. Naive interpretations in this sense will not be discussed in the present paper.

More refined methods of introducing an interpretation are based upon certain theories of meaning. In this paper I shall briefly discuss two such theories. According to the first theory the interpretation of an expression is determined by its 'use'. Applying this to our problem and using our own terminology we arrive at the result that the interpretation of an observation language is uniquely and completely determined by its characteristic. This result we shall call the *principle of pragmatic meaning*. According to the second theory the interpretation of an observational term is determined by what is

---

[8] There is little doubt that assumption (c) is silently made by nearly every positivist.

[9] The fact that any language (and everyday language in particular) has ontological consequences taken together with the stability-thesis (expressed in some form of conceptual realism, e.g. Platonism) was amply utilized for metaphysical speculations by the Peripatetics and by their followers. Cf. e.g. J. Gredt, *Die Aristotelisch–Thomistische Philosophie* (Freiburg, 1935), I.

'given' (or 'immediately given') immediately before either the acceptance or the rejection of any observation sentence containing that term. This we shall call the *principle of phenomenological meaning*. Within positivism (in the sense defined in section 1) these two principles play an all-important role. It will be our task to show that they are both untenable.

Take first the principle of pragmatic meaning. Combined with the (empirical) fact (if it is a fact; of section 6) that the characteristic of the everyday language is fairly stable, this principle implies the stability thesis. The stability thesis will be refuted in section 6. At the same time we shall explain how it is possible for the interpretation of a language to change without any perceptible effect upon its characteristic. This amounts to a refutation of the principle of pragmatic meaning.

A more general objection is this: the four conditions in section 2 can be satisfied by human beings and their verbal utterances as well as by machines and their reactions. It is quite obvious that, however well behaved and useful a physical instrument may be, the fact that in certain situations it consistently reacts in a well-defined way does not allow us to infer (logically) what those reactions mean: first, because the existence of a certain observational ability (in the sense elucidated at the end of section 2) is compatible with the most diverse interpretations of the things observed;[10] and secondly, because no set of observations is ever sufficient for us to infer (logically) any one of those interpretations (problem of induction). It should then be equally obvious that, however well behaved and useful a human observer may be, the fact that in certain situations he (consistently) produces a certain noise, does not allow us to infer what this noise means.

As an example of an (implicit) application of the principle of pragmatic meaning I shall now discuss Bohr's idea of complementarity. This idea which has greatly contributed towards the understanding of microscopic phenomena employs some philosophical assumptions which cannot be accepted without criticism. Bohr has repeatedly emphasized, and here I am quite prepared to follow him, that 'no content can be grasped without a form' (*E*, 240)[11] and that, more especially, 'any experience . . . makes its appearance within the frame of our customary points of view and forms of perception' (*A*, 1). He has also pointed out – and here it will be necessary to criticize him – that 'however far the phenomena transcend the scope of classical physical *explanation*, the account of all evidence must be expressed in classical *terms*' (*E*, 209; cf. also *A*, 77, 53, 94, etc.) which implies that the 'forms of perception' referred to above are, and will be, those of classical physics: 'We can by no means dispense with those forms which colour our whole language and in terms of which all experience must ultimately be

---

[10] Cf. the end of section 2 as well as section 5.
[11] The letters refer: *E*, to P. A. Schilpp (ed.), *Albert Einstein: Philosopher-Scientist* (Evanston, 1953); *A*, to Bohr, *Atomic Theory*.

described' $(A, 5)$. To sum up: the observation language of physics is a positivistic observation language whose interpretation is the same as the interpretation of classical physics *before* the advent of quantum mechanics. How can this be reconciled with the fact that classical physics is contradicted by the quantum of action?

According to Bohr it can be reconciled by restricting the application of classical terms in a way which (a) 'provides room for new physical laws'[12] and especially for the quantum of action; which (b) still allows us to describe any possible experiment in classical terms; and which (c) leads to correct predictions. Any set of rules satisfying (a), (b) and (c) is called by Bohr a 'natural generalization of the classical mode of description' $(A, 56)$. He emphasizes that the laws (or rather rules of prediction) employed by such a generalization 'cannot be included within the frame formed by our accustomed modes of perception' $(A, 12; 22, 87)$ for they impose restrictions upon this very frame. Or, to put it in different words: the laws of quantum mechanics do not admit of a coherent and universal interpretation in intuitive terms. Bohr seems to assume that this will hold for any future theory of microscopic entities.

Now it may be conceded that the laws of quantum mechanics do not admit of a straightforward interpretation on the basis of a classical model, as such a model would be incompatible either with the principle of superposition or with the individuality of the microscopic entities. It may also be conceded that *as a matter of fact* we do find it difficult (though by no means impossible) to form an intuitive picture of processes which are not dependent upon the classical framework. But from this psychological predicament we can by no means infer (assumption 1) that such intuitive understanding will never be possible. And it would be even less correct to assume on that basis that the concept of a non-classical process cannot be formed (assumption 2); for it is well known that we can form and handle concepts even of those things which we cannot readily visualize. Yet these two assumptions play an important role in Bohr's philosophy: according to Bohr the laws of matrix mechanics (or of wave mechanics) and, indeed, the laws of any future quantum theory are symbolic 'expedients which enable us to express in a consistent manner essential aspects of the phenomena' $(A, 12)$, i.e. of classical situations; he emphasizes that they do not form a 'new conceptual scheme' $(A, 111$, against Schrödinger's interpretation of wave mechanics) for the description of universal features of the world different from those of classical physics. And according to Bohr it would even be a 'misconception to believe that the difficulties of the atomic theory may be evaded by eventually replacing the concepts of classical physics by new conceptual forms' $(A, 16)$, as there exist 'general limits of man's

---

[12] Niels Bohr, 'Can the Quantum Mechanical Description of Physical Reality be considered Complete?', *Phys. Rev.*, 48 (1936), 701.

capacity to create concepts' (*A*, 96). How can this defeatist attitude be understood?

I think that it can be understood if we explain more thoroughly the ideas upon which Bohr's interpretation is based. The first idea is that the belief in classical physics has influenced not only our thinking but also our experimental procedures and even our 'forms of perception'. This idea gives a correct description of the effect which the continued use of a fairly general physical theory may have upon our practices and upon our perceptions: it will become increasingly difficult to imagine an alternative account of the facts. The second idea is inductivism. According to inductivism we invent only such theories as are suggested by our observations. Combined with the first idea inductivism implies that it is psychologically impossible to create non-classical concepts and to invent a non-classical 'conceptual scheme'. The third idea is the principle of pragmatic meaning. According to this idea, the use of classical methods and the existence of classical 'forms of perception' imply that the observation language possesses a classical interpretation (see above). As a non-classical picture of the world would lead to an interpretation which is inconsistent with this classical interpretation, such a non-classical picture, apart from being psychologically impossible, would even involve a logical absurdity. I think that Bohr's defeatist attitude expressed in the quotation at the end of the last paragraph is due to his implicit belief in the principle of pragmatic meaning and his explicit adoption of the inductivistic doctrine (cf. *A*, 18 as well as the quotation at the beginning of this paper).

As opposed to this it is sufficient to point out that even in a situation where all *facts* seem to suggest a theory which cannot any longer be maintained to be universally true, that even in such a situation the invention of new conceptual schemes need not be psychologically impossible so long as there exist abstract pictures of the world (metaphysical or otherwise) which may be turned into alternative interpretations.[13] And our foregoing criticism of the principle of pragmatic meaning shows that such alternative interpretations need not lead into logical absurdity either (for this cf. also nn.20 and 21). It follows that the permanence of the classical 'forms of perception' can be accounted for without adopting a positivistic philosophy of science; and that it leads to positivism only if two philosophical ideas are used (inductivism; the principle of pragmatic meaning) which can easily be shown to be incorrect.

5. PHENOMENOLOGICAL MEANING

The principle of phenomenological meaning takes over where the principle of pragmatic meaning seems to fail. It admits that behaviour does not

[13] Cf. for this point section 7 of the present paper.

determine interpretations. But apart from behaving in a certain way, man has also feelings, sensations and more complex experiences. The principle of phenomenological meaning assumes that interpretations are determined by what is experienced: in order to explain to a person what 'red' means one need only create circumstances in which red is experienced. The things experienced (or 'immediately perceived') in those circumstances completely determine the meaning of the word 'red' (theory of ostensive definition). Or, to put it in more general terms: the meaning of an observational term is determined by what is 'immediately given' at the moment of the acceptance of any observational sentence containing that term.

In order to get some insight into the implications of this principle let us first take the phrase 'immediately given' in its widest sense. The properties of the things which are 'immediately given' in this wide sense and their relations can be 'read off' the experiences without any difficulty being felt,[14] i.e. the *acceptance* (or the rejection) of any description of those things is uniquely determined by the observational situation. The question arises (and is answered in the affirmative by the principle of phenomenological meaning) whether this amounts to a determination of the *meaning* of the description accepted (rejected).

Our answer to this question (which is negative) will be given in three steps. This answer amounts to a refutation of the principle of phenomenological meaning.

(a) First, consider the relation between an immediately given object or a *phenomenon P* (this phenomenon may include a set of leading questions) and (the acceptance of) a sentence *S* assumed to be uniquely determined by that phenomenon. This relation I shall call the relation of phenomenological adequacy.

I shall first show that at the moment of the utterance of *S* this relation cannot be immediately given in the same sense in which *P* is immediately given, i.e. it cannot be a phenomenon. My argument will be by *reductio ad absurdum*. Indeed, assume that the observer *O* utters *S* (or thinks that *S* is the case) because (and only after) he has discovered that *S* is phenomenologically adequate or that it 'fits' *P*. This would mean that *O* (1) not only attends to *P* and *S*, but also to a third phenomenon *P'* (the relation between *P* and *S*); and (2) that he has identified *P'* as the relation of phenomenological adequacy. According to the idea we are investigating at the moment he could have done the latter only by confronting *P'* with a further phenomenon *S'* (either a thought, or a sentence) to the effect that *P'* was the relation of phenomenological adequacy, and by discovering that *S'* fits *P'*.

---

[14] For the problems of phenomenological description cf. E. Tranekjaer-Rasmussen, *Bevidsthedsliv og Erkendelse* (Copenhagen, 1956), ch. 2. With respect to the usefulness of phenomenological analysis for philosophy I learned much from discussions with Professor Tranekjaer-Rasmussen as well as from his book.

This discovery in its turn pre-supposes $(1')$ that he not merely attends to P, S, P', S', but also to a further phenomenon P'' (the relation between P' and S'); and $(2')$ that he has indentified P'' as the relation of phenomenological adequacy; and so on *ad infinitum*. Hence, the observer will have to perform infinitely many acts of introspection before ever being able to utter an observation sentence. This means that the conditions of adequate report which we are considering at the moment are such that no observer will ever be able to say anything – which is patently absurd. And as it would be equally absurd to assume that there are infinitely many distinct phenomena in our mind but that we attend only to some of them, we have to conclude that at the moment of the utterance of an observational sentence S by an observer O only those phenomena exist and are attended to which are adequately described by S. The relation of phenomenological adequacy is not part of the experience of O.

From this it follows at once that the utterance of a certain observation sentence cannot be justified by saying that it 'fits' the phenomena. For if by appealing to the relation of phenomenological adequacy we make it part of our experience, we have thereby changed the original phenomenon. And our description of the new phenomenon will still be in need of justification. It is no good repeating 'but I *experience* P'', for the question discussed is not what is experienced, but whether what is experienced has been described adequately. And we have shown that this question cannot be answered by appealing to the relation of phenomenological adequacy. This refutes the contention, implicit in the principle of phenomenological meaning, that questions of meaning can be decided by introspection or by attendance to what is immediately given. The phenomenon which appears at the moment of observation can at most be regarded as a (phenomenological) *cause* of the acceptance (or rejection) of S.

(b) The idea that it can be more, for example that it can also provide us with an *interpretation* of the sentence produced, altogether puts the cart before the horse. It is of course true that some of the phenomena which can be brought into the relation of phenomenological adequacy with other phenomena do also possess an interpretation. But this interpretation is not conferred upon them because they 'fit', but it is an essential presupposition of the 'fitting'. This is easily seen when considering signs whose interpretation has been forgotten; they no longer fit the phenomena which previously evoked their acceptance. It follows that the principle of phenomenological meaning would in most cases either lead to interpretations which are different from the ones considered by its champions (see also the next paragraph); or it would be inapplicable. And it would be inapplicable in exactly those cases in which it is supposed to provide us with an interpretation – i.e. in the cases of signs which have not yet been given any meaning.

(c) But does introspection perhaps play a selective role? That is, is it perhaps possible that, given a phenomenon P and a class of *interpreted* sentences, the relation of phenomenological adequacy allows us to select those sentences which correctly describe P (which possess the 'correct' interpretation)? I believe that introspection cannot even play this more modest role of a selector. One reason is the existence of 'secondary interpretations':[15] I may feel a strong inclination to call the vowel e 'yellow'. The important thing is now that I feel this inclination only if 'yellow' carries its usual meaning. But according to this usual meaning 'yellow' is not applicable to sounds. A second reason, which I consider to be a very decisive one, is the existence of phenomenological situations whose phenomenologically adequate descriptions are self-contradictory. An example of such a situation has been described by E. Tranekjaer-Rasmussen.[16] A third reason is that, given some phenomenon, one can always construct an infinite series of descriptions, all of them fitting this particular phenomenon. One possible method of constructing this series which has achieved some importance in epistemological discussions consists in reducing the infinitely many consequences of the usual descriptions one by one. However, the assumption that one interpretation of a sign S could be more 'correct' than another one must be criticized also for more general reasons: if we consider signs in isolation, then any interpretation which we confer upon them is a matter of convention (cf. n.6). The same applies if we do not consider them in isolation, but as parts of a complicated linguistic machinery – unless we invoke the principle of pragmatic meaning which has already been criticized.

To sum up: the meaning of an observational term and the phenomenon leading to its application are two entirely different things.[17] Phenomena cannot determine meaning, although the fact that we have adopted a certain interpretation may (psychologically) determine the phenomena. That is, the strict adherence to an interpretation and the rejection of all accounts which are different from it may lead to a situation where the relation between phenomena and propositions will be one to one. In such a situation a distinction between phenomena and interpretations on the one hand and phenomena and objective facts on the other cannot readily be drawn; the principle of phenomenological meaning as well as the principle that descriptions are uniquely determined by facts will appear to be correct and Bacon's philosophy will appear to be the only reasonable one. It is

[15] The term 'secondary meaning' and the example are both due to Wittgenstein, *Philosophical Investigations* (Oxford, 1953), 216, para. 3ff.

[16] 'Perspectoid Distances', *Acta Psychologica*, 11 (1955), 297. Cf. also E. Rubin, 'Visual Figures Apparently Incompatible with Geometry', *Acta Psychologica*, 7 (1950), 365ff. These two papers deserve far more attention than they have so far received from philosophers.

[17] This distinction has been emphasized with great clarity by E. Kaila. See his article 'Det fraemmande sjaelslivets kunksapsteoretiska problem', *Theoria*, 2 (1933), 144ff as well as his essay 'Ueber das System der Wirklichkeitsbegriffe', *Acta Phil. Fenn.*, 2 (1936), 17ff (containing a polemic against Russell similar to the one later in this section).

important to realize that there are no factual reasons which would allow us to conclude that such a situation will never arise. This point will be further elaborated in section 7.

So far we have used the term 'introspection' in the wide sense of 'attendance to what is easily described'. But our analysis applies also if a more sophisticated idea is used of what is immediately given, for example if it is assumed that the 'given' is not directly accessible but must be found either by a special effort, or appears only under special conditions (e.g. when the reduction-screen is used). For the result of the special effort, or the things appearing under the special conditions mentioned, will again be phenomena. And we have already shown that phenomena cannot determine interpretations.

As in section 4 we shall conclude this section also by examining a philosophical argument which employs (silently) the principle of phenomenological meaning. The argument is Russell's. It is based upon the assumption (which is a consequence of the principle of phenomenological meaning) that phenomenologically simple objects must possess simple logical properties. Russell[18] considers statements of everyday language such as 'there is a dog' uttered in the presence of a dog. This statement is logically complex in the sense that if it is true many other statements will be true as well (such as the statement 'if a cat enters the room I shall hear a bark'). The more complex a statement is, the more easily it can be refuted. Hence, it is natural to suspect, as Russell does, that a more 'modest' statement (which implies 'less' consequences) will stand a greater chance of remaining true. Such a statement would also be logically simpler that 'there is a dog'. Russell seems to assume that the statement 'there is a canoid patch of colour', while being true whenever 'there is a dog' is true, satisfies the condition of being logically simpler than 'there is a dog' *because it is about a simpler phenomenon* (a patch of colour is two-dimensional, does not bark; a dog is three-dimensional, barks etc.) But he is thrice mistaken. First of all, a sentence of an observation language should be phenomenologically adequate. Now in the example taken above what is seen is a dog. Hence, 'there is a canoid patch of colour' is phenomenologically inadequate since the phenomenon 'canoid patch of colour' (realized e.g. by looking at the picture of a dog) is definitely different from the phenomenon 'dog' (how else would we distinguish between dogs and pictures of dogs?). Secondly, if there is a dog 'there is a canoid patch of colour' is also false; for a picture of a dog is not a dog. Thirdly, 'there is a canoid patch of colour' is by no means logically simpler than 'there is a dog' – the statement is about a physical object (a patch of colour) of a certain shape (doglike) and hence belongs to the same category as 'cat', 'dog', etc. Of course, I can confer upon this statement an interpretation which makes it less pretentious e.g. by omitting

[18] *Inquiry Into Meaning and Truth* (New York, 1940), 139.

from its consequences those concerning touch – but if the procedure can be carried out at all – section 6 (4) – it can also be applied to 'there is a dog' and thereby shown to be independent of the phenomenological character of the objects described.

The content of the last two sections can now be summed up by saying that neither the 'use' of observation sentences, nor the phenomena which accompany their application in observational situations can determine their interpretation. As positivists have not produced any further attempt to justify the interpretations of their observation languages we arrive at the result that fundamentally all those interpretations are naive in the sense explained at the beginning of section 4.

But is it possible to introduce interpretations in a more reasonable way? If we want to answer this question, we must first take a closer look at the stability thesis.

6. REFUTATION OF THE STABILITY THESIS: 'EVERYDAY LANGUAGE'

Consider for this purpose a language L which ascribes colours to self-luminescent objects. The predicates of this language P_i ($i = 1, 2, 3, \ldots$) are colour-predicates. We shall assume them to be observable. We shall also assume (1) that the characteristic of L has been defined; and (2) that the methods of observation implied in this characteristic involve only such velocities, masses, etc. as are met on the everyday level and can be produced and handled with relative ease.

Human beings using L will interpret the descriptive signs of this language in a way which depends upon their 'prejudices' (in Bacon's sense) i.e. upon their general ideas about things and their properties. A view frequently met is that the P_i designate properties of objects and that the objects possess those properties whether they are observed or not. We shall adopt this interpretation.

Now assume that a theory is formulated according to which the wavelength of light, as measured by an observer B in accordance with the characteristic of L, depends (among other things) upon the relative velocity of B and the light source (Doppler effect). Combined with the statement (of psychology) that an observer who watches a self-luminescent body a emitting light of the wavelength $\lambda_i < \lambda < \lambda_i'$, when using L will accept '$P_i(a)$' ($i = 1, 2, 3 \ldots$), this theory leads to the following result: what is asserted of a on the basis of the operations (described in the characteristic L) which terminate in ascribing to a (or withholding from a) the sign P_i, is that it is (or is not) an instance of a relation rather than an instance of a predicate. But this means that adopting the foregoing theory leads to an interpretation of L which is different from the one originally used.[19]

[19] An example which is slightly more technical but at the same time more straightforward is

In this new interpretation the expression '$P_i(a)$' is no longer complete and unambiguous. It will depend upon a parameter p (the relative velocity of a and the coordinate system of the observer – which may, or may not, be observable). An unambiguous description of the state of affairs it now refers to will be given by some such expression as '$P_i(a, p)$'.

It does not follow that the *use* of '$P_i(a)$' as defined in the characteristic of L will have to be discontinued. As stated above, this characteristic restricts the application of '$P_i(a)$' to the everyday level. On the everyday level the dependence of colour upon velocity remains unnoticed. Hence, no difficulty will arise if we go on using '$P_i(a)$' as we used it before the Doppler effect was discovered. Of course, we must not conclude that therefore '$P_i(a)$' still possesses its former *interpretation*: an expression does not cease to designate a relation even if for all situations inside the domain of its applicability this relation depends upon one term only and remains completely insensitive towards all changes of the remaining terms.[20]

At this stage is seems appropriate to make a few remarks about the role of everyday language in scientific practice. It has been frequently asserted that the language in which we describe our surroundings, chairs, tables and also the ultimate results of experiment (pointer-reading) is fairly insensitive

the following one: the magnitudes (properties) of classical physics can be determined at any time with any required precision. On the other hand, quantum-mechanical entities are complementary in the sense that at a given time they are able to possess only some of their possible properties. Now classical mechanics is a special case of quantum mechanics which means that all the objects of the macroscopic level obey the laws of quantum mechanics. Hence, we must re-interpret the signs of classical physics as designating properties which apply to their objects (the objects of the macroscopic level) in almost all circumstances (whereas according to classical physics proper they apply strictly in all circumstances). This means that *having adopted quantum mechanics we must drop the classical interpretation of classical physics*. For a formal discussion of the same point cf. G. Temple, 'The Fundamental Paradox of Quantum Theory', *Nature*, 135 (1957), 957ff and the discussion following that note, as well as G. Ludwig, *Die Grundlagen der Quantenmechanik* (Berlin, 1954), 49.

[20] It is important to realize that after the advent of a new theory T the so-called 'everyday level' (1) will be defined by physical conditions which are stated in terms of T; and (2) will be characterized as the totality of observable phenomena which do not contain the new phenomena, predicted by T. It frequently happens that those physical conditions are compatible with the characteristic of the 'everyday language'. If they are not it will even be necessary to change the characteristic of that language.

Applying the consideration in the text to our second example, n.19, we may say that if quantum mechanics is correct, then we must interpret all physical magnitudes, classical magnitudes included, as elements of a ring of non-commuting entities. This means that even the familiar properties of objects, such as their position, their momentum, their colour, etc., must be interpreted as Hermitian entities not all of which commute. Now there is *no practical need* to reformulate the language by means of which we describe our experiments or to change its characteristic since the error, committed on the macroscopic level, by the identification of the Hermitian entities of quantum mechanics and the classical properties, can be shown to be negligible. But although the *smallness* of the error allows us to continue the *use* of the classical practices and the classical 'forms of perception' on the macroscopic level, the *existence* of the error forbids us to regard this is an indication of the persistence of the classical *interpretation* of those forms. Cf. also the discussion at the end of section 4.

towards changes in the theoretical 'superstructure'. It seems somewhat doubtful whether even this modest thesis can be defended; first, because a uniform 'everyday language' does not exist. The language used by the 'everyday man' (whoever that may be) is a mixture of languages, as it were, i.e. it is a means of communication which has received its interpretation from various and often incompatible and obsolete theories. Secondly, it is not correct that this mixture does not undergo important changes: terms which at some time were regarded as observational elements of 'everyday language' (such as the term 'devil') are no longer regarded as such. Other terms, such as 'potential', 'velocity', etc., have been included in the observational part of everyday language, and many terms have assumed a new use. The fact that the pragmatic properties of some parts of the everyday language have remained unchanged may well be due to the fact that the people using these particular sections are not interested in science and do not know its results; after all, theories as such cannot influence linguistic habits. What can influence those habits is the *adoption* of theories by certain people.

Yet it must be conceded that even the scientist who employs parts of everyday language for the purpose of giving an account of his experiments does not introduce a new use for familiar words such as 'pointer', 'red', 'moving', etc. whenever he changes his theories. Does it follow, as is asserted in the stability thesis, that he is always talking about the same things and that he always employs the same interpretation of his observation language? Our foregoing analysis shows that this need not be the case. At the same time it is explained how the interpretation of a language can change without any perceptible effect upon its characteristic. This amounts to a refutation of the principle of pragmatic meaning. It also becomes clear that the analysis of everyday language cannot provide us with an interpretation either.

On the basis of the foregoing discussion we may now tentatively put forward our *thesis I: the interpretation of an observation language is determined by the theories which we use to explain what we observe, and it changes as soon as those theories change*.

In the remainder of this section I shall indicate some of the consequences of this thesis. Its logical basis will be investigated in the next, and last section.

Consider first the following objection:[21] the idea that interpretations depend upon theories makes nonsense of crucial experiments. A crucial experiment is a case where we want to decide by observation which of two given theories has to be abandoned. Hence, the meaning of the observation

---

[21] This difficulty was pointed out to me by Professor H. Feigl. I should like to add at this point that discussions with Professor Feigl and the members of his *Centre* have greatly helped me to clarify my ideas.

sentence (which sentence is supposed to be the impartial judge between the two theories) must be independent of these theories. To this I reply that, in the same way in which the acceptance (or the rejection) of a particular sentence in an observational situation is a pragmatic event whose result is interpreted independently of and sometimes only after its occurrence, so the acceptance (or the rejection) of a theory on the basis of a crucial experiment is a pragmatic event (a psychological event) which is afterwards interpreted as a theoretical decision in terms of those theories which survived the test.

The consequences in whose light our thesis I should be judged, are (*inter alia*) the following:

(1) According to thesis I, we must distinguish between appearances (i.e. phenomena) and the things appearing (the things referred to by the observational sentences in a certain interpretation). This distinction is characteristic of realism.

(2) The distinction between observational terms and theoretical terms is a pragmatic (psychological) distinction which has nothing to do with the logical status of the two kinds of term. On the contrary, thesis I implies that the terms of a theory and the terms of an observational language used for the tests of that theory give rise to exactly the same logical (ontological) problems. *There is no special 'problem of theoretical entities'.* And the belief in the existence of such a problem is due to the adoption of either the principle of pragmatic meaning or of the principle of phenomenological meaning.[22]

(3) This has implications with respect to such problems as the mind–body problem: it may happen that two observation languages with different characteristics are united and jointly interpreted by one and the same theory. Maxwell's electrodynamics plays this role with respect to the phenomena of light and electricity. Any application of either the principle of pragmatic meaning or of the principle of phenomenological meaning will in such a case tend to regard the unification either as illegitimate, or as 'purely formal'. The mind–body problem owes its existence to exactly this situation: *phenomenologically* pains and warts are different entities – hence, no unification is possible. But as our discussion of the principle of phenomenological meaning should have made clear (cf. especially item b), the assertion that pains and bodily affairs are different entities cannot be based upon introspection *unless we use already a certain interpretation* which implies this assertion. The point of thesis I is then that there may exist other and more satisfactory interpretations in which the difference does not any more exist.

[22] This shows, by the way, that the existential problems of observable entities are identical with the existential problems of the so-called 'theoretical terms'. An example is the problem of the existence of the devil which was decided *on the plane of theories* and not on the plane of observational practices. Cf. for this last point Lecky's admirable *History of the Rise of Rationalism in Europe* (New York, 1872), i, 9 and *passim*. Logically speaking all terms are 'theoretical'.

A very instructive model of such an interpretation has been discussed by J. O. Wisdom[23]

(4) But if the interpretation of a statement (such as of 'I am in pain now') depends upon the theories used (in this case upon psycho-physiological theories) then we cannot determine the logical complexity of the statement independently of those theories, even if the statement should belong to an observational language; more especially, we cannot stipulate that 'I am in pain now' should have no consequence apart from the phenomenon which leads to its production – unless psychology leaves room for such a stipulation.

(5) A theory (such as electrodynamics) may be understood even by a blind person. The only difference between a blind person and a seeing person consists in the fact that the first one uses a different part of the theory (or of the consequences of the theory) as his observation language. Hence, even a blind person may understand 'red' and similar terms (of his *theoretical* language) and there is no reason why he should not be able to explain 'red' to a seeing person 'by ostension'. This being so we cannot assume that when ceasing to be blind he automatically improves his knowledge of redness. It is to be admitted that he will now possess a new (and very effective) method of deciding (in the pragmatic sense of section 2) whether or not a given object is red. But just as the invention of a new microscope will change our notion of certain microscopic organisms only if it leads to new theories about them, in the same way the fact that our observer is now able to see red will lead him to a new notion of redness only if it leads him to new theories about red – and this need not be the case.

7. THE LOGICAL BASIS OF THE ARGUMENTS IN SECTION 6

The arguments against the stability thesis which were developed in the last section do not yet go to the root of the matter. They consist in the assertion that *as a matter of fact* scientists re-interpret their observation language L as soon as a new theory is devised which has consequences within L. This assertion is neither true nor sufficient to establish that the stability thesis is incorrect. That it is not true may be seen from the example which we discussed at the end of section 4. But it is not sufficient either, as we would like to attack positivism even if it were generally accepted. This means that scientific method, as actually practised, cannot show that positivism is false. What we are referring to when discussing the issue between positivism and realism (as exemplified in our discussion of the stability thesis) are certain *ideals* concerning the form of our knowledge. In short: *the issue between positivism and realism is not a factual issue which can be decided by pointing to certain*

[23] 'A New Model for the Mind–Body Relationship', *Br. J. Phil. Sci.*, 2 (1952), 295ff.

actually existing things, procedures, forms of language, etc., it is an issue between different ideals of knowledge.[24]

There seem to exist two objections against this characterization of the situation. The first objection is that it makes the resolution of the issue an arbitrary one. The second objection is that different ideals of knowledge cannot be realized equally easily. Taking the second objection first, we are prepared to admit that there may be *psychological* difficulties in inventing theories of a certain type, especially if metaphysical views are held which seem to recommend radically different theories. However, a stronger variant of the second objection is frequently used. According to this stronger variant, all our theoretical knowledge is (uniquely) determined by the facts and cannot be chosen at will. Against this objection we repeat (cf. section 5) that what is determined by the 'facts' is the acceptance (or rejection) of sentences *which are already interpreted* and which have been interpreted independently of the phenomenological character of what is observed. The impression that every fact suggests one and only one inter-pretation and that therefore our views are 'determined' by the facts, this impression will arise only when (with respect to the language used) the relation of phenomenological adequacy is a one–one relation. As has been pointed out before (see section 5 (c)), such a situation arises whenever a fairly general point of view was held long enough to influence our expectations, our language and thereby our perceptions, and when during that period no alternative picture was ever seriously considered. We may prolong such a situation either by explaining away adverse facts with the help of *ad hoc* hypotheses which are framed in terms of the points of view to be conserved; or by reducing more successful alternatives to 'instruments of prediction' which, being devoid of descriptive meaning, cannot clash (in the phenomenological sense) with any experience (cf. n.7); or by devising a criterion of significance according to which such alternatives are meaning-less. The important thing is that *such a procedure can always be carried out* (although it may at times require some ingenuity to devise *ad hoc* hypotheses which, while explaining away some distressing facts, do not at the same time contradict a different part of the theory to be preserved). This means that we can always arrange matters in such a way that either the principle of phenomenological meaning or the principle of pragmatic meaning will seem to be correct and that the stability thesis correctly describes the relation of our knowledge to experience. But we can also choose the opposite procedure, i.e. we can take refutations seriously and regard alternative theories, in spite of their unusual character, as descriptive of really existing things, properties, relations etc. In short, *although the truth of a*

---

[24] The normative character of epistemology has been stressed by V. Kraft in his paper 'Der Wissenschaftscharakter der Erkenntnislehre', *Actes du congrés de l'union internationale de philo-sophie des sciences* (Zürich, 1954), 85ff.

theory may not depend upon us, its form (and the form of our theoretical knowledge in general) *can always be arranged so as to satisfy certain demands.* With this the second objection collapses.

But does the fact that the form of our knowledge can be adjusted to meet our demands, make the rejection of positivism an arbitrary act (first objection)? It does not, as we judge an ideal by the consequences which its realization may or may not imply. In what follows I want to discuss the consequences of the positivistic procedure and show why I think that realism is to be preferred.

The positivistic procedure has just been explained. We have also explained some of its consequences. The first consequence is that the stability thesis will give a correct account of the interpretation of the observational language. In section 3 we have pointed out that this leads to a metaphysical ontology.[25] Now we are able to identify the basis of this ontology: it is a theory or a general point of view which has been conserved because it appears to be phenomenologically adequate. The price we have to pay if we proceed in this way is that the chosen theory will finally be completely void of empirical content. The second consequence is that owing to the one to one nature of the relation of phenomenological adequacy, a distinction between thought and imagination on the one hand and sensation on the other cannot readily be drawn. We may even say that in this respect positivism leads to a restriction on the argumentative use of our language, and perhaps to its complete elimination. This means that positivistic knowledge is connected with a more primitive and naturalistic stage of human development than its alternative. A third consequence is this: as the example of section 6 shows, it may turn out that some of the elements chosen (like redness) exist only under certain conditions which involve a relation to the (physical situation of the) observer. If the theory expressing this relation is regarded as a mere means of prediction, then we cannot explain the conditioned existence of the elements by saying that what we thought to be a property was in fact a relation: for we are not allowed to describe objective existence to this relation. We are forced to say (as is admitted by all positivists from Berkeley to Ayer) that our elements turned out to be subjective: positivism sooner or later leads to subjectivism.

As opposed to positivism, a realistic position does not admit any dogmatic and incorrigible statement into the field of knowledge. Hence, also, our knowledge of what is observed is not regarded as unalterable and this in

---

[25] If the belief in the devil had been combined with a strong belief in the stability thesis it would never have been possible to substitute for it a new and more reasonable account of the phenomena which constituted its observational core. Luckily enough, mankind was not prevented by positivistic prejudices from abandoning this belief. It is different with some more recent views about the constitution of matter which by virtue of an implicit belief in the principle of pragmatic meaning seem to be regarded as the *conditio sine qua non* of physical understanding. Cf. also the end of section 4.

spite of the fact that it may have a counterpart in the phenomena themselves. This means that at times interpretations will have to be considered which do not 'fit' the phenomena and which clash with what is immediately given. Interpretations of this kind could not possibly emerge from close attention to the 'facts'. It follows that we need a non-observational source for interpretations. Such a source is provided by (metaphysical) speculation which is thus shown to play an important role within realism. However, the results of such speculation must be made testable, and having been transformed in this way they must be interpreted as descriptive of general features of the world (otherwise we are thrown back upon the old account of what is observed). This procedure (a) allows us to draw a clear boundary line between objective states of affair and the states of the observer, though it admits that we may be mistaken with regard to the exact position of the boundary line; it (b) is empirical in the sense that no dogmatic statements are allowed to become elements of knowledge; it (c) is liable to encourage progress by admonishing us to adapt even our sensations to new ideas; and it (d) allows for the *universal* application of the argumentative function of our language, and not only for its application *within* a given frame which itself can only be either described, or expressed (e.g. in our 'forms of perception').

These are some of the consequences of both positivism and realism with respect to the nature of our experiences. The presentation of these consequences does not yet amount to a definite decision in favour of either positivism or realism. After all, such a decision is a practical act which cannot *follow* from any theoretical consideration, although it may be *motivated* by theoretical considerations. What I intended to do in this paper was to provide some motives. And I also wanted to show that the apparently innocuous statements which I quoted at the beginning of this essay lead to consequences which may prove to be an irritation to at least some positivists.

3
On the interpretation of scientific theories

1. According to positivism the interpretation of scientific theories is a function of either experience, or of some observational language. There are various views about the nature of this function and we may therefore distinguish a great many varieties of positivism: (a) theoretical terms are explicitly definable on the basis of observational terms; (b) theoretical terms are extensionally reducible to observational terms; (c) theoretical terms are intensionally reducible to observational terms; (d) theoretical terms are implicitly definable with the help of interpretative systems which either may, or may not, contain probability statements; and so on. In the present paper I shall discuss two objections which may be raised against all these varieties. Indeed, it seems to me that the difficulties of positivism as revealed by these objections cannot be overcome by inventing a new and ingenious connection between theoretical terms and observational terms, but only by altogether dropping the idea that the meaning of theoretical terms depends upon such a connection.

2. My first objection against positivism is that it implies that statements describing causally independent situations may yet be semantically dependent. In order to understand this objection consider the attempt to explicate statements about material objects in terms of what is seen, heard, felt, etc. by observers of a certain kind. It is well known that what is seen, heard, or felt by observers depends upon the object as well as upon the physiological status of the observers themselves. This status may be changed by influences (e.g. drugs, or hypnosis) which act independently of the influences of the observed objects. Any attempt to explain the properties of material objects on the basis of experience will have to take such additional influences into account. An explication will therefore be of the form

$$F (M, S, O) \tag{1}$$

where F is a complicated logical constant, not necessarily extensional, M a (general or particular) situation pertaining to material objects, O a (general or particular) observable situation, and S a situation which we shall call the mediating situation and which is identical with the conditions of observation. In a special case these conditions may include a reference to the intensity of the light irradiating the object, to the absence of obstacles between the object and the observer at the time of observation, to the properties of the retina and of the brain of the observer, and to many other situations which, although causally independent of the observed object (turning out the light does not influence the material object, although it makes it invisible), do yet contribute to its observable effect. Those terms of '*S*' which are neither observational terms, nor terms of '*M*', will be called the *mediating terms*. In our above example the mediating terms are 'light', 'intervening obstacle', 'retina', 'brain', and others. Now if we adopt the principle, characteristic of the positivistic approach, that the descriptive terms of '*M*' 'obtain only an indirect, and incomplete interpretation by the fact that some of them are connected . . . with observational terms',[1] then we shall have to assume that their interpretation is implicitly defined by statement (1), and therefore dependent upon all the terms of '*S*', although we know at the same time that the situation *M* is causally dependent only upon part of the situation *S,* and possibly altogether independent of it.

As a second example, consider the attempt to explain the theoretical terms of celestial mechanics on the basis of observational terms referring to bright dots as seen either through a telescope, or on a photographic plate. In this case the mediating situation consists in the optical properties of the planets, the properties of the light which is reflected by them, the properties of the atmosphere of the earth, the properties of telescopes, and so on. Again, the interpretation of sentences containing the terms to be explicated will depend upon the interpretation of other sentences referring to states of affairs which are in no causal relation whatever to the states of affairs referred to by the former. For example, the interpretation (the 'meaning') of 'mass of the sun' will partly depend upon the interpretation of 'refractive index of the atmosphere of the earth'.

We shall now discuss three methods which one might adopt in order to overcome this difficulty. The first method consists in denying that situations described with the help of theoretical terms only either exist, or can be regarded as causes, or elements of other situations not so describable. This move is hardly appropriate for a philosopher who has set himself the task of analysing physics, as it completely disregards the existential character of general scientific theories.[2] The second method consists in the attempt to eliminate the mediating terms from the rules of correspondence which connect theoretical terms and observational terms. It is obvious that simple omission of the mediating terms will not do. For a necessary criterion of the adequacy of any logical reconstruction of science is that is translates true sentences into true sentences, and it can easily be shown that the statements

[1] R. Carnap, 'The Methodological Character of Theoretical Concepts', in *Minnesota Studies in the Philosophy of Science* (Minneapolis, 1956), i, 47.
[2] H. Feigl, 'Existential Hypotheses', *Phil. Sci.* 17 (1950), 35ff.

resulting from the original rules of correspondence by the omission of the mediating terms will in general be empirically false ('table in x at time $t =$ anybody inspecting region x at time t perceives a table' is empirically false; in a dark room no table will be seen by anybody). But the attempt to give an observational account of the mediating terms cannot succeed either, as any such account involves further mediating terms and will therefore never come to an end. The third method, which has not yet led to any concrete suggestion, would have to consist in devising semantical rules which make the interpretation of the theoretical terms dependent upon the interpretation of the observational terms only, and this in spite of the fact that the rules of correspondence contain the mediating terms as well. It is difficult to see how such a procedure would account for differences in the interpretation of theoretical terms without dropping the principle that they do not possess an independent meaning: we may safely assume that 'Ax' means something different in 'If a colourblind man inspects x and sees grey, then Ax', and in 'If a man with perfect coloursight inspects x and sees grey, then Ax', and yet the method discussed at the present moment does not allow us to explain this difference by pointing to a difference in the observational terms employed. It is also difficult to see how a change of the logical constant F in formula (1) could do the trick. For although such a change may influence the kind of dependence existing between S and M, it will not, and cannot (see our discussion of the third method, immediately above) eliminate this influence.

3. My second objection is closely related to the first. It may be stated by saying that given two situations, S' and S'', which are causally independent of each other, a change of our theories about S' which preserves this causal independence may yet imply a change of the interpretation of S''. I shall explain this second objection with the help of an example consisting of a formalism T expressing the state of affairs T', which is interpreted by an interpretative system J:

   T is a formalization of celestial mechanics T'. The descriptive terms of that formalism are functors, such as 'mass', 'force', 'acceleration', 'heliocentric coordinate', etc., whose variables range over particles of matter. I shall assume that the observation terms are again functors, such as 'declination', 'high ascension' whose variables range over luminescent points in the sky. The interpretative system used will be fairly complicated as it must take into account refraction, aberration, etc. Now the behaviour of the planets and, more especially, the properties of the force acting between them may be safely said to be causally independent of the thermal and optical properties of the atmosphere of the earth. Yet, if the 'meaning' of the theoretical terms is to depend upon their connection with observation and upon nothing else (see the quotation from Carnap above), then any change of our assumptions about these latter properties will necessitate a change of the interpretative system used, and thereby affect the interpretation of the theoretical terms of T.

To sum up: the first objection is that according to positivism statements describing causally independent situations will nevertheless be semantically dependent; the second objection is that a change in our knowledge concerning a situation S^1 which is causally independent of another situation S^2 will yet imply a change in the interpretation of the terms of 'S^2'. Both objections are based upon the principle (which I shall call the *principle of semantic independence*) that the interpretation of a statement describing a situation, which is causally independent of the situation described by another statement should be independent of the interpretation of this latter statement.

4. The above objections cannot be removed by pointing out that the logical constant F in formula (1) can be chosen in such a way that a complete interpretation of theoretical terms is never obtained. For our quarrel is not with the fact that, given a certain interpretative system, the positivistic method of interpretation does not leave room for a further specification of meaning; it is with the fact that any specification, however incomplete and 'open', which is based upon formula (1) contradicts the principle of semantic independence and must therefore be regarded as inadequate.

Another attempt to escape our two objections consists in advocating probabilistic rules of correspondence. This move, which has been suggested by A. Pap, seems to be inspired by the realization that a more 'liberal' account of the connection between the meaning of theoretical terms and the meaning of observational terms is required. Pap seems to agree with my criticism as far as non-probabilistic Fs are concerned.[3] He seems to assume, however, that a probabilistic F will solve the difficulties. My criticism of Pap's proposal will proceed in two steps. I consider first his assertion that different probabilistic interpretative systems need not contradict each other (whereas different non-probabilistic interpretative systems for the same theory will frequently contradict each other). Now this is a point which I would hardly contest and I do not see how it can be regarded as a solution to the difficulties I have pointed out. For I have not attacked the positivistic theory of meaning specification on the grounds that two inconsistent interpretative systems lead to different meanings of the theoretical terms, but rather on the grounds that any two different interpretative systems lead to different meanings of the theoretical terms, and this in spite of the fact that the difference may be one of situations which are causally independent of whatever theoretical situation may obtain. Hence, it is sufficient for me that

[3] Private communication.

$P(T/A) = p$ and $P(T/A\&B) = p'$ are different, the second implying that T depends upon B, the first not implying any such assertion.

This last example leads at once to our decisive second attack against probabilistic meaning specification. For assume T to be a state of affairs described in theoretical terms only, O an observational situation, and M the mediating situation. Now if, as is frequently the case, T and M are causally independent, then $P(T/M\&O) = P(T/O)$, i.e. mediating situations, although necessary as conditions of observation (cf. our objections against the second method discussed in section 2) can be eliminated from any probabilistic statement about the relation of theoretical terms to observational terms. This quite obviously shows that a probabilistic F is not only inadequate as a meaning rule, but also as a statement expressing which tests are relevant and which are not. This finishes the probabilistic rescue manoeuvre.

5. It is instructive to trace the origin of the difficulty of positivism as expressed in our two objections. According to physical theory any observable state of affairs, i.e. any state of affairs which is big enough to be accessible to inspection by human beings, is the result of a superposition of many influences. According to the positivistic account of these theories every single influence contributing to the observational state of affairs is to be described with the help of theoretical terms. Briefly and crudely we may say that according to physics any observational situation is the result of a superimposition of many theoretical entities which are partly independent of each other, and it is therefore a complicated and intricate affair. Positivism turns the situation upside down.[4] For a positivist the observational situations are the primitive and unanalysable elements in terms of which the theories must be understood. Now if we realize that the theoretical entities of a given theory make only a very small contribution to any observational situation, then the attempt to explain them exclusively on the basis of observation will at once be recognized as absurd. It is absurd because it regards as simple what is complicated, and because it attempts to explain a state of affairs (the state of some theoretical entity) in terms of other states of affairs to which it makes an occasionally very insignificant contribution. It may be pointed out that for Aristotelian physics this absurdity does not exist. For the Aristotelians were much more inclined to take observable states of affairs at their face value. But it is impossible to believe at the same time in physics and in the above account of the interpretation of its terms.

6. Having criticized the positivistic theory of the interpretation of general

---

[4] This has been pointed out by Feigl in his 'Existential Hypotheses'. His point of view is in many respects similar to the one defended here, although it may not be quite as radical.

empirical terms, we now turn to positive suggestions. The result of our criticism was the interpretation of scientific terms must be independent of their occurrence in statements of the form (1), or what amounts to the same, it must be independent of the connection with experience. As a theory is testable only insofar as it is connected with experience via these very statements, it follows that if a theory is to be meaningful at all, its interpretation must go beyond whatever counts as its 'empirical content': *the interpretation of any physical theory contains metaphysical elements*, the term 'metaphysical' here being used as synonymous with 'non-empirical'. If this statement seems surprising it is partly because of the new and technical sense which the word 'metaphysical' has received in the hands of the positivists. In the next section we shall show that this result which we have here derived on the basis of the principle of semantic independence, has a natural place within realism.

7. Realism asserts that there exist states of affairs which are causally independent of the states of observers, measuring instruments and the like, but which may influence these instruments and these observers. It also admits that whatever influence a real state of affairs exerts upon an observer, it will not be the only influence, but will have to interact with many other influences, some of them known, some of them unknown. According to the realistic interpretation, a scientific theory aims at a description of states of affairs, or properties of physical systems, which transcends experience not only insofar as it is general (whereas any description of experience can only be singular), but also insofar as it *disregards all the independent causes which, apart from the situations described by the theory, may influence the observer or his measuring instrument*. For example, Newtonian astronomy describes the structure of the planetary system – the mutual interaction of the planets and their behaviour – without taking into account all the disturbances experienced by the light which, having left the sun, having been reflected and diffracted by the atmosphere of the earth as well as in the lens of some telescope, reports this structure only in a more or less distorted way. Of course, any attempt to test Newtonian astronomy will have to take these distortions into account; for a test of a situation consists in connecting a cause C provided by it with an effect to which other causes have also contributed; and it presupposes that all these causes are known as well. But this must of course not be taken to imply that the 'meaning' of the statement that C obtains, depends upon the meaning of statements describing those other causes. *The interpretation of a scientific theory depends upon nothing but the state of affairs it describes*.[5] This is an immediate consequence of the principle of semantic independence.

This result has consequences with respect to the interpretation of

[5] For this cf. also Feigl, 'Existential Hypotheses'.

metaphysics. A metaphysical theory does not contain any indication as to how we are to test it, a scientific theory contains some such indications without thereby making its whole content accessible to test. On the basis of what we said above it is no longer possible to distinguish 'metaphysical meaning' (or 'nonsense') and 'scientific meaning' by referring to testability (although testability may of course be regarded as a useful criterion for separating scientific theories from metaphysical theories). For it may happen that a scientific theory and a metaphysical theory describe exactly the same state of affairs, the one in a testable way, the other in a way which is inaccessible to test (in a similar way to that in which a sentence which is decidable in one theory, and undecidable in another may yet in both cases refer to the same state of affairs) and that they therefore possess identically the same meaning. But the discussion of this possibility would already transcend the scope of the present paper which was to criticize positivism.

4

Explanation, reduction and empiricism

The main contention of the present paper is that a formal account of reduction and explanation is impossible for general theories, or non-instantial theories,[1] as they have also been called. More especially, it will be asserted and shown that wherever such theories play a decisive role both Nagel's theory of reduction[2] and the theory of explanation associated with Hempel and Oppenheim[3] cease to be in accordance with actual scientific practice and with a reasonable empiricism. It is to be admitted that these two 'orthodox' accounts fairly adequately represent the relation between sentences of the 'All-ravens-are-black' type, which abound in the more pedestrian parts of the scientific enterprise.[4] But if the attempt is made to extend these accounts to such comprehensive structures of thought as the Aristotelian theory of motion, the impetus theory, Newton's celestial mechanics, Maxwell's electrodynamics, the theory of relativity, and the quantum theory, then complete failure is the result. What happens here when a transition is made from a theory T' to a wider theory T (which, we shall assume, is capable of covering all the phenomena that have been covered by T') is something much more radical than incorporation of the *unchanged* theory T' (unchanged, that is, with respect to the meanings of its main descriptive terms as well as to the meanings of the terms of its observation language) into the context of T. What does happen is, rather, a

[1] In what follows, the usual distinction will be drawn between empirical generalizations, on the one side, and theories, on the other. Empirical generalizations are statements, such as "All A's are B's' (the A's and B's are not necessarily observational entities), which are tested by inspection of instances (the A's). Universal theories, such as Newton's theory of gravitation, are not tested in this manner. Roughly speaking their test consists of two steps: (1) derivation, with the help of suitable boundary conditions, of empirical generalizations and (2) tests, in the manner indicated above, of these generalizations. One should not be misled by the fact that universal theories too can be (and usually are) put in the form 'All As are Bs'; for whereas in the case of generalizations this form reflects the test procedure in a very direct way, such an immediate relation between the form and the test procedures does not obtain in the case of theories. Many thinkers have been seduced by the similarity of form into thinking that the test procedures will be the same in both cases.

[2] Nagel has explained his theory in [60]. I shall quote from the reprint of the article in [20], 288ff.

[3] For the theory of Hempel and Oppenheim see [47]. I shall quote from the reprint in [23], 319ff.

[4] For important exceptions, see n.72.

replacement of the ontology (and perhaps even of the formalism) of T' by the ontology (and the formalism) of T, and a corresponding change of the meanings of the descriptive elements of the formalism of T' (provided these elements and this formalism are still used). This replacement affects not only the theoretical terms of T' but also at least some of the observational terms which occurred in its test statements. That is, not only will description of things and processes in the domain in which T' had been applied be infiltrated, either with the formalism and the terms of T, or if the terms of T' are still in use, with the meanings of the terms of T, but the sentences expressing what is accessible to direct observation inside this domain will now mean something different. In short, introducing a new theory involves changes of outlook both with respect to the observable and with respect to the unobservable features of the world, and corresponding changes in the meanings of even the most 'fundamental' terms of the language employed. This is the position which will be defended here.

The position may be said to consist of two ideas. The first idea is that the influence, upon our thinking, of a comprehensive scientific theory, or of some other general point of view, goes much deeper than is admitted by those who would regard it as a convenient scheme for the ordering of facts only. According to this first idea scientific theories are ways of looking at the world and their adoption affects our general beliefs and expectations, and thereby also our experiences and our conception of reality. We may even say that what is regarded as 'nature' at a particular time is our own product in the sense that all the features ascribed to it have first been invented by us and then used for bringing order into our surroundings. As is well known, it was Kant who most forcefully stated and investigated this all-pervasive character of theoretical assumptions. However, Kant also thought that the very generality of such assumptions and their omnipresence would forever prevent them from being refuted. As opposed to this, the second idea implicit in the position to be defended here demands that our theories be testable and that they be abandoned as soon as a test does not produce the predicted result. It is this second idea which makes science proceed to better and better theories and which creates the changes described in the introductory paragraphs of this essay.

Now, it is easily seen that the mere statement of the second idea will not do. What we need is a guarantee that despite the all-pervasive character of a scientific theory as it is asserted in the first idea, it is still possible to specify facts that are inconsistent with it. Such a possibility has been denied by some philosophers. These philosophers started out by reacting against the claim that scientific theories are nothing but predictive devices; they recognized that their influence goes much deeper; however, they then doubted that it would be possible ever to get outside any such theory and therefore, either they became apriorists (Poincaré, Eddington), or they returned to instrumentalism. For these thinkers there seemed to exist only a choice between two evils – instrumentalism and apriorism.

A closer look at the arguments leading up to this dilemma shows that they all proceed from a test model in which a single theory is confronted with the facts. As soon as this model is replaced by a model in which we make use of at least two factually adequate but mutually inconsistent theories, the first idea becomes compatible with the demand for testability which must now be interpreted as a demand for crucial tests, either between two explicitly formulated theories, or between a theory and our background knowledge. In this form, however, the test model turns out to be inconsistent with the orthodox theory of explanation and reduction. It is one of my aims here to exhibit this inconsistency.

It will be necessary, for this purpose, to discuss two principles which underlie the orthodox approach: (A) the principle of deducibility, and (B) the principle of meaning invariance. According to the principle of deducibility, explanation is achieved by deduction in the strict logical sense. This principle leads to the demand, which is incompatible with the test model just outlined, that all successful theories in a given domain must be mutually consistent. According to the principle of meaning invariance, an explanation must not change the meanings of the main descriptive terms of the explanandum. This principle, too, will be found to be inconsistent with empiricism.

It is interesting to note that (A) and (B) play a role within both modern empiricism and some very influential school philosophies. Thus it is one of the basic assumptions of Platonism that the key terms of sentences expressing knowledge (epistēmē) refer to unchangeable entities and must therefore possess a stable meaning. Similarly, the key terms of Cartesian physics – i.e. 'matter', 'space', 'motion' – and the terms of Cartesian metaphysics – such as 'god' and 'mind' – are supposed to remain unchanged in any explanation involving them. Compared with these similarities, between the school philosophies on the one hand and modern empiricism on the other, the differences are of very minor importance.[5] These differences lie in the terms of which stability of meaning is required. A Platonist will direct his attention to numbers and other ideas, and he will demand that words referring to these entities retain their (Platonic) meanings. Modern empiricism, on the other hand, regards empirical terms as fundamental and demands that their meanings remain unchanged.

It will turn out in the course of this essay that any form of meaning invariance is bound to lead to difficulties when the task arises either of giving a proper account of the growth of knowledge, and of discoveries contributing to this growth, or of establishing correlations between entities

[5] Concerning these similarities, see Popper's discussion of essentialism in [66], ch. 3 and *passim*, as well as Dewey's very different account in [21], especially ch. 2.

which are described with the help of what we shall later call incommensurable concepts. It will also turn out that these are exactly the difficulties we encounter in trying to solve such age-old problems as the mind–body problem, the problem of the reality of the external world, and the problem of other minds. That is, it will usually turn out that a solution of these problems is deemed satisfactory only if it leaves unchanged the meanings of certain key terms and that it is exactly this condition, the condition of meaning invariance, which makes them insoluble. It will also be shown that the demand for meaning invariance is incompatible with empiricism. Taking all this into account, we may hope that once contemporary empiricism has been freed from the elements which it still shares with its more dogmatic opponents, it will be able to make swift progress in the solution of the above problems. It is the purpose of the present paper to develop and to defend the outlines of such a disinfected empiricism.[6]

1. TWO ASSUMPTIONS OF CONTEMPORARY EMPIRICISM

Nagel's theory of reduction is based upon two assumptions. The first assumption concerns the relation between the secondary science, the discipline to be reduced, on the one side, and the primary science, the

---

[6] As will be shown in section 2, the empiricism of the thirties was disinfected in the sense desired here. However, later on modern empiricism re-adopted some very undesirable principles of traditional philosophy. *Added 1980.* This paper owes much to discussions with D. Bohm, H. Feigl, S. Körner, T. Kuhn, G. Maxwell, H. Putnam, E. Tranekjaer-Rasmussen. Popperians, who would gladly assert that they invented the multiplication table, if they could get away with it, have described it as a rehash of Popper's own ideas. Thus in his *Objective Knowledge* (Oxford, 1972), 205, Karl Popper sadly notes that after 1962 people started referring to me instead of him when discussing matters of explanation and theory comparison while Imre Lakatos proclaimed, with his usual propagandistic flair, that Popper's theses 'were given wider circulation' by my paper which, he added, was 'not as good (and definitely not as clear) as Popper's and Duhem's original exposition' (*Philosophical Papers* (Cambridge, 1978), I, 109, n.4 and text). Now the ideas which Popper and Lakatos claim for the Popperian school (Popper: 'the same idea of *mine* . . .' *Objective Knowledge*, my italics) are neither Popper's own (as Lakatos soon realized) nor are they the ideas I had in mind when writing my paper. Popper correctly remarks that his *Aim of Science* was a 'starting point' of my paper: he overlooks that it was a starting point of criticism, not of repetition: Popper, repeating Duhem, emphasizes that high level theories often conflict with established laws and thus *inspire* crucial experiments. I, on the other hand, was interested in cases where alternatives not only inspire such experiments *but are necessary for producing the corresponding evidence* (cf. the Brownian motion example in section 6 and the more detailed discussion of the case in ch. 8, and ch. 3 of *A M*; the Brownian motion example was suggested to me by David Bohm). Lakatos (*Philosophical Papers,* I, 109, n.5) is aware of the purely psychological (or, as he calls it, 'catalytic') function of alternatives in Popper's methodology, regards it as insufficient, includes me in his criticism and ascribes to himself the view I actually hold in the present paper (and explain in greater detail in ch. 6.1) viz. that 'alternatives are *necessary* parts of the falsifying process'. He also overlooks that the paper introduces incommensurability and ends up by arguing that explanations inevitably contain 'subjective' elements. Small wonder that a paper with such unpopperian tendencies should look 'not so clear' to an author who reads it as a repetition of Popper's repetition of Duhem. Or do Popperians claim to have invented subjectivism and incommensurability as well?

discipline to which reduction is made, on the other. It is asserted that this relation is the relation of deducibility. Or, to quote Nagel:

> The objective of the reduction is to show that the laws, or the general principles of the secondary science, are simply logical consequences of the assumptions of the primary science.[7]   (1)

The second assumption concerns the relation between the meanings of the primitive descriptive terms of the secondary science and the meanings of the primitive descriptive terms of the primary science. It is asserted that the former will not be affected by the process of reduction. Of course, this second assumption is an immediate consequence of (1), since a derivation is not supposed to influence the meanings of the statements derived. However, for reasons which will become clear later, it is advisable to formulate this invariance of meaning as a separate principle. This is also done by Nagel, who says: 'It is of the utmost importance to observe that the expressions peculiar to a science will possess meanings that are fixed by its own procedures, and are therefore intelligible in terms of its own rules of usage, *whether or not the science has been, or will be, reduced to some other disciplines'*.[8] Or, to express it in a more concise manner:

> Meanings are invariant with respect to the process of reduction.   (2)

(1) and (2) admit of two different interpretations, just as does any theory of reduction and explanation: such a theory may be regarded either as a description of actual scientific practice, or as a prescription which must be followed if the scientific character of the whole enterprise is to be guaranteed. Similarly, (1) and (2) may be interpreted as assertions concerning actual scientific practice, or as demands to be satisfied by the theoretician who wants to follow the scientific method. Both of these interpretations will be scrutinized below.

Two very similar assumptions, or demands, play a decisive role in the orthodox theory of explanation, which may be regarded as an elaboration of suggestions that were first made, in a less definite form, by Popper.[9] The first assumption (demand) concerns again the relation between the explanandum, or the facts to be explained, on the one side, and the explanans, the discipline which functions as the basis of explanation, on the other. It is again asserted (required) that this relation is (be) the relation of deducibility. Or, to quote Hempel and Oppenheim:

> The explanandum must be a logical consequence of the explanans; in other words, the explanandum must be logically deducible from the information contained in the explanans, for otherwise the explanans would not constitute adequate grounds for the explanation.[10]   (3)

[7] [20], 301. A more elaborate form of this condition is called the 'condition of derivability' on [61], 354.

[8] [20], 301. My italics. See also [61], 345, 352.   [9] [69], section 12.   [10] [47], 321.

Considering what has been said in the case of reduction one would expect the assumption (demand) concerning meanings to read as follows:

Meanings are invariant with respect to the process of explanation. (4)

However, despite the fact that (4) is a trivial consequence of (3), this assumption has never been expressed in as clear and explicit a way as (2).[11] There was even a time when a consequence of (4), the assertion that *observational* meanings are invariant with respect to the process of explanation, seemed to be in doubt. It is for this reason that I have separated (2) from (1), and (4) from (3).

It is not difficult to show that, with respect to observational terms, (4) is consistent with the earlier positivism of the Vienna Circle. Their main thesis, that all descriptive terms of a scientific theory can be explicitly defined on the basis of observation terms, guarantees the stability of the meanings of observational terms (unless one assumes that an explicit definition changes the meaning of the definiens, a possibility that to my knowledge has never been considered by empiricists). And as the chain of definitions leaves unchanged those terms which are already defined, (4) turns out to be correct as well.

However, since these happy and carefree days of the *Aufbau*, logical empiricism has been greatly modified. The changes that took place were mainly of two kinds. On the one side, new ideas were introduced concerning the relation between observational terms and theoretical terms. On the other side, the assumptions made about the observational language itself were modified. In both cases the changes were quite drastic, but for our present purpose a brief outline must suffice. The early positivists assumed that observational terms refer to subjective impressions, sensations and perceptions of some sentient being. Physicalism for some time retained the idea that a scientific theory should be based upon experiences, and that the ultimate constituents were sensations, impressions and perceptions. Later, however, a behaviouristic account was given of these perceptions to make them accessible to intersubjective testing. Such a theory was held, for some time, by Carnap and Neurath.[12] Soon afterwards the idea that it is experiences to which we must refer when trying to interpret our observation

---

[11] An exception is Nagel who, in [61], 338, defines reduction as 'the explanation of theory or a set of experimental laws established in one area of inquiry by a theory usually, though not invariably, formulated for some other domain'. This implies that the condition of meaning invariance formulated by him for the process of reduction is supposed to be valid in the case of explanation also. On pp. 86–7, meaning invariance for observational terms is stated quite explicitly: an experimental law 'retains a meaning that can be formulated independently of [any] theory . . . [It] has . . . a life of its own, not contingent on the continued life of any particular theory that may explain the law'.

[12] For this and the following, see Carnap's account in [13], especially the passages in small print on pp. 223–4.

statements was altogether abandoned.[13] According to Popper, who has been responsible for this decisive turn, we must *'distinguish sharply between objective science on the one hand, and "our knowledge" on the other'*. It is conceded that 'we can become aware of facts only by observation'; but it is denied that this implies an interpretation of observation sentences in terms of experiences, whether these experiences are explained subjectivistically or as features of objective behaviour.[14] For example, we may admit that the sentence 'This is a raven' uttered by an observer who points at a bird in front of him is an observational sentence and that the observer has produced it because of the impressions, sensations and perceptions he possesses. We may also admit that he would not have uttered the sentence had he not possessed these impressions. Yet the sentence is not therefore about impressions; it is about a bird which is neither a sensation nor the behaviour of some sentient being. Similarly, it may be admitted that the observation sentences which a scientific observer produces are prompted by his impressions. However, their content will again be determined not by these impressions, but by the entities allegedly described. Therefore, in the case of classical physics, 'every basic statement must either be itself a statement about relative positions of physical bodies . . . or it must be equivalent to some basic statement of this 'mechanistic' . . . kind'.[15]

The descriptive terms of Carnap's 'thing-language', too, no longer refer to experiences. They refer to properties of objects of medium size which are accessible to observation, i.e. which are such that a normal observer can quickly decide whether or not an object possesses such a property.[16] 'What we have called observable predicates', says Carnap, 'are predicates of the thing-language (they have to be clearly distinguished from what we have called perception terms . . . whether these are now interpreted subjectivistically, or behavioristically).'[17]

The characterization of observation statements implicit in the above quotations is a causal characterization, or if one wants to use more recent terminology, a pragmatic characterization:[18] an observation sentence is distinguished from other sentences of a theory, not, as was the case in earlier positivism, by its content, but by the cause of its production, or by the fact that its production conforms to certain behavioural patterns.[19] This being the case, the fact that a certain sentence belongs to the observation language does not allow us to make any inference concerning the kind of entities described in it.

[13] *Ibid.*, p. 223: 'It is stipulated that under given circumstances any concrete statement may be regarded as a protocol statement.'   [14] Popper [69], 98, original italics.
[15] Popper [69], 103. Popper himself does not restrict his characterization to the observation statements of classical physics.
[16] Carnap [14], 63, Explanation 1. Page references are to the reprint of this article in [22], 47ff.
[17] *Ibid.*, p. 69.   [18] For this terminology see Morris [59], 6ff.
[19] See again Explanation 1 of [14], as well as my elaboration of this explanation in [31].

It is worthwhile to dwell a little longer on the features of this *pragmatic theory of observation*, as I shall call it. In the case of measuring instruments, the pragmatic theory degenerates into a triviality: nobody would ever dream of asserting that the way in which we interpret the movements of, say, the hand of a voltmeter is uniquely determined either by the character of this movement itself or by the processes inside the instrument; a person who can see and understand only these processes will be unable to infer that what is indicated is voltage, and he will be equally unable to understand what voltage is. Taken by themselves the indications of instruments do not mean anything unless we possess a theory which teaches us what situations we are to expect in the world, and which guarantees that there exists a reliable correlation between the indications of the instrument and such a particular situation. If one theory is replaced by another with a different ontology, then we may have to revise the interpretation of all our measurements, however self-evident such a particular interpretation may have become in the course of time. According to the phlogiston theory, measurements of weight before and after combustion are measurements of the amount of phlogiston added or lost in the process. Today we must give a completely different interpretation of the results of these measurements. Again, Galileo's thermoscope was initially supposed to measure an intrinsic property of a heated body; however, with the discovery of the influence of atmospheric pressure, of the expansion of the substance of the thermoscope (which, of course, was known beforehand) and of other effects (non-ideal character of the thermoscopic fluid), it was recognized that the property measured by the instrument was a very complicated function of such an instrinsic property, of the atmospheric pressure, of the properties of the particular enclosure used, of its shape, and so on.[20] Indeed, the point of view outlined in the beginning of the present paper gives an excellent account of the way in which results of measurement, or indications of instruments, are reinterpreted in the light of fresh theoretical insight. Nobody would dream of using the insight given by a new theory for the readjustment of some general beliefs only, leaving untouched the interpretation of the results of measurement. And nobody would dream of demanding that the meanings of observation statements as obtained with the help of measuring instruments remain invariant with respect to the change and progress of knowledge. Yet precisely this is done when the measuring instrument is a human being, and the indication is the behaviour of this human being, or the sensations he has, at a particular time.

It is not easy to set down in a few lines the reasons for this exceptional treatment of human observers. Nor is it possible to criticize them thoroughly and thereby fully pave the way for the acceptance of the pragmatic theory

[20] For historical references, see [18], especially the articles on the phlogiston theory (J. B. Conant) and on the early development of the concept of temperature (D. Roller).

of observation. However, such a comprehensive criticism is not really necessary here. It was partly given by those very same philosophers who are responsible for the formulation of the pragmatic theory (which most of them dropped later on, their own excellent arguments in favour of it notwithstanding).[21] For example it was pointed out that the attempt to derive observational meanings from phenomena obliterates the distinction between (psychological and sociological) facts and (linguistic) conventions.[22] It is assumed that the urge we feel under certain circumstances to say 'I am in pain' and the peculiar character of this urge (it is different from the urge we feel when we say 'I am hungry') already determines the meaning of the main descriptive term of the sentence uttered, i.e. 'pain or hunger'. Conversely, the attempt to uphold this distinction between facts and conventions leads at once to the separation, characteristic of the pragmatic theory, of the observational character of a statement from its meaning: according to the pragmatic theory, the fact that a statement belongs to the observational domain has no bearing upon its meaning. Even if its production is accompanied by very forceful sensations and related to them in a manner that makes substitution by a different sentence psychologically very difficult or perhaps even impossible, even then we are free to interpret the sentence in whatever way we like. It is very important to point out that this freedom of interpretation obtains also in psychology, where our sentences are indeed about subjective events. Whatever restrictions of interpretation we accept are determined by the language we use, or by the theories or general points of view whose development has led to the formulation of this language.

The freedom of interpretation admitted by the pragmatic theory did not exist in the earlier positivism. Here sensations were thought to be the objects of observation. According to it, whether or not a statement is a sense-datum statement and, therefore, part of the observation language could be determined by logical analysis. Conversely, the assertion that a certain statement belongs to the observation language there implied an assertion about the kind of entities described (for example sense data). The ontology of the observational domain was therefore fixed independently of theorizing. This being the case, the demand for a unified ontology (which was still retained) could be met only by adopting one or other of the following two procedures: it could be met either by denying a descriptive function to the sentences of a theory and by declaring that these sentences are nothing but parts of a complicated prediction machine (*instrumentalism*), or by conferring upon these sentences an interpretation that completely depends upon their connection with the observational language as well as upon the (fixed) interpretation of the latter (*reductionism*). It is important to realize that it is

[21] Cf. Carnap [11] and [12].
[22] For a very clear presentation of this distinction, see Popper [66], ch. 5.

the clash between realism, on the one side, and the combination of the theory of sense data with the demand for a unified ontology, on the other, which necessitates this transition to either instrumentalism or reductionism.

One of the most surprising features of the development of contemporary empiricism is that the very articulate formulation of the pragmatic account of observation was not at once followed by an equally articulate formulation of a realistic interpretation of scientific theories. After all, realism had been abandoned mainly because the theory of sense data had made it incompatible with the demand for a unified ontology. The arrival of the pragmatic theory of observation removed this incompatibility and thereby opened the way for a hypothetical realism of the kind outlined earlier. Yet, in spite of this possibility, the actual historical development was in a completely different direction. The pragmatic theory was retained for a while (and is still retained, in footnotes, by some empiricists[23]), but it was soon combined either with instrumentalism or with reductionism. As the reader can verify for himself, such a combination in effect amounts to abandoning the pragmatic theory, a more complicated language with a more complicated ontology taking the place of the sense-datum language of the earlier point of view. How close the most recent offspring of this development is to the old sense-datum ideology may be seen in a paper by Rudolf Carnap.

There Carnap analyses scientific theories with the help of his well-known double-language model consisting of an observational language, L_O and a theoretical language, L_T, the latter containing a postulate system, T. The languages are connected to each other by correspondence rules, i.e. by sentences containing observational terms and theoretical terms. With respect to such a system, Carnap asserts that 'there is no independent interpretation for L_T. The system T is itself an uninterpreted postulate system. The terms $[L_T]$ obtain only an indirect and incomplete interpretation by the fact that some of them are connected by correspondence rules with observational terms, and the remaining terms of $[L_T]$ are connected with the first ones by the postulates of T.'[24]

This procedure quite obviously presupposes that the meaning of the observational terms is fixed independently of their connection with theoretical systems. If the pragmatic theory of observation were still retained by Carnap, then the interpretation of an observational statement would have to be independent of the behavioural pattern exhibited in the observational situation as well. It is not clear how, then, the observation sentence could be given any meaning at all. Now, Carnap is very emphatic about the fact that incorporation into a theoretical context is not sufficient for providing an interpretation, since no theoretical context possesses an 'independent

[23] See Hempel [46], especially n.10.   [24] See Carnap [15], 47.

interpretation'.[25] We must, therefore, suspect that, for Carnap, incorporation of a sentence into a complicated behavioural pattern has implications for its meaning, i.e. we must suspect that he has silently dropped the pragmatic theory. This is indeed the case. He asserts that 'a complete interpretation of L_O' is given since 'L_O is used by a certain language community as a means of communication',[26] adding in a later passage that if people use a term in such a fashion that for some sentences containing the term 'any possible observational result can never be absolutely conclusive evidence, but at best evidence yielding a high probability, then the appropriate place for [the term] in a dual language system . . . is in L_T rather than in L_O'.[27] These two passages together seem to imply that the meaning of an observational statement is already fixed by the way in which the sentence expressing it is handled in the immediate observational situation (note the emphasis upon absolute confirmability for observational sentences), i.e. they seem to imply the rejection of the pragmatic theory.

As I said above, this tacit withdrawal from the pragmatic theory of observation is one of the most surprising features of modern empiricism. It is responsible for the fact that this philosophy, despite the apparent progress that has been made since the thirties, still relies on the assumption that observational meanings are invariant with respect to the process of explanation and perhaps even with full meaning invariance. (The behaviouristic criterion of observability will be satisfied by any language that has been used for a long time; a long history and the observational plausibility brought about by it are the best preconditions for the petrification of meanings. This applies to Platonism as well as to modern empiricism.)

This completes my comments on (4). Two points remain: first, that the unwitting and partial return to the ideology of sense data is responsible for many of the 'inner contradictions' which are so characteristic of contemporary empiricism as well as for the pronounced similarity of this philosophy to the 'school philosophies' it has attacked; second, that (4) has been accepted, not only by philosophers, but also by many physicists who believe in the so-called Copenhagen Interpretation of microphysics. It is one of Niels Bohr's most fundamental ideas that 'however far the new phenomena' found on the microlevel 'transcend the scope of classical physical *explanation*, the account of all evidence must be expressed in classical terms'.[28] I shall not discuss, in the present section, the arguments which Bohr has developed in favour of this idea. Let me only say that it leads immediately to the invariance of the meanings of the descriptive terms of the observation language, the classical signs now playing the role of the observational vocabulary.

---

[25] For a detailed criticism of this assertion, see my [27] and [39].
[26] Carnap [15], 40.   [27] *Ibid.*, 69.
[28] [6], 209ff. For a more detailed account of Bohr's philosophy of science, see [32].

To sum up: two ideas which are common to both the modern empiricist's theory of reduction and to his theory of explanation are:

(A) reduction and explanation is (or should be) by derivation;
(B) the meanings of (observational) terms are invariant with respect to both reduction and explanation.

In the sections to follow it will be my task to scrutinize these two basic principles. I shall begin with (A).

2. CRITICISM OF REDUCTION OR EXPLANATION BY DERIVATION

The task of science, so it is assumed by those who hold the theory about to be criticized, is the explanation, and the prediction, of known singular facts and regularities with the help of more general theories. In what follows we shall assume T' to be the totality of facts and regularities to be explained, D' the domain in which T' makes correct predictions, and T (domain $D' \subset D$) the theory which functions as the basis of explanation.[29] Considering (3), we shall have to demand that T be either strong enough to contain T' as a logical consequence, or at least compatible with T' (inside D', that is). Only theories which satisfy one or the other of the two demands just stated are admissible as explanantia. Or, taking the demand for explanation for granted,

> only such theories are admissible (for explanation and prediction) in a given domain which either contain the theories already used in this domain, or are at least consistent with them. (5)

It is in this form that (A) will be discussed in the present section and in the sections to follow.

As has just been shown, condition (5) is an immediate consequence of the logical empiricist's theory of explanation and reduction, and it is therefore adopted – at least by implication – by all those who defend that theory. However, its correctness has been taken for granted by a much wider circle of thinkers, and it has also been adopted independently of the problem of explanation. Thus, in his essay 'Studies in the Logic of Confirmation', C. G. Hempel demands that 'every logically consistent observation report' be 'logically compatible with the class of all the hypotheses which it confirms', and more especially, he has emphasized that observation reports do 'not confirm any hypotheses which contradict each other'.[30] If we adopt this

---

[29] In what follows it will not be necessary explicitly to distinguish between 'T' and T, and this distinction will therefore not be made. Also terms such as 'consistent', 'incompatible' and 'follows from' will be applied to pairs of theories, $[T, T']$, and they will then mean that T *taken together with the conditions of validity of T', or the boundary conditions characterizing D'*, is compatible with, consistent with, or sufficient to derive, T'.

[30] [45], 105, condition (8.3). It was J. W. N. Watkins who drew my attention to this property of Hempel's theory.

principle, then a theory T will be confirmed by the observations confirming a more narrow theory T' only if it is compatible with T'. Combining this with the principle that a theory is admissible only if it is confirmed to some degree by the evidence available, we at once arrive at (5).

Outside philosophy, (5) has been taken for granted by many physicists. For instance, Ernst Mach in his *Waermelehre* makes the following remark: 'Considering that there is, in a purely mechanical system of absolutely elastic atoms, no real analogue for the *increase of entropy*, one can hardly suppress the idea that a violation of the second law . . . should be possible if such a mechanical system were the *real* basis of thermodynamic processes.'[31] And he insinuates that, for this reason, the mechanical hypotheses must not be taken too seriously.[32] More recently, Max Born has based his arguments against the possibility of a return to determinism upon (5) and the assumption, which we shall here take for granted,[33] that the theory of wave mechanics is incompatible with determinism. 'If any future theory should be deterministic', he says, 'it cannot be a modification of the present one, but must be entirely different. How this should be possible without sacrificing a whole treasure of well-established results I leave the determinist to worry about.'[34]

The use of (5) is not restricted to such general remarks, however. A decisive part of the quantum theory itself, the so-called quantum theory of measurement, is the immediate result of the postulate that the behaviour of macroscopic objects, such as measuring instruments, must obey some classical laws precisely and not just approximately. For example, macroscopic objects must always dwell in a well-defined classical state, and this despite the fact that their microscopic constituents exhibit very different behaviour. It is this postulate which leads to the introduction of abrupt jumps in addition to the continuous changes that occur in accordance with Schrödinger's equation.[35] An account of measurement which very clearly exhibits this feature has been given by Landau and Lifshitz. These authors point out that

> the classical nature of the apparatus means that . . . the reading of the
> apparatus . . . has some definite value . . . This enables us to say that the

[31] [53], 364.

[32] For a much more explicit statement of what appears in [53] only as an insinuation, see [54].

[33] Born believes that this assumption has been established by von Neumann's proof. In this he is mistaken; see [29]. However, there exist different and quite plausible arguments for the incompatibility of determinism and wave mechanics, and it is for this reason that I take the assumption for granted. An outline of these plausible arguments is given in [37]. It should be noted that von Neumann himself did not share Born's inductivism. See [62], 327.

[34] [7], 109. In his treatment of the relation between Kepler's laws and Newton's theory, which, he thinks, applies to all pairs of theories which overlap in a certain domain and are adequate in this domain. Born explicitly accepts (5). For an analysis of Born's inductivism, see Popper [68].

[35] See [30].

state of the system apparatus + electron after the measurement will in actual fact be described, not by the entire sum $[\Sigma A_n(q)\Phi_n(\zeta)$ where q is the coordinate of the electron, ζ the apparatus coordinate] but by only the one term which corresponds to the 'reading' g_n of the apparatus, $A_n(q)\Phi_n(\zeta)$.[36]

Moreover, most of the arguments against suggestions such as those put forth by Bohm, de Broglie and Vigier make more or less explicit use of (5).[37] A discussion of this condition is therefore very topical and leads right into the centre of contemporary arguments about microphysics.

This discussion will be conducted in three steps. It will first be argued that most of the cases which have been used as shining examples of scientific explanation *do not* satisfy (5) and that it is not possible to adapt them to the deductive scheme. It will then be shown that (5) *cannot* be defended on empirical grounds and that it leads to very unreasonable consequences. Finally, it will turn out that once we have left the domain of empirical generalizations (5) *should not* be satisfied either. In connection with this last, methodological step, the elements of a positive methodology for theories will be developed, and the historical, psychological, and semantical aspects of such a methodology will be discussed. Altogether the three steps will show that (A) is in disagreement both with actual scientific practice and with reasonable methodological demands. I start now with the discussion of the actual inadequacy of (5).

3. THE FIRST EXAMPLE

A favourite example of both reduction and explanation is the reduction of what Nagel calls the Galilean science to the physics of Newton,[38] or the explanation of the laws of the Galilean physics on the basis of the laws of the physics of Newton. By the Galilean science (or the Galilean physics) is meant, in this connection, the body of theory dealing with the motion of material objects (falling stones, penduli, balls on an inclined plane) near the surface of the earth. A basic assumption here is that the vertical accelerations involved are constant over any finite (vertical) interval. Using T' to express the laws of this theory, and T to express the laws of Newton's celestial mechanics, we may formulate Nagel's assertion to the effect that the one is reducible to the other (or explainable on the basis of the other) by saying that

$$T \ \& \ d \vdash T' \tag{6}$$

[36] [52], 22. See also von Neumann's treatment of the Compton-effect in [62], 211–15.
[37] See [32], [36], [38].
[38] [20], 291. I am aware that, from a historical point of view, the discussion to follow is not adequate. However, I am here interested in the systematic aspect, and I have therefore allowed myself what could only be regarded as great liberties if the main interest were historical.

where d expresses, in terms of T, the conditions valid inside D'. In the case under discussion d will include a description of the earth and its surroundings (supposed to be free from air; we shall also abstract from all those phenomena which are due to the rotation of the earth and whose inclusion would strengthen, rather than weaken our case), and reference will be made to the fact that the variation H of the height above ground level in the process described is very small if compared with the radius R of the earth.

As is well known, (6) cannot be correct: as long as H/R has some finite value, however small, T' will not follow (logically) from T and d. What will follow will rather be a law, T''', which, while being experimentally indistinguishable from T' (on the basis of the experiments which formed the inductive evidence for T' in the first place), is nevertheless inconsistent with T'. If, on the other hand, we want to derive T' precisely, then we must replace d by a statement which is patently false, as it would have to describe the conditions in the close neighbourhood of the earth as leading to a vertical acceleration that is constant over a finite interval of vertical distance. It is therefore impossible, for quantitative reasons, to establish a deductive relationship between T and T', or even to make T and T' compatible. This shows that the present example is not in agreement with (5) and is, therefore, also incompatible with (A), (1), and (3).

In this situation, we may adopt one or other of two procedures. We may either declare that the Galilean science can neither be reduced to, nor explained in, terms of Newton's physics;[39] or we may admit that reduction and explanation are possible, but deny that deducibility, or even consistency (on the basis of suitable boundary conditions), is a necessary condition of either. It is clear that the question as to which of these two procedures is to be adopted is of subordinate importance (after all, it is purely a matter of terminology that is to be settled here!) compared with the question of whether newly invented theories should be consistent with, or contain, those of their predecessors with which they overlap in empirical content. We shall therefore defer settlement of the terminological problem and concentrate on the question of consistency, or derivability. And we shall use the terms 'explanation' and 'reduction' either in a vague and general sense, awaiting further explication, or in the manner suggested by Nagel and by Hempel and Oppenheim. The usage adopted should always be clear from the context.

The objection which has just been developed – so it is frequently pointed out – cannot be said to endanger the correct theory of explanation, since everybody would admit that explanation may be by approximation only. This is a curious remark. It criticizes us for taking seriously, and objecting to, a criterion which has either been universally stated as a necessary condition of explanation, or which plays a central role in some theories of

[39] This suggestion was made to me by my teacher Viktor Kraft.

confirmation: condition (3). Dropping (3) means giving up altogether the orthodox theory, for (3) formed the very core of this theory.[40] On the other hand, the remark that we explain 'by approximation' is much too vague and general to be regarded as the statement of an alternative theory. As a matter of fact, it will turn out that the idea of approximation can no longer be incorporated into a formal theory, since it contains elements which are essentially subjective. However, before dealing with this aspect of explanation we shall inquire a little more closely into the reasons for the failure of (3). Such an inquiry will lead to the result not only that (3) is false, but also that it is very unreasonable to assume that it could be true.

4. REASONS FOR THE FAILURE OF (5) AND (3)

The basic argument is really very simple, and it is very surprising that it has not been used earlier. It is based upon the fact that *one and the same set of observational data is compatible with very different and mutually inconsistent theories.* This is possible for two reasons: first, because theories, which are universal, always go beyond any set of observations that might be available at any particular time; second, because the truth of an observation statement can always be asserted within a certain margin of error only.[41] The first reason allows for theories to differ in domains where experimental results are not yet available. The second reason allows for such differences even in those domains where observations have been made, provided the differences are restricted to the margin of error connected with the observations.[42] Both reasons taken together give us considerable freedom in the construction of our theories.

However, this freedom which experience grants the theoretician is nearly always restricted by conditions of an altogether different character. These additional conditions are neither universally valid, nor objective. They are connected partly with the tradition in which the scientist works, with the beliefs and the prejudices which are characteristic of that tradition; and they are connected partly with his own personal idiosyncrasies. The formal apparatus available, and the structure of the language he speaks, will also strongly influence the activity of the scientist. Whorff's assertion to the effect that the properties of the Hopi language are not very favourable for the development of a physics like the one with which we are acquainted may very well be correct.[43] Of course, it must not be overlooked that man is

[40] This was emphasized, in private communication, by Viktor Kraft and David Rynin.
[41] As J. W. N. Watkins has pointed out to me, this invalidates Hempel's conditions 9.1 and 9.2 (in [45]). An attempt to bring logical order into the relation between observation statements and the more precise statements derived from a theory has been made by S. Körner [50], 140.
[42] Even this condition is too strong, as will be shown below.
[43] See [74].

capable not only of applying, but also of inventing, languages.[44] Still, the influence of the language from which he starts should never be underestimated. Another factor which strongly influences theorizing is metaphysical beliefs. The Neoplatonism of Copernicus was at least a contributing factor in his acceptance of the system of Aristarchus.[45] Also, the issue between the followers of Niels Bohr and the realists, being still undecidable on the basis of experimentation, is mainly metaphysical in character.[46] That the choice of theories may be influenced even by aesthetic motives can be seen from Galileo's reluctance to accept Kepler's ellipses.[47]

Taking all this into account we see that the theory which is suggested by a scientist will also depend, apart from the facts at his disposal, on the tradition in which he participates, on the mathematical instruments he accidentally knows, on his preferences, on his aesthetic prejudices, on the suggestions of his friends, and on other elements which are rooted, not in facts, but in the mind of the theoretician and which are therefore subjective. This being the case it is to be expected that theoreticians working in different traditions, in different countries, will arrive at theories which, although in agreement with all the known facts, are mutually inconsistent. Indeed, any consistency over a long period of time would have to be regarded not, as is suggested by (3), (A) and (5), as a methodological virtue, but as an alarming sign that no new ideas are being produced and that the activity of theorizing has come to an end. Only the doctrine that theories are uniquely determined by the facts could have persuaded people that a lack of ideas is praiseworthy and that its consequences are an essential feature of the development of our knowledge.

At this point it is worth mentioning what will be explained in great detail later: that the freedom of theorizing granted by the indeterminateness of facts is of great methodological importance. It will turn out that many test procedures presuppose the existence of a class of mutually incompatible, but factually adequate, theories. Any attempt to reduce this class to a single theory would decrease the empirical content of this remaining theory and would therefore be undesirable from the point of view of empiricism. The freedom granted by the indeterminateness of facts is not only psychologically important (it allows scientists of different temperament to follow their different inclinations and thereby gives them satisfaction which goes beyond the satisfaction derived from the exclusive consideration of facts); it is also needed for methodological reasons.

---

[44] As is done by Bohr, Heisenberg and von Weizsaecker in their philosophical writings as well as by some Wittgensteinians. For the point of view of these physicists, see [34] and [38], as well as the end of section 7.

[45] See T. S. Kuhn [51], 128ff.

[46] See [36].   [47] See E. Panofsky [63].

So far I have assumed that the experimental evidence which inside D' confirms T and T' is the same for both theories. Although this may be so in the specific example discussed, it is certainly not true in general. Experimental evidence does not consist of facts pure and simple, but of facts analysed, modelled, and manufactured according to some theory.

The first indication of this manufactured character of the evidence is seen in the corrections which we apply to the readings of our measuring instruments, and in the selection which is made among those readings. Both the corrections and the selection made depend upon the theories held, and they may be different for the theoretical complex containing T, and for the theoretical complex containing T'. Usually T will be more general and more sophisticated than T', and it will also be invented a considerable time after T'. New experimental techniques may have been introduced in the meantime. Hence, the 'facts' within D' which count as evidence for T will be different from the 'facts' within D' which counted as evidence for T' when the latter theory was first introduced. An example is the very different manner in which the apparent brightness of stars was determined in the seventeenth century and is determined now. This is another important reason why T usually will not satisfy (5) with respect to T': not only are T and T' connected with different theoretical ideas leading to different predictions, even in the domain where they overlap and are both confirmed, but the better experimental techniques and the improved theories of measurement will usually provide evidence for T which is different from the evidence for T' even within the domain of common validity. In short, introducing T very often leads to recasting the evidence for T'. The demand that T should satisfy (5) with respect to T' would in this case imply the demand that new and refined measurements not be used, which is clearly inconsistent with empiricism.[48]

A further indication of the 'manufactured' character of the experimental evidence is seen in the fact that observable results, and indeed anything conveyed with the help of a language, are always expressed in some theory or other. Because this fact will also be of importance in connection with my criticism of (B), and because it leads to a further criticism of (A), I shall discuss at length the example I have chosen for its elucidation.

---

[48] Against the argument in the last paragraph it might be pointed out that results of measurement which are capable of improvement, and which therefore change, do not belong to the observational domain, but must be formulated with the help of singular statements of the theoretical language. (For this move see, Carnap [15], 40.) Observational statements proper are such qualitative statements as 'Pointer A coincides with mark B', or 'A is greater than B' – and these statements will not change, or be eliminated, whatever the development of the theory, or of the methods of measurement. This point will be dealt with, and refuted, in section 7.

5. SECOND EXAMPLE: THE PROBLEM OF MOTION[49]

From its very beginning, rational cosmology, the creation of the Ionian 'physiologists', was faced with the problem of change and motion (in the general sense in which it includes locomotion, qualitative alteration, quantitative augmentation and diminution, as well as generation and corruption). The problem arose in two forms. The first was the possibility of change and motion. This form of the problem had to be solved by the invention of a cosmology which allowed for change, i.e. which was not such that the occurrence of change was (unwittingly) excluded from it by the very nature of the assumptions upon which it was based. The second form, which arose once the first had been solved in a satisfactory manner, was the cause of change. As was shown by Parmenides, the early monistic theories of Thales, Anaximander and others could not solve the first form of the problem. For Parmenides himself, this did not refute monism; it refuted the existence of change.

The majority of thinkers went a different path, however. They regarded monism as refuted and started with pluralistic theories. In the case of the atomic theory, which was one of these pluralistic theories, the relation between Parmenides' arguments and pluralism is very clear. Leucippus who 'had associated with Parmenides in philosophy',[50] 'thought he had a theory which was in harmony with the senses, and did not do away with coming into being and passing away, nor motion, nor with the multiplicity of things'.[51] This is how the atomic theory arose, as an attempt to solve problems created by the empirical inadequacy of the early monism of the Ionians.

However, the theory which was most influential in the Middle Ages and which also tried to solve what I have above called the second form of the problem was Aristotle's theory of motion as the actualization of potentiality. According to Aristotle,

> motion is a process arising from the continuous action of a source of motion, or a 'motor', and a 'thing moving.'[52]   (7)

This principle, according to which any motion (and not only accelerated motion) is due to the action of some kind of force, can be easily supported by such common observations as a cart drawn by a horse and a chair pushed

---

[49] For a more detailed account of the theories mentioned in this section, see M. Clagett [17]. Concerning the first part of the present section, see J. Burnet [10], as well Clagett [16] and Popper [68].

[50] Aristotle [2], A, 8 324b35.

[51] Theophrastus quoted from Burnet [10], 333.   [52] Clagett [17], 425.

around by an angry husband. It gets into difficulties when one considers the motion of things thrown: stones continue to move despite the fact that contact with the motor apparently ceases when they leave the hand. Various theories were suggested to eliminate this difficulty. From the point of view of later developments, the most important one of these theories is the impetus theory. The impetus theory retains (7) and the general background of the Aristotelian theory of motion. Its distinction lies in the specific assumptions it makes concerning the causes that are responsible for the motion of the projectile. According to the impetus theory, the motor (for example the hand) transfers to the projectile an inner moving force which is responsible for its continued motion, and which is continually decreased by the resisting air and by the gravity of the projectile. A stone in empty space would therefore either remain at rest or move (along a straight line) with constant speed, depending on whether its impetus is zero or possesses a finite value.[53]

At this point a few words must be said about the characterization of locomotion. The question as to its proper characterization was a matter of dispute. To us it seems quite natural to characterize motion by space transversed, and, as a matter of fact, one of the suggested characterizations did just this: it defined motion kinematically by reference to space transversed. This apparently very simple characterization needs further specification if an account is to be given of non-uniform movements where the distinction becomes relevant between average velocity and instantaneous velocity. Compared with the actual space transversed by a given body, the instantaneous velocity is a rather abstract notion since it refers to the space that would be transversed if the velocity were to remain constant over a finite interval of time.

Another characterization of motion is the dynamical. It defines motion in terms of the forces which bring it about in accordance with (7). Adopting the impetus theory the motion of a stone thrown would have to be characterized by its inherent impetus, which pushes it along until it is exhausted by the opposing forces of friction and gravity.

Which characterization is the better one to take? From an operationalist point of view (and we shall adopt this point of view, since we want to follow the empiricist as far as possible), the dynamical characterization is definitely to be preferred: while it is fairly easy to observe the impetus enclosed in a moving body by bringing it to a stop in an appropriate medium (such as soft wax) and then noting the effect of such a manoeuvre, it is much more difficult, if not nearly impossible, to arrange matters in such a way that from a given moment on, a non-uniformly moving object assumes a

---

[53] I have added the parentheses because of the absence from the earlier forms of the impetus theory, of an explicit consideration of direction.

constant speed with a value identical with the value of the instantaneous velocity of the object at that moment, and then to watch the effect of this procedure.

With the use of the dynamical characterization, the 'inertial law' pronounced above reads as follows:

> The impetus of a body in empty space which is not under the influence of any
> outer force remains constant. (8)

Now, in the case of inertial motions, (8) gives correct predictions about the behaviour of material objects. According to (3), explanation of this fact will involve derivation of (8) from a theory and suitable initial conditions. Disregarding the demand for explanation, we can also say, on the basis of (5), that any theory of motion that is more general than (8) will be adequate only if it contains (8) which, after all, is a very basic law. According to (2), the meanings of the key terms of (8) will be unaffected by such a derivation. Assuming Newton's mechanics to be the primary theory, we shall therefore have to demand that (8) be derivable from it *salva significatione*. Can this demand be satisfied?

At first sight it would seem that it is much easier to derive (8) from Newton's theory than it is to establish the correctness of (6): as opposed to Galileo's law (8) is not in quantitative disagreement with anything asserted by Newton's theory. Even better, (8) seems to be identical with Newton's first law so that the process of derivation seems to degenerate into a triviality.[54]

In the remainder of the present section, it will be shown that this is not so and that it is impossible to establish a deductive relationship between (8) and Newton's theory. Later on this will be the starting point of our criticism of (B).

Let me repeat, before beginning the argument, that (8), taken by itself, cannot be attacked on empirical grounds. Indeed, we have indicated a primitive method of measuring impetus, and the attempt to confirm (8) by using this method will certainly show that within the domain of error connected with such crude measurements, (8) is perfectly all right. It is, therefore, quite in order to ask for the explanation or the reduction of (8), and the failure to arrive at a satisfactory solution of this task cannot be blamed upon the empirical inadequacy of (8).

We now turn to an analysis of the main terms of (8). According to Nagel the meaning of these terms is to be regarded as 'fixed' by the procedures and assumption of the impetus theory, and any one of them is 'therefore

---

[54] There existed theories, among them the theory of *mail* by Abu'l-Barakat, where quantitative disagreement with Newton's laws was to be expected: in these theories, the impetus decreased with time in the same manner in which a hot poker that is removed from the fire gradually loses the heat stored in it. See Clagett [17], 513.

intelligible in terms of its own rules of usage'.[55] What are these meanings, and what are the rules which establish them?

Take the term 'impetus'. According to the theory of which (8) is a part, the impetus is the force responsible for the movement of the object that has ceased to be in direct contact, by push, or by pull, with the material mover. If this force did not act, if the impetus were destroyed, then the object would cease to move and fall to the ground (or simply remain where it is, in case the movement were on a frictionless horizontal plane). A moving object which is situated in empty space and which is influenced neither by gravity nor by friction is not outside the reach of any force. It is pushed along by the impetus, which may be pictured as a kind of inner principle of motion (similar, perhaps, to the vital force of an organism, which is the inner principle of *its* motion).

We now turn to Newton's celestial mechanics and the description in terms of this theory, of the movement of an object in empty space. (Newton's theory still retains the notion of absolute space and allows therefore for such a description to be formed.) Quantitatively, the same movement results. But can we discover in the description of this movement, or in the explanation given for it, anything resembling the impetus of (8)? It has been suggested that the momentum of the moving object is the perfect analogue of the impetus. It is correct that the measure of this magnitude (mv) is identical with the measure that has been suggested for the impetus.[56] However, it would be very mistaken if we were, on that account, to identify impetus with momentum. For whereas the impetus is supposed to be something that pushes the body along,[57] the momentum is the result rather than the cause of its motion. Moreover, the inertial motion of classical mechanics is a motion which is supposed to occur by itself, and without the influence of any causes. After all, it is this feature which, according to most historians, radical empiricists included, constitutes one of the main differences between the Aristotelian theory and the celestial mechanics of the seventeenth, eighteenth and nineteenth centuries: in the Aristotelian theory, the natural state in which an object remains without the assistance of any causes is the state of rest. A body at rest (in its natural place, we should add) is not under the influence of any forces. In Newtonian physics it is the state of being at rest or in uniform motion which is regarded as the natural state. This means, of course, the explicit denial of a force such as the impetus is supposed to represent.

This denial need not mean that the concept of such a force cannot be formed within Newton's mechanics. After all, we deny the existence of

---

[55] [20], 301.

[56] See Clagett [17], 523.

[57] For an elaborate discussion of the difference between momentum and impetus, see Anneliese Maier [58]. For what follows, see also M. Bunge [9], ch. 4.4.

unicorns and use in this denial the very concept of a unicorn. Is it then possible to define a concept such as impetus in terms of the theoretical primitives of Newton's theory? The surprising fact is that any attempt to arrive at such a definition leads to disappointment (which shows, by the way, that theories such as Newton's are expressed in a language that is much more tightly knit than is the language of everyday life). I have already pointed out that the momentum, which would give us the correct numerical value is not what we want. What we want is a *force* that acts upon the isolated object and is responsible for its motion. The concept of such a force can of course be formed within Newton's theory. But given that the movement under review (the inertial movement) occurs with constant velocity, and Newton's second law, we obtain in all relevant cases zero for the value of this force, which is not the measure we want. A positive measure is obtained only if we assume that the movement occurs in a resisting medium (which is of course, the original Aristotelian assumption), an assumption which is inconsistent with another feature of the case considered, the fact that the inertial movement is supposed by Newton's theory to occur in empty space. I conclude that the concept of impetus, as fixed by the usage established in the impetus theory, cannot be defined in a reasonable way within Newton's theory. And this is not surprising. For this usage involves laws, such as (7), which are inconsistent with Newtonian physics.

In the last argument, the assumption that the concept of force is the same in both theories played an essential role. This assumption was used in the transition from the assertion, made by the impetus theory, that inertial motions occur under the influence of forces, to the calculation of the magnitude of these forces on the basis of Newton's second law. Its legitimacy may be derived from the fact that both the impetus theory and Newton's theory apply the concept of force under similar circumstances (paradigm-case argument). Still, meaning and application are not the same thing, and it might well be objected that the transition performed is not legitimate, since the different contexts of the impetus theory and of Newton's theory confer different meanings upon one and the same word 'force'. This being the case, our last argument is based upon a *quaternio terminorum* and is, therefore, invalid. In order to meet this objection, we may repeat our argument using the word 'cause' instead of the word 'force' (the latter has a somewhat more specific meaning). But if someone again retorts that 'cause' has a different meaning in Newton's theory from the one it has in the impetus theory, then all I can say is that continuing the objection will in the end establish what I wanted to show in a more simple manner: the impossibility of defining the notion of an impetus in the descriptive terms of Newton's theory. To sum up: the concept of impetus is not 'explicable in terms of the theoretical primitives of the primary science'.[58] And this is

[58] Nagel [20], 302.

exactly as it should be, considering the conflict between some very basic principles of these two theories.

However, explication in terms of the primitives of the primary science is not the only method which was considered by Nagel in his discussion of the process of reduction. Another way to achieve reduction, which he mentions immediately after the above quotation, 'is to adopt a material, or physical hypothesis according to which the occurrence of the properties designated by some expression in the premises of the primary science is a sufficient, or a necessary and sufficient, condition for the occurrence of the properties designated by the expressions of the secondary discipline'. Both procedures are in accordance with (4), or with (2), or at least Nagel thinks that they are: 'in this case', he says, referring to the procedure just outlined, 'the meaning of the expressions of the secondary science as *fixed by the established usage of the latter,* is not declared to be analytically related to the meanings of the corresponding expressions of the primary science'.[59] Let us now see what this second method achieves in the present case.

To start with, this method amounts to introducing a hypothesis of the form

$$\text{impetus} = \text{momentum} \qquad (9)$$

where each side retains the meaning its possesses in its respective discipline. The hypothesis then simply asserts that wherever momentum is present, impetus will also be present (see the above quotation of Nagel's), and it also asserts that the measure will be the same in both cases. Now this hypothesis, although acceptable within the impetus theory (after all, this theory permits the incorporation of the concept of momentum), is incompatible with Newton's theory. It is therefore not possible to achieve reduction and explanation by the second method.

To sum up: a law such as (8) which, as I have argued, is empirically adequate and in quantitative agreement with Newton's first law, is nevertheless incapable of being reduced to Newton's theory and therefore incapable of explanation in terms of the latter. Whereas the reasons we have so far found for irreducibility were of a quantitative nature, this time we met a qualitative reason, as it were: the incommensurable character of part of the conceptual apparatus of (8), on the one side, with part of Newton's theory, on the other.

Taking together the quantitative as well as the qualitative argument, we are now presented with the following situation: there exist pairs of theories, T and T', which overlap in a domain D' and which are apparently incompatible (though experimentally indistinguishable) in this domain. Outside D', T has been confirmed, and it is also more coherent, more general and less *ad hoc* than T'. But the conceptual apparatus of T and T' is

[59] *Ibid.,* my italics.

such that it is possible neither to define the primitive descriptive terms of T' on the basis of the primitive descriptive terms of T, nor to establish correct empirical relations involving both these terms (correct, that is, from the point of view of T). This being the case, explanation of T' on the basis of T, or reduction of T' to T, is clearly impossible if both explanation and reduction are to satisfy (A) and (B). Altogether, the use of T necessitates the elimination both of the conceptual apparatus of T' and of the laws of T'. The conceptual apparatus has to be eliminated because its use involves principles, such as (7) in the example above, which are inconsistent with the principles of T; and the laws have to be eliminated because they are inconsistent with what follows from T for events inside D'. The demand for explanation and reduction clearly cannot arise if this demand is interpreted as the demand for the explanation, or reduction, of T', rather than of a set of laws that is in some respect similar to T' but in other respects (meanings of fundamental terms included) very different from it. For such a demand would imply the demand to derive from correct premises what is false, and to incorporate what is incommensurable.

The effect of the transition from T' to T is rather to be described in the manner indicated in the introductory remarks above, where I said that what happens when a transition is made from a restricted theory T' to a wider theory T (which is capable of covering all the phenomena which have been covered by T') is something much more radical than incorporation of the unchanged theory T' into the wider context of T. It is rather a replacement of the ontology of T' by the ontology of T, and a corresponding change in the meanings of all descriptive terms of T' (provided these terms are still employed). Let me add here that the not-too-well-known example of the impetus theory versus Newton's mechanical theory is not the only instance where this assertion holds. As I shall show a little later, more recent theories also correspond to it. Indeed, it will turn out that the principle correctly describes the relation between the elements of *any* pair of non-instantial theories satisfying the conditions which I have just enumerated.

This finished step one of the argument against the assumption that reduction and explanation are by derivation. What I have shown (and shall show in later sections) is that some very important cases which have been, or could be, used as examples of reduction (and explanation) are not in agreement with the condition of derivability. It will be left to the reader to verify that this holds in almost all cases of explanation by theories: assumption (A) does not give a correct account of actual scientific practice. It has also been shown that in this respect the thesis formulated at the beginning of this paper is much more adequate.

Against this result it may be pointed out, with complete justification, that scientific method, as well as the rules for reduction and explanation connected with it, is not supposed to describe what scientists are actually doing. Rather, it is supposed to provide us with normative rules which should be followed, and to which actual scientific practice will correspond only more or less closely. Adopting this point of view, one cannot regard the arguments of the last few sections as ultimately decisive. They are satisfactory insofar as they show that the 'orthodox' are wrong when asserting that (A), (B) and (5) reflect actual scientific practice. But they do not dispose of these principles if they are interpreted as demands to be followed by the scientist (although, of course, they provide ample material for such disproof). I therefore proceed now to a methodological criticism of the demands of the orthodox. The first move in this criticism will be the examination of an argument which has sometimes been used to defend (5).

6. METHODOLOGICAL CONSIDERATIONS

The argument runs as follows: (α) a good theory is a summary of facts; (β) the predictive success of T' (I will continue to use the notation introduced in section 2) has shown T' to be good theory inside D'; hence (γ), if T, too, is to be successful inside D', then it must either give us all the facts contained in T', i.e. it must give us T', or at least it must be compatible with T'.

It is easily seen that this very popular argument will not do.[60] We can show this by considering its premises. Premise (α) is acceptable if it is not taken in too strict a sense (e.g. if it is not interpreted as implying an ontology of mutually independent 'facts' as has been suggested by the early Wittgenstein). Interpreted in such a loose manner (α) simply says that a good theory not only will be able to answer many questions, but will also answer them correctly. Now if this is to be the interpretation of (α), then (β) cannot possibly be correct: in (β) the predictive success of T' is taken to indicate that T' will give a correct account of *all* the facts inside its domain. However, one must remember that because of the general character of statements expressing laws and theories, their predictive success can be established only with respect to part of their content. Only part of a theory can at any time be known to be in agreement with observation. From this limited knowledge nothing can be inferred (logically!) with respect to the remainder.[61]

We must also consider the margin of error involved in every single test. Hence, from a purely logical point of view, new theories will be restricted only to the extent to which their predecessors have been tested and

---

[60] A sloppy version of this argument occurs frequently in arguments by physicists. It ought to be mentioned, by the way, that Hempel's condition 8 leads to the very same result, viz. to the demand that new theories be consistent with their confirmed predecessors.

[61] This point derives from Hume. That Hume's arguments are still not understood by many thinkers and are therefore still in need of repetition has been emphasized by Popper [69], Reichenbach [70], Goodman [42], and others.

confirmed.[62] Only to this extent will it be necessary for them to agree with their predecessors. In domains where tests have not yet been carried out, or where only very crude tests have been made, we have complete freedom as to how to proceed, and this quite independently of which theories were originally used here for the purpose of prediction. Clearly this last condition, which is in agreement with empiricism, is much less restrictive than either (3) or (5).

One might hope to arrive at more restrictive conditions by adding inductive argument to logical reasoning. True, from a logical point of view we can only say that part of T' has been found to be in agreement with observation and that T need agree only with that part and not, as is demanded in (5), with the whole of T'. However, if inductive reasoning is used as well, then we shall perhaps have to admit that this partial confirmation has established T', and that therefore the whole of T' should be covered by T. Does this help us to strengthen the condition mentioned at the end of the last paragraph and to demonstrate (5) after all?

It is clear that inductive reasoning cannot establish (5). For let us assume that T agrees with T' only where T' has been confirmed and is different from T' in all other instances, without having as yet been refuted. In this case T will satisfy our own condition of the last paragraph, and it will not satisfy any stronger condition (except accidentally). Can inductive reasoning prompt us to eliminate T? It is not easily seen how far this could be the case, since T shares all its confirming instances with T'. If T' is established by these instances, then so is T – unless we use formal considerations (which I shall discuss later). Again we arrive at the result that, considering facts, there is not much to choose between T and T', and that (5) cannot be defended on empirical grounds.

It is worthwhile to inquire a little more closely into the effect which the adoption of (5), and, incidentally, also of Hempel's condition 8,[63] would have upon the development of scientific knowledge. It would lead to the elimination of a theory, not because it is inconsistent with the facts, but because it is inconsistent with another, and as yet unrefuted, theory whose confirming instances its shares. This is a strange procedure to be adopted by thinkers who claim above all to be empiricists. However, the situation becomes even worse when we inquire why the one theory is retained and the other rejected. The answer (which is, of course, not the answer given by the empiricist) can only be that the theory which is retained was there first. This shows that in practice the allegedly empirical procedure (5) leads to the preservation of the old theories and to the rejection of the new theories

---

[62] As was mentioned in section 4, it is hardly ever the case that two theories which have been discussed in very different historical periods will be based upon exactly the same observations. The condition is therefore still too strict.

[63] See [45], 105.

even before these new theories have been confronted with the facts. That is, it leads to the same result as transcendental deduction, intuitive argumentation, and other forms of *a priori* reasoning, the only difference being that now it is in the name of experience that such results are obtained. This is not the only instance where, on closer scrutiny, a rather close relation emerges between some versions of modern empiricism and the 'school philosophies' it attacks.

We must now consider the argument that formal criteria may provide a principle of choice between T and T' that is independent of fact. Such formal criteria can indeed be given.[64] However, while usually a more general and coherent theory is preferred to a less general collection of laws, because of being less *ad hoc*, (5) tends to reverse this, as general theories of a high degree of coherence usually violate (5). Again this principle is seen to be incompatible with reasonable methodology.

Two things have been shown so far. First, the invalidity of an argument used for establishing (5). Second, the undesirability, from an empirical point of view, of some consequences of this argument. However, all this has little weight when compared with the following most important consideration.

Within contemporary empiricism, discussions of testing and of empirical content are usually carried out in the following manner: it is asked how a theory is related to its empirical consequences and what these consequences are. True, in the derivation of these consequences reference will have to be made to principles or theorems which are borrowed from other disciplines and which then occur in the correspondence rules. However, these principles and theorems play a subordinate role when compared with the theory under review; and it is, of course, also assumed that they are mutually consistent and consistent with the theory. One may therefore say that, for the orthodox procedure, the natural unit to which discussions of empirical content and of test methods are referred is always a single theory taken together with those of its consequences that belong to the observation language.

This manner of discussion does not allow us to give an adequate account of crucial experiments which involve more than one theory, none of which are expendable or only of psychological importance. A very good example of the structure of such crucial tests is provided by the more recent development of thermodynamics. As is well known, the Brownian particle is a perpetual-motion machine of the second kind, and its existence refutes the (phenomenological) second law. However, could this fact have been discovered in a direct manner, by direct investigation of the observational consequences of thermodynamics? Consider what such a refutation would have required: the proof that the Brownian particle is a perpetual-motion

[64] See Popper [69], ch. 6.

machine of the second kind would have required (a) measurement of the exact motion of the particle in order to ascertain the changes of its kinetic energy plus the energy spent on overcoming the resistance of the fluid, and (b) precise measurements of temperature and heat transfer in the surrounding medium in order to ascertain that any loss occurring here was indeed compensated by the increase of the energy of the moving particle and the work done against the fluid as mentioned in (a). Such measurements, however, are beyond experimental possibilities.[65] Hence, a direct refutation of the second law, based upon an investigation of the testable consequences of thermodynamics alone, would have had to wait for one of those rare, not repeatable, and therefore *prima facie* suspicious, large fluctuations in which the transferred heat is indeed accessible to measurement. This means that such a refutation would have never taken place, and, as is well known, the actual refutation of the second law was brought about in a very different manner. It was brought about via the kinetic theory and Einstein's utilization of it in the calculation of the statistical properties of the Brownian motion. In the course of this procedure the phenomenological theory (T') was incorporated into the wider context of statistical physics (T) in such a manner that (5) was violated; and then a crucial experiment was staged (Perrin's investigations).

It seems to me that the more general our knowledge becomes the more important it will be to carry out tests in the manner indicated, not by comparing a single theory with experience, but by staging crucial experiments between theories which, although in accordance with all the known facts, are mutually inconsistent and give widely different answers in unexplored domains. This suggests that outside the domain of empirical generalizations *the methodological unit to which we refer when discussing questions of test and empirical content consists of a whole set of partly overlapping, factually adequate, but mutually inconsistent theories.* To the extent to which utilization of such a set provides additional tests which for empirical reasons could not have been carried out in a direct manner, the use of a set of this kind is demanded by empiricism. For the basic principle of empiricism is, after all, to increase the empirical content of whatever knowledge we claim to possess.

On the other hand, the fact that (5) does not allow for the formation of such sets now proves this principle to be inconsistent with empiricism. By excluding valuable tests it decreases the empirical content of the theories that are permitted to remain (and which, as indicated above, will usually be the theories which were there first). This last result of a consistent application of (5) is of very topical interest: it may well be, as has been pointed out by Bohm and Vigier,[66] that the refutation of the quantum-mechanical

---

[65] Concerning the extreme difficulties of following the motion of the Brownian particle in all its details, see R. Fuerth [41]. See also below ch. 8, n.15 and text as well as ch. 12.

[66] See the discussion remarks of these two physicists in [49], as well as those of Bohm [4].

uncertainties presupposes just an incorporation of the present theory into a wider context, which is no longer in accordance with the idea of complementarity and which therefore suggests new and decisive experiments. And it may also be that the insistence on the part of the majority of contemporary physicists upon (5) will, if successful, forever protect these uncertainties from refutation. This is how modern empiricism may finally lead to a situation where a certain point of view petrifies into dogma by being, in the name of experience, completely removed from any conceivable criticism.

To sum up the arguments of the present section: it has been shown that neither (5) nor (A) can be defended on the basis of experience. On the contrary, a strict empiricism will admit theories which are factually adequate and yet mutually inconsistent. An analysis of the character of tests in the domain of theories has revealed, moreover, that the existence of sets of partly overlapping, mutually inconsistent, and yet empirically adequate theories is not only possible, but also required. I shall conclude the present section by discussing in a little more detail the logical and psychological consequences of the use of such a set.

Increase of testability will not be the only result. The use of a set of theories with the properties indicated above will also improve our understanding of each of its members by making it very clear what is denied by whichever theory happens to be accepted in the end. Thus, it seems to me that our understanding of Newton's somewhat obscure notion of absolute space and of its merits is greatly improved when we compare it with the relational ideas of Berkeley, Huyghens, Leibniz and Mach, and when we consider the failure of the latter ideas to give a satisfactory account of the phenomenon of inertial forces. Also, the study of general relativity will lead to a deeper understanding of this notion than could be obtained from a study of the *Principia* alone.[67] This is not meant to be understood in a

---

[67] This, by the way, is one of the reasons why an axiomatic exposition of physical principles, such as Newton's, is inferior by far to a dialectic exposition where many ideas are considered, and the pros and cons discussed, until finally one theory is pronounced the most satisfactory. Of course, if one holds that, concerning theories the only relation of interest is the relation between a single theory and 'the facts', and if one also believes that these facts single out a certain theory more or less uniquely, then one will be inclined to regard discussion of alternatives as a matter of history, or of psychology, and one will even wish to hide, with some embarrassment, the situation at the time when the clear message of the facts had not yet been grasped. However, as soon as it is recognized that the refutation (and thereby also the confirmation) of a theory necessitates its incorporation into a family of mutually inconsistent alternatives, the discussion of these alternatives becomes of paramount importance for methodology and should be included in the presentation of the theory that is accepted in the end. For the same reason, adherence either to the distinction between a context of discovery (where alternatives are considered, but given a psychological function only) and a context of justification (where they are not mentioned any more), or strict adherence to the axiomatic approach must be regarded as an arbitrary and very misleading restriction of methodological discussion: much of what has been called 'psychological', or 'historical', in past discussions of method is a very relevant part of the theory of test procedures. Considering all this, the increased attention paid to the historical aspects of a subject, and

psychological sense only. For just as the meaning of a term is not an intrinsic property of it, but is dependent upon the way in which the term has been incorporated into a theory, in the same manner the content of a whole theory (and thereby again the meaning of the descriptive terms which it contains) depends upon the way in which it is incorporated into both the set of its empirical consequences and the set of all the alternatives which are being discussed at a given time.[68] Once the contextual theory of meaning has been adopted, there is no reason to confine its application to a single theory, or a single language, especially as the boundaries of such a theory or of such a language are almost never well defined. The considerations above have shown, moreover, that the unit involved in the test of a specific theory is not this theory taken together with its own consequences, but is a whole class of mutually incompatible and factually adequate theories. Hence, both consistency and methodological considerations suggests such a class as the context from which meanings are to be made clear.[69]

Also, the use of such a class rather than of a single theory is a most potent antidote against dogmatism. Psychologically speaking, dogmatism arises, among other things, from the inability to imagine alternatives to the point of view in which one believes. This inability may be due to the fact that such alternatives have been absent for a considerable time and that, therefore, certain ways of thinking have been left undeveloped; it may also be due to the conscious elimination of such alternatives. However that may be, persistence of a single point of view will lead to the gradual establishment of well-circumscribed methods of observation and measurement; it will lead to codification of the ways in which these results are interpreted; it will lead to a standardized terminology and to other developments of a similarly conservative kind. This being the case, the gradual acceptance of the theory by an ever-increasing number of people must finally bring about a trans-

the attempts to break down the distinction between the synthetic and the analytic must be welcomed as steps in the right direction. However, even here there are drawbacks. Only very few of the enthusiastic proponents of an increased study of the history of a subject realize the methodological importance of their investigations. The justification they give for their interest is either sentimental or psychological ('it gives me ideas'), or based upon some very implausible notions concerning the 'growth' of knowledge. What these thinkers need in order not to fall victims to all sorts of quasi-philosophies is a methodological backbone, and I hope that the theory of test which has been sketched above in its merest outlines will provide such a backbone.

[68] In the twentieth century, the contextual theory of meaning has been defended most forcefully by Wittgenstein; see [75] as well as my summary in [28]. However, it seems that Wittgenstein is inclined to confine this theory to the inside of his language games: Platonism of concepts is replaced by Platonism of (theories or) games. For a brief criticism of this attitude see [35].

[69] Textbooks and historical presentations very often create the impression either that such classes never existed and that physicists (at least the 'great' ones) at once arrived at the one good theory, or that their existence must not be taken too seriously. This is quite understandable. After all, historians have been just as much under the influence of inductivist ideas as the physicists and the philosophers.

formation of even the most common idiom that is taught in very early youth. In the end, all the key terms will be fixed in an unambiguous manner, and the idea (which may have led to such a procedure in the first place) that they are copies of unchanging entities, and that change of meaning, if it should happen, is due to human mistake – this idea will now be very plausible. Such plausibility reinforces all the manoeuvres which may be used for the preservation of the theory (elimination of opponents included).[70]

The conceptual apparatus of the theory having penetrated nearly all means of communication, such methods as transcendental deduction and analysis of usage, which are further means of solidifying the theory, will be very successful. Altogether it will seem that at last an absolute and irrevocable truth has been arrived at. Disagreement with facts may of course occur, but, being now convinced of the truth of the existing point of view, its proponents will try to save it with the help of *ad hoc* hypotheses. Experimental results that cannot be accommodated, even with the greatest ingenuity, will be put aside for later consideration. The result will be absolute truth, but, at the same time, it will decrease in empirical content to such an extent that all that remains will be no more than a verbal machinery which enables us to accompany any kind of event with noises (or written symbols) which are considered true statements by the theory.

The picture painted above is by no means exaggerated. The way in which, for example, the theory of witchcraft and demonic influence crept into the most common ways of thinking, and could be preserved for quite a considerable time, offers a vivid illustration of each point mentioned in the last paragraph. Moreover, the story of its overthrow furnishes another illustration of our thesis that comprehensive theories cannot be eliminated by a direct confrontation with 'the facts'.

Let us compare such a dogmatic procedure with the effects of the use of a class of theories rather than a single theory. First of all, such a procedure will encourage the building of a great variety of measuring instruments. There will be no *one* way of interpreting the results, and the theoretician will be trained to switch quickly from one interpretation to another.[71] Intuitive appeal will lose its paralysing effect, transcendental deduction which, after all, presupposes uniformity of usage, will be impossible, and the question of agreement with the facts will assume a very prominent position. Experimental results which are inconsistent with one theory may be consistent with a different theory; this eliminates the motives for *ad hoc* hypotheses, or

---

[70] Today, of course, the elimination takes the more refined form of a refusal to publish (or to read) what is not in agreement with the accepted doctrine. However, this liberalism applies to physical theories only. It does not seem to apply to political theories.

[71] As Joseph Agassi pointed out to me, this method was consciously used by Faraday in order to escape the influence of prejudice. Concerning its role in modern discussions about the microlevel, see [37].

at least reduces them considerably. Nor will it be necessary to use instrumentalism as a means of getting out of trouble, since a coherent account may be provided by an alternative to the theory considered. The likelihood that empirical results will be left lying around will also be smaller; if they do not fit one theory, they will fit another. It is not at all superfluous to mention the tremendous development of human capabilities encouraged by such a procedure and the antidotes it contains against the wish to set up, and to obey, all-powerful regimes, be they political, religious, or scientific. Taking all this into account, we are inclined to say that *whereas unanimity of opinion may be fitting for a church, or for the willing followers of a tyrant, or some other kind of 'great man', variety of opinion is a methodological necessity for the sciences and, a fortiori, for philosophy.* Neither (A), nor (B), nor (5) allows for such variety. It follows that, to the extent to which both principles (and the philosophy behind them) delimit variety and demand future theories to be consistent with theories already in existence, they contain a theological element (which lies, of course, in the worship of 'facts' that is so characteristic of nearly all empiricism).

This finishes my criticism of (A) and (5). (A) has been shown to be in disagreement not only with scientific practice but also with the principles of a sound empiricism. The account of theorizing given in the introduction has been shown to be superior to the hierarchy of axioms and theorems which seems to be the favourite model of contemporary empiricism. The use of a set of mutually inconsistent and partially overlapping theories has been found to be of fundamental importance for methodology. The desideratum mentioned in connection with what has here been called the second idea has thereby been fulfilled. Serious doubt has been thrown upon the correctness and the desirability of (B). I now turn to the refutation of (B).

7. CRITICISM OF THE ASSUMPTION OF MEANING INVARIANCE

In section 5 it was shown that the 'inertial law' (8) of the impetus theory is incommensurable with Newtonian physics in the sense that the main concept of the former, the concept of impetus, can neither be defined on the basis of the primitive descriptive terms of the latter, nor related to them via a correct empirical statement. The reason for this incommensurability was also exhibited: although (8), taken by itself, is in quantitative agreement both with experience and with Newton's theory, the 'rules of usage' to which we must refer in order to explain the meanings of its main descriptive terms contain law (7) and, more especially, the law that constant forces bring about constant velocities. Both of these laws are inconsistent with Newton's theory. Seen from the point of view of this theory, any concept of a force whose content is dependent upon the two laws just mentioned will possess zero magnitude, or zero denotation, and will therefore be incapable of expressing features of actually existing situations. Conversely, it will be capable of being used in such a manner only if all connections with Newton's theory have first been severed. It is clear that this example refutes (B) if we interpret that thesis as a description of how science actually proceeds.

We may generalize this result in the following fashion: consider two theories, T' and T, which are both empirically adequate inside D', but which differ widely outside D'. In this case the demand may arise to explain T' on the basis of T, i.e. to derive T' from T and suitable initial conditions (for D'). If we assume T and T' to be in quantitative agreement inside D', such derivation will still be impossible if T' is part of a theoretical context whose rules of usage involves laws inconsistent with T.[72]

It is my contention that the conditions just enumerated apply to many pairs of theories which have been used as instances of explanation and reduction. Many (if not all) such pairs on closer inspection turn out to consist of elements which are incommensurable and therefore incapable of mutual reduction and explanation. However, the above conditions admit of still wider application and lead then to very important consequences with regard to the structure and development both of our knowledge and of the language used for its expression. After all, the principles of the context of which T' is a part need not be explicitly formulated, and as a matter of fact they rarely are. To bring about the situation described above (sets of mutually incommensurable concepts), it is sufficient that they govern the use of the main terms of T'. In such a case T' is formulated in an idiom of whose implicit rules of usage are inconsistent with T (or with some consequences of T in the domain where T' is successful). Such inconsistency will not be obvious at a glance; it will take considerable time before the incommensurability of T and T' can be demonstrated. However, as soon as this demonstration has been carried out, the idiom of T' must be given up and replaced by the idiom of T. Of course, one need not go through the laborious and very uninteresting task of analysing the context of which T' is part.[73] All that is needed is the adoption of the terminology and the 'grammar' of the most detailed and most successful theory throughout the domain of its application.[74] This automatically takes care of whatever

---

[72] Since this difficulty can arise even in the domain of empirical generalizations, the orthodox account may be inappropriate for them as well.

[73] There are many philosophers (including my friends in the Minnesota Center) who would admit that the importance of linguistic analysis is very limited. However, they would still hold that its application is necessary in order to find out to what extent the advent of a new theory modifies the customary idiom. The considerations above would show that even this is granting too much and that one travels best without any linguistic ballast.

[74] One hears frequently that a *complete* replacement of the grammar and the terminology of the old language is impossible because this old language will be needed for introducing the new language and will, therefore, infect at least part of the new language. This is curious reasoning indeed if we consider that children learn languages without the help of

incommensurabilities may arise, and it does so without any linguistic detective work (which therefore turns out to be entirely unnecessary for the progress of knowledge).

What has just been said applies most emphatically to the relation between (theories formulated in) some commonly understood language and more abstract theories. That is, languages such as the 'everyday language', that notorious abstraction of contemporary linguistic philosophy, frequently contain (not explicity formulated, but implicit in the way in which its terms are used) principles which are inconsistent with newly introduced theories, and they must therefore either be abandoned and replaced by the language of the new and better theories even in the most common situations, or be completely separated from these theories (which would lead to a situation where it is possible to believe in various kinds of 'truth'). It is far from correct to assume that everyday languages are so widely conceived, so tolerant, indefinite and vague that they will be compatible with any scientific theory, that science can at most fill in details, and that a scientific theory will never run against the principles implicitly contained in them. The very opposite is the case. As will be shown later, everyday languages, like languages of highly theoretical systems, have been introduced in order to give expression to some theory or point of view, and they therefore contain a well-developed and sometimes very abstract ontology. It is very surprising that the champions of ordinary language should have such a low opinion of its descriptive power.

However, before turning to this part of the argument, I shall briefly discuss another example where the questionable principles of T' have been explicitly formulated, or can at least be easily unearthed.

The example which is dealt with by Nagel is the relation between phenomenological thermodynamics and the kinetic theory. Employing his own theory of reduction and, more especially, the condition I have quoted in n.11 above, Nagel claims that the terms of statements which have been derived from the kinetic theory (with the help of correlating hypotheses similar to (9)) will have the meanings they originally possessed within the phenomenological theory, and he repeatedly emphasizes that these meanings are fixed by 'its own procedures' (the procedures of the phenomenological theory) 'whether or not [this theory] has been, or will be, reduced to some other discipline'.[75]

As in the case of the impetus theory, we shall begin our study of the correctness of this assertion with an examination of these 'procedures' and 'usages'. More especially, we shall start with an examination of the usage of

---

a previously known idiom. Is it really asserted that what is possible for a small child will be impossible for a philosopher, a linguistic philosopher at that?

[75] [20], 301.

the term 'temperature', 'as fixed by the established procedures' of thermo-dynamics.

Within thermodynamics proper,[76] temperature ratios are defined by reference to reversible processes of operating between two levels, L' and L'', each of these levels being characterized by one and the same temperature throughout. The definition,

$$T':T'' = Q':Q'' \qquad (10)$$

identifies (after a certain arbitrary choice of units) the ratio of the tempera-tures with the ratio between the amount of heat absorbed at the higher level and the amount of heat rejected at the lower level. Closer inspection of the 'established usage' of the temperature thus defined shows that it is sup-posed to be

independent of the material of the substance chosen for the cycle, and unique.

$$(11)$$

This property can be inferred from the extension of the concept of tempera-ture thus defined to radiation fields, and from the fact that the constants of the main laws in this domain are universal, rather than dependent upon either the thermometric substance or the substance of the system investi-gated.

It can be shown by an argument not to be presented here that (10) and (11) taken together imply the second law of thermodynamics in its strict (phenomenological) form: the concept of temperature as 'fixed by the established usages' of thermodynamics is such that its application to concrete situations entails the strict (i.e. non-statistical) second law.

The kinetic theory, however, does not give us such a concept, whatever procedure is adopted. First of all, there does not exist any dynamical concept that possesses the required property. The statistical account, on the other hand, allows for fluctuations of heat back and forth between two levels of temperature and, therefore, again contradicts one of the laws implicit in the 'established usage' of the thermodynamic temperature. The relation between the thermodynamic concept of temperature and what can be defined in the kinetic theory, therefore, can be seen to conform to the pattern that has been described at the beginning of this section: we are again dealing with two incommensurable concepts. The same applies to the relation between the purely thermodynamic entropy and its statistical counterpart; whereas the latter admits of very general application, the former can be measured by infinitely slow reversible processes only. Taking all this into consideration we must admit that it is impossible to relate the

[76] See Fermi [22], section 9. I am talking about classical thermodynamics, not about Prigogine's thermodynamics of open systems.

kinetic theory and the phenomenological theory in the manner described by
Nagel, or to explain all the laws of the phenomenological theory in the
manner demanded by Hempel and Oppenheim on the basis of the statistical theory. Again replacement rather than incorporation, or derivation (with
the help, perhaps, of premises containing statistical as well as phenomenological concepts), is seen to be the process that characterizes the transition
from a less general theory to a more general one.

It ought to be pointed out that the discussion is very idealized. The
reason is that a purely kinetic account of the phenomena of heat does not
exist. What exists is a curious mixture of phenomenological and statistical
elements, and it is this mixture which has received the name 'statistical
thermodynamics'. However, even if this is admitted, it remains that the
concept of temperature as it is used in this new and mixed theory is different
from the original, purely phenomenological concept. To our point of view,
according to which terms change their meanings with the progress of
science, Nagel raises the following objection:

> The redefinition of expressions with the development of inquiry [so it is
> noted], is a recurrent feature in the history of science. Accordingly, though it
> must be admitted that in an earlier use the word 'temperature' had a meaning
> specified exclusively by the rules and procedures of thermometry and classi-
> cal thermodynamics, it is *now* so used that temperature is 'identical by
> definition' with molecular energy. The deduction of the Boyle–Charles law does
> not therefore require the introduction of a further postulate, whether in the
> form of a coordinating definition or a special empirical hypothesis, but simply
> makes use of this definitional identity. This objection illustrates the unwitting
> double talk into which it is so easy to fall. It is certainly possible to redefine the
> word 'temperature' so that it becomes synonymous with 'mean kinetic en-
> ergy'. But it is equally certain that on this redefined usage the word has a
> different meaning from the one associated with it in the classical science of
> heat, and therefore a meaning different from the one associated with the word
> in the statement of the Boyle–Charles law. However, if thermodynamics is to
> be reduced to mechanics, it is temperature in the sense of the term in the
> classical science of heat which must be asserted to be proportional to the
> mean kinetic energy of gas molecules. Accordingly, if the word 'temperature'
> is redefined as suggested by the objection, the hypothesis must be invoked
> that the state of bodies described as 'temperature' (in the classical thermo-
> dynamic sense) is also characterized by 'temperature' in the redefined sense
> of the term. This hypothesis, however, will then be one that does not hold as a
> matter of definition. . . . Unless this hypothesis is adopted, it is not the
> Boyle–Charles law which can be derived from the assumptions of the kinetic
> theory of gases. What is derivable without the hypothesis is a sentence similar
> in syntactical structure to the standard formulation of the law, but possessing
> a sense that is unmistakably different from what the law asserts.[77]

Let me at once admit the correctness of the last assertion. After all, it has
been my contention throughout this paper that extension of knowledge

[77] [61], 357–8.

leads to a decisive modification of the previous theories both as regards the quantitative assertions made and as regards the meanings of the main descriptive terms used. Applying this to the present case I shall therefore at once admit that incorporation into the context of the statistical theory is bound to change the meanings of the main descriptive terms of the phenomenological theory. The difference between Nagel and myself lies in the following. For me, such a change to new meanings and new quantitative assertions is a natural occurrence which is also desirable for methodological reasons (the last point will be established later in the present section). For Nagel such a change is an indication that reduction has not been achieved, for reduction in Nagel's sense is supposed to leave untouched the meanings of the main descriptive terms of the discipline to be reduced (cf. his 'if thermodynamics is to be reduced to mechanics, it is temperature in the sense of the term in the classical science of heat which must be asserted to be proportional to the mean kinetic energy of gas molecules'). 'Accordingly', he continues, quite obviously assuming that reduction in his sense can be carried through 'if the word "temperature" is redefined as suggested by the objection, the hypothesis must be invoked that the state of bodies described as "temperature" (in the classical thermodynamic sense) is also characterized by "temperature" in the redefined sense of the term. *This* hypothesis ... will then be one that does not hold as a matter of definition.' It will also be a false hypothesis because the conditions for the definition of the phenomenological temperature are never satisfied in nature (see the arguments above in the text and compare also the arguments in connection with formula (9)), which is only another sign of the fact that reduction, in Nagel's sense, of the phenomenological theory to the statistical theory is not possible (obviously the additional premises used in the reduction are not supposed to be false). Once more arguments of meaning have led to quite unnecessary complications.

Further examples exhibiting the same features can be easily provided. Thus in classical, pre-relativistic physics the concept of mass (and, for that matter, the concept of length and the concept of duration) was absolute in the sense that the mass of a system was not influenced (except perhaps causally) by its motion in the coordinate system chosen. Within relativity, however, mass has become a relational concept whose specification is incomplete without indication of the coordinate system to which spatiotemporal descriptions are all to be referred. Of course, the values obtained on measurement of the classical mass and of the relativistic mass will agree in the domain D', in which the classical concepts were first found to be useful. This does not mean that what is measured is the same in both cases: what is measured in the classical case is an *intrinsic property* of the system under consideration; what is measured in the case of relativity is a *relation* between the system and certain characteristics of D'. It is also impossible to define the exact classical concepts in relativistic terms or to relate them with the help of an empirical generalization. Any such procedure would imply the false assertion that the velocity of light is infinitely large. It is therefore again necessary to abandon completely the classical conceptual scheme once the theory of relativity has been introduced; and this means that it is imperative to use relativity in the theoretical considerations put forth for the explanation of a certain phenomenon as well as in the observation language in which tests for these considerations are to be formulated; for the empirically untenable consequences of the attempts above to give a reduction of classical terms to relativistic terms emerges whether or not the elements of the definition belong to the observation language.

Many more examples can be added to those discussed in the present paper (including, the impetus theory, phenomenological thermodynamics and the classical conception of mass). They show that the postulate of meaning invariance is incompatible with actual scientific practice. That is, it has been shown that in most cases it is impossible to relate successive scientific theories in such a manner that the key terms they provide for the description of a domain D', where they overlap and are empirically adequate, either possess the same meanings or can at least be connected by empirical generalizations. It is also clear that the methodological arguments against meaning invariance will be the same as the arguments against the derivability condition and the consistency condition, since the demand for meaning invariance implies the demand that the laws of later theories be compatible with the principles of the context of which the earlier theories are part, and this demand is a special case of condition (5). Using our earlier arguments against (5) we may now infer the untenability, on methodological grounds, of meaning invariance as well. And as our argument is quite general we may also infer that it is undesirable that the 'ordinary' usage of terms be preserved in the course of the progress of knowledge. Wherever such preservation is observed, we shall feel inclined to think that the suggested new theories are not as revolutionary as they perhaps ought to be, and we shall have the suspicion that some *ad hoc* procedures have been adopted. Violation of ordinary usage, and of other 'established' usages, on the other hand, is a sign that real progress has been made, and it is to be welcomed by anybody interested in such progress (provided of course that this violation is connected with the suggestion of a new point of view or a new theory and is not just the result of linguistic arbitrariness).

Our argument against meaning invariance is simple and clear. It proceeds from the fact that usually some of the principles involved in the determination of the meanings of older theories or points of view are inconsistent with the new, and better, theories. It points out that it is natural to resolve this contradiction by eliminating the troublesome and unsatisfactory older principles and to replace them by principles, or theorems, of the new theory. And it concludes by showing that such a procedure will also lead to the elimination of the old meanings and thereby to the violation of meaning invariance.

The most important method used for escaping the force of this clear and simple argument is the transition to instrumentalism. Instrumentalism maintains that the new theory must not be interpreted as a series of statements, but that it is rather to be understood as a predictive machine whose elements are tools rather than statements and therefore cannot be incompatible with any principle already in existence. This very popular move (popular, that is, because it is also used by scientists) admittedly cuts the ground from beneath our argument and makes it inapplicable. However, it has never been explained why a new and satisfactory theory should be interpreted as an instrument, whereas the principles behind the established usage, which can easily be shown to be empirically inadequate, are not so interpreted. After all, the only advantage of the latter is that they are familiar – an advantage which is a psychological and historical accident and which should therefore not have any influence upon questions of interpretation and of reality. One may try to answer this criticism by ascribing an instrumental function to all principles, old or new, and not only to those contained in the most recent theory. Such a procedure means acceptance of a sense-datum account of knowledge. Having shown elsewhere that such an account is impossible,[78] I can now say that this consequence of a universal instrumentalism is tantamount to its refutation. The result is that neither a restricted nor a universal instrumentalism can be carried through in a satisfactory manner. This disposes of the instrumentalist move.

While instrumentalism possesses at least a semblance of plausibility the arguments to be discussed now are devoid even of this feature. Indeed, I am very hesitant to apply the world 'arguments' to these expressions of confused thinking, their wide acceptance and asserted self-evidence notwithstanding. Consider for example the following question (which is supposed to be a criticism of our suggestion that after the acceptance of the kinetic theory the word 'temperature' will be in need of re-interpretation):[79] 'If the meaning of "temperature" is [now] the same as that of "mean kinetic energy of molecular motion", what are we talking about when milk is said to have a temperature of 10° Cels? Surely not the kinetic energy of the molecular constituents of the liquid, for the uninstructed layman is able to understand what is said here without possessing any notion about the molecular composition of the milk.'

[78] See [33].
[79] The argument in connection with this question can be found in [20], 293. It is not clear to me whether or not Nagel would be prepared to support the argument.

It may be quite correct that the 'uninstructed layman' does not think of molecules when speaking about the temperature of his milk and that he has not the slightest notion of the molecular constitution of the liquid either.[80] However, what has he to do with our argument, according to which a person who has already accepted and understood the theory of the molecular constitution of gases, liquids and solids cannot at the same time demand that the pre-molecular concept of temperature be retained? It is not at all denied by our argument that the uninstructed layman may possess a concept of temperature that is very different from the one connected with the molecular theory (after all, some uninstructed laymen, intelligent clergymen included, still believe in ghosts and in the devil). What is denied is that anybody can consistently continue using this more primitive concept *and at the same time believe in the molecular theory*. This does not mean that a person may not, on different occasions, use concepts which belong to different and incommensurable frameworks. The only thing that is forbidden by him is the use of both kinds of concepts in the same argument; for example, he may not use the one kind of concept in his observation language and the other kind in his theoretical language. Any such combination – and this is the gist of our considerations in this section – would introduce principles which are mutually inconsistent and thereby destroy the argument in which it is supposed to occur. It is evident that this position is not at all endangered by the objection implied by the question above.

However, quite apart from being so obviously irrelevant to our thesis the objection reflects an attitude that must appear quite incredible to anybody who possesses even the slightest acquaintance with the history of knowledge. The question insinuates that the laymen's ability to handle the word 'temperature' according to the rules prescribed for it in some simple idiom indicates his understanding of the thermal properties of bodies. It insinuates that the existence of an idiom allows us to infer the truth of the principles which underlie this idiom. Or, to be more specific, it insinuates that what is being used is, on that account alone, already exhibited as adequate, useful and perhaps irreplaceable. After all, the reference to the layman's understanding of the world 'temperature' is not made without purpose. It is made with the purpose of preserving the common meaning of

[80] By the way, who *is* this uninstructed layman? From the purpose for which he is being employed in many arguments, it would seem to emerge that he is not supposed to know much science, or much politics, or much religion, or much of anything. This means that in these times of mass communication and mass education he must be very careful not to read the wrong parts of his newspaper, he must be careful not to leave his television set on for too long a time, and he must also not allow himself to converse too much with his friends, his children etc. That is, he must be either a savage or an idiot. I really wonder what are the motives which lead to a philosophy where the most interesting language is the language of savages or of idiots.

this word since, it is alleged, this common meaning can be understood and is not in need of replacement. The discussion of a specific example will at once show the detrimental effect of any such procedure.

The example chosen now brings us to the second part of this section where we investigated the relation, not between explicitly formulated theories, but between a theory and the implicit principles that govern the usage of the descriptive terms of some idiom. As was said a little earlier, it is our conviction that everyday languages, far from being so widely and generally conceived that they can be made compatible with any scientific theory, contain principles that may be inconsistent with some very basic laws. It was also pointed out that these principles are rarely expressed in an explicit manner (except, perhaps, in those cases where there is an attempt to defend the corresponding idiom against replacement or change) but that they are implicit in the rules that govern the use of its main descriptive terms. And our point was that, once these principles are found to be empirically inadequate, they must be given up and with them the concepts that are obtained by using terms in accordance with them. Conversely, the attempt to retain these concepts will lead to the conservation of false laws and to a situation where every connection between concepts or facts is severed.

The example which I have chosen to show this involves the pair 'up–down'. There existed a time when this pair was used in an absolute fashion, without reference to a specified centre, such as the centre of the earth. That it was used in such a manner can be easily seen from the 'vulgar' remark that the antipodes would 'fall off' the earth if the earth were spherical,[81] as well as from the more sophisticated attempts of Thales, Xenophanes and others to find support for the earth as a whole, assuming that it would otherwise fall 'down'.[82] These attempts, as well as that remark about the antipodes, employ two assumptions: first, that any material object is under the influence of a force; second, that this force acts in a privileged direction in space which must therefore be regarded as anisotropic. It is this privileged direction to which the pair 'up–down' refers. The second assumption is not explicitly made; it can only be derived from the way in which the pair 'up–down' is used in arguments such as those mentioned above.[83] We have here an example of a cosmological assumption (the anisotropic character of space) implicit in the common idiom.

This example refutes the thesis which has been defended by some

---

[81] For a discussion of this remark and of a related 'vulgar' remark concerning the shape and arrangement of the terrestrial waters, see Pliny [64], II, 161–6, quoted in Cohen and Drabkin [19], 159–61.

[82] For a description and criticism of these attempts, see [1], 294a 12ff; also quoted in [19], 143–8.

[83] For the atomist's conception of space, which, at least since Epicurus, seems to be influenced by the popular ideas discussed above, see M. Jammer [48], 11.

philosophers that everyday languages are fairly free from hypothetical elements and therefore ideally suited as observational languages.[84] It refutes the thesis by showing that even the most harmless part of a common idiom may rest upon very far-reaching assumptions and must therefore be regarded as hypothetical to a very high degree.

Another remark concerns the changes of meaning needed once the Newtonian (or perhaps even the Aristotelian) explanation of the fall of heavy bodies is adopted. Newtonian space is isotropic and homogeneous. Hence, accepting this theory, one can no longer use the pair 'up–down' in the previous fashion and at the same time assume that one is describing actual features of physical situations. More especially, one cannot retain the absolute use of this pair for the description of observable features, since such features are quite obviously assumed to exist. Any person accepting Newton's physics and the conception of space it contains must, therefore, give a new meaning even to such a familiar pair of terms as the pair 'up–down', and he must now interpret it as a relation between the direction of a motion and a centre that has been fixed in advance. And as Newton's theory is preferable, on empirical grounds, to the older and 'absolute' cosmology, it follows that the relational usage of the pair 'up–down' will be preferable too. Conversely, the attempt to retain the old usage amounts to retaining the old cosmology, and this despite the discoveries which have shown it to be obsolete.

To this argument it may be, and has been, objected that the vulgar usage of the pair 'up–down' was never supposed to be so general as to be applicable to the universe as a whole. This may be the case (although I do not see any reason for assuming that 'ordinary' people are so very cautious as to apply the pair to the surface of the earth only; all the passages referred to in the above quotations contradict this assumption and so does the fact that at all times real ordinary people – and not only their Oxford substitutes – were very much interested in celestial phenomena.[85] However, even such a restriction would not invalidate our argument. It would rather show that the pair was used for singling out an absolute direction near the surface of the earth and that it did not assume such a direction to exist throughout the universe. It is clear that even this modest position is incompatible with the ideas implicit in the Newtonian point of view, which does not allow for local anisotropies either.

Consider now, after this example, the following argument in favour of the thesis that what is being used is, on that account alone, already exhibited as

---

[84] This thesis was introduced by Herbert Feigl in discussions with me. For my own position, see also Phillip Frank [40].

[85] The reason why Oxford philosophers so rarely discuss the influence of astronomy upon everyday languages may perhaps be found in the weather of their favourite discussion place. However, this reason unfortunately does not explain their ignorance in physics, theology, mythology, biology and even linguistics.

adequate, useful, and perhaps irreplaceable. The argument is Austin's, and it has been repeated by G. J. Warnock.[86]

> Language [writes Warnock] is to be used for a vast number of highly important purposes; and it is at the very least unlikely that it should contain either much more, or much less, than those purposes require. If so, the existence of a number of different ways of speaking is very likely indeed to be an indication that there is a number of different things to be said. . . . Where the topic at issue really is one that does constantly concern most people in some practical way – as for example perception, the ascription of responsibility, or the assessment of human character or conduct – then it is certain that everyday language is as it is for some extremely good reasons; its verbal variety is certain to provide clues to important distinctions.[87]

If I understand this passage correctly, it means that the existence of certain distinctions in a language may be taken as an indication of similar distinctions in the nature of things, situations and the like. And the reason for this is that people who are in constant contact with things and situations will soon develop the correct linguistic means for describing their properties. In short, human beings are good inductive machines in domains of concentrated interested, and their inductive ability will be the better the greater their concern, or the greater the practical value of the topic treated. Consequently, languages containing distinctions of practical interest are very likely to be adequate and irreplaceable.

There are many objections against this train of reasoning. First of all, it would seem to be somewhat arbitrary to restrict interests to those which can be derived from the immediate necessities of the physical life of the human race. From history we learn that the motives emerging from abstract considerations such as those found in a myth, or in a theo-astronomical system, are at least as strong as the more pedestrian motives connected with the immediate fulfilment of material needs (after all, people have killed and died for their convictions). If a language can be trusted because of the commitment of those who use it, and if commitment is found to range over a much wider area than had first been imagined – over physics, astronomy (think of Giordano Bruno), and biology – then the result will be that the principle we are discussing (that what is being used for a purpose is on that account alone already useful and irreplaceable) must be applied to any language and any theory that has ever been developed and seriously tested. However – and this is the second point – there exist many theories and languages which have been found to be inadequate, and this despite their usefulness and despite the zeal of those

---

[86] [72], 150–1.
[87] Astronomy is again omitted. It would seem to me that problems of astronomy had a much greater influence upon the formation of our language than problems of perception, which are of a very ephemeral nature and also are very technical. The skies and the stars (which, after all, were assumed to be gods) were everyone's concern.

who developed them. This applies to the language of Aristotelian physics, which had to be introduced into medieval thinking under very great difficulties and whose influence went much further than is sometimes realized; it applies to the language of the physics of Newton; and it applies to many other languages. Of course, this is the result one would expect; success under even very severe tests does not guarantee infallibility; no amount of commitment or success can guarantee the perennial reliability of inductions.

The principle which we have been discussing does not occur only in philosophy. Bohr's contention that the account of all quantum-mechanical evidence must forever 'be expressed in classical terms' has been defended in a very similar manner.[88] According to Bohr, we need our classical concepts not only if we want to give a summary of facts; without these concepts the facts to be summarized could not even be stated. Like Kant before him, Bohr observes that our experimental statements are always formulated with the help of certain theoretical terms and that the elimination of these terms would lead, not to the 'foundations of knowledge' as a positivist would have it, but to complete chaos. 'Any experience', he asserts, 'makes its appearance within the frame of our customary points of view and forms of perception'[89] – and at the present moment the forms of perception are those of classical physics.

Does it follow, as is asserted by Bohr, that we can *never* go beyond the classical framework and that therefore all our future microscopic theories will have to use the notion of complementarity as a fundamental notion?

It is quite obvious that the actual use of classical concepts for the description of experiments within contemporary physics can never justify such an assumption, even if these concepts happen to have been very successful in the past (Hume's problem). For a theory may be found whose conceptual apparatus, when applied to the domain of validity of classical physics, would be just as comprehensive and useful as the classical apparatus without coinciding with it. Such a situation is by no means uncommon. The behaviour of the planets, of the sun, and of the satellites can be described both by the Newtonian concepts and by the concepts of general relativity. The order introduced into our experiences by Newton's theory is retained and improved upon by relativity. This means that the concepts of relativity are sufficiently rich for the formulation of all the facts which were stated before with the help of Newtonian physics. Yet the two sets of concepts are completely different and bear no logical relation to each other.

Other examples of the same kind can be provided very easily. What we are dealing with here is, of course, again the old problem of induction. No

---

[88] See p. 22 above.
[89] [5], 1. For a more detailed account of what follows, see [32], [34], [36], [37], [38]. But see [36] and ch. 1 for a defence of this procedure.

number of examples of the usefulness of an idiom is ever sufficient to show that the idiom will have to be retained forever. And if it is objected, as it has been in the case of the quantum theory, that the language of classical physics is the only actual language in existence for the description of experiments,[90] then the reply must be that man is not only capable of using theories and languages but that he is also capable of inventing them.[91] How else could it have been possible, to mention only one example, to replace the Aristotelian physics and the Aristotelian cosmology with the new physics of Galileo and Newton? The only conceptual apparatus then available was the Aristotelian theory of change with its opposition of actual and potential properties, the four causes, and the like. Within this conceptual scheme, which was also used for the description of experimental results, Galileo's (or rather Descartes') law of inertia does not make sense, nor can it be formulated. Should, then, Galileo have followed Heisenberg's advice and have tried to get along with the Aristotelian concepts as well as possible, since his 'actual situation . . . [was] such that [he did] use the Aristotelian concepts'[92] and since 'there is no use discussing what could be done if we were other beings that we are'? By no means. What was needed was not improvement or delimitation of the Aristotelian concepts; what was needed was an entirely new theory. This concludes our argument against the principle that a useful language is to be regarded as adequate and irreplaceable and, thereby, fully restores the force of our attack against meaning invariance, as well as reinforces the positive suggestions made in connection with this attack and especially the idea that conceptual changes may occur anywhere in the system that is employed at a certain time for the explanation of the properties of the world we live in.

As I indicated in the introductory discussion, this transition from a point of view which demands that certain 'basic' terms retain their meaning, come what may, to a more liberal point of view which allows for changes anywhere in the system employed is bound to influence profoundly our attitude with respect to many philosophical problems and will also facilitate their solution.[94] Let me take the mind–body problem as an example. It seems to me that the difficulties of this problem are to be sought precisely in the fact that meaning invariance is regarded as a necessary condition of its satisfactory solution. That is, it is assumed, or even demanded, that the meanings of at least some terms of the problem must remain constant throughout the discussion of the problem and further that these terms must retain their meanings in the solution as well.

Of course, different schools will apply the demand for meaning

[90] See Heisenberg [44], 56, and von Weizsaecker [73], 110.

[91] See also n.74, above.

[92] This is a paraphrase of a passage in Heisenberg [44], 56.

[94] Note, however, that ethical arguments may restrict the change of our ideas concerning man: we may not want a superman.

invariance to different concepts. A Platonist will demand that terms such as 'mind' and 'matter' remain unchanged, whereas an empiricist will require that some observational terms, such as the term 'pain', or the more abstract term 'sensation', retain their (common) meaning. Now a closer analysis of these key terms will, I think, reveal that they are incommensurable in exactly the sense in which this term was defined at the beginning of this section. This being the case, it is of course completely impossible either to reduce them to each other, or to relate them to each other with the help of an empirical hypothesis, or to find entities which belong to the extension of both kinds of terms. That is, the conditions under which the mind–body problem has been set up as well as the particular character of its key terms are such that a solution is forever impossible: a solution of the problem would require combining what is incommensurable without allowing for a modification of meanings which would eliminate this incommensurability.

All these difficulties disappear if we are prepared to admit that, in the course of the progress of knowledge, we may have to abandon a certain point of view and the meanings connected with it – for example, if we are prepared to admit that the mental connotation of mental terms may be spurious and in need of replacement by a physical connotation according to which mental events, such as pains, states of awareness, and thoughts, are complex physical states of either the brain or the central nervous system, or perhaps the whole organism. I personally happen to favour this idea that at some time sensations will turn out to be fairly complex central states which therefore possess a definite location inside the human body (which need not coincide with the place where the sensation is *felt* to be). I also hope that it will be possible to carry out a similar analysis of all so-called mental states.

Whatever the merit of this belief of mine, it cannot be refuted by reference to the fact that what we 'mean' by 'a sensation', or by 'a thought', is nothing that could have a location, an internal structure, or physical ingredients. For if my belief is correct, and it it is indeed possible to develop a materialistic theory of human beings, then we shall of course be forced to abandon the mental connotations of the mental terms, and we shall have to replace them by physical connotations. According to the point of view which I am defending here, the only legitimate way of criticizing such a procedure would be to criticize this new materialistic theory by either showing that it is not in agreement with experimental findings or pointing out that it possesses some undesirable formal features (e.g. by pointing out that it is *ad hoc*). Linguistic counter-arguments have, I hope, been shown to be completely irrelevant.

The considerations in these last paragraphs are of course very sketchy. Still, I hope that they give the reader an indication of the changes implied by the renunciation of the principle of meaning invariance as well as of the nefarious influence this principle has had upon traditional philosophy (modern empiricism included).

8. SUMMARY AND CONCLUSION

Two basic assumptions of the orthodox theory of reduction and explanation have been found to be in disagreement with actual scientific practice and with reasonable methodology. The first assumption was that the explanandum is *derivable* from the explanans. The second assumption was that *meanings are invariant* with respect to the process of reduction and explanation. We may sum up the results of our investigation in the following manner:

Let us assume that T and T' are two theories satisfying the conditions outlined at the beginning of section 3. Then, from the point of view of scientific method, T will be most satisfactory if it is *inconsistent* with T' in the domain where they both overlap;[95] and if it is *incommensurable* with T'.

It is clear that a theory which is satisfactory according to the criterion just pronounced will not be capable of functioning as an explanans in any explanation or reduction that satisfies the principles put forth by Hempel and Oppenheim or Nagel. Paradoxically speaking: *Hempel–Oppenheim explanations cannot use satisfactory theories as explanantia. And satisfactory theories cannot function as explanantia in Hempel–Oppenheim explanations.* How is the theory of explanation and reduction to be changed in order to eliminate this very undesirable paradox?

It seems to me that the changes that are necessary will make it impossible to retain a formal theory of explanation, because these changes will introduce pragmatic or 'subjective' considerations into the theory of explanation. This being the case, it seems perhaps advisable to eliminate altogether considerations of explanation from the domain of scientific method and to concentrate upon those rules which enable us to compare two theories with respect to their formal character and their predictive success and which guarantee the constant modification of our theories in the direction of greater generality, coherence and comprehensiveness. I shall now give a more detailed outline of the reasons which have prompted me to adopt this pragmatic point of view.

Consider again T and T' as described above. Under these circumstances, the set of laws T'' following from T inside D' will either be inconsistent with T' or incommensurable with it. In what sense, then, can T be said to explain T'? This question has been answered by Popper for the case of the inconsistency of T' and T''.

---

[95] This condition has been discussed with great clarity by Popper in [67]. It was this discussion (as well as dissatisfaction with [60]) that was the starting point of the present analysis of the problem of explanation.

Newton's theory [he says] unifies Galileo's and Kepler's. But far from being a mere conjunction of these two theories – which play the part of *explicanda* for Newton – *it corrects them while explaining them*. The original explanatory task was the deduction of the earlier results. It is solved, not by deducing them, but by deducing something better in their place: new results which, under the special conditions of the older results, come numerically very close to these older results, and at the same time correct them. Thus the empirical success of the old theory may be said to corroborate the new theory; and in addition, the corrections may be tested in their turn. . . . . What is brought out strongly by [this] . . . situation . . . is the fact that the new theory cannot possibly be *ad hoc*. . . . . Far from repeating its *explicandum*, the new theory contradicts it and corrects it. In this way, even the evidence of the *explicandum* itself becomes independent evidence for the new theory.[96]

J. W. N. Watkins suggested to me that this theory may be summarized as follows. Explanation consists of two steps. The first step is a derivation, from T, of those laws which obtain under the conditions characterizing D'. The second step is a comparison of T''' and T' and the realization that both are empirically adequate, i.e. fall within the domain of uncertainty of the observational results: T explains T' satisfactorily only if T is true and there exists a consequence T''' of T for the conditions of validity of T' such that T''' and T' are at least equally strong and also experimentally indistinguishable.

The first question that arises in connection with Mr Watkins' formulation is this: experimentally indistinguishable on the basis of which observations? T' and T''' may be indistinguishable by the crude methods used at the time when T was first suggested, but they may well be distinguishable on the basis of later and more refined methods. Reference to an observational method will therefore have to be included in the clause of experimental indistinguishability. The notion of explanation will be relative to this observational material. It will not make sense any longer to ask whether or not T explains T'. The proper question will be whether T explains T' given the observational material, or the observational methods O. Using this new mode of speech we are forced to deny that Kepler's laws are explained by Newton's theory relative to the present observations – and this is perfectly in order; for these observations refute Kepler's laws and thereby eliminate the demand for explanation. It seems to me that this theory can well deal with all the problems that arise when T and T' are commensurable, but inconsistent inside D'. It does not seem to me that it can deal with the case where T' and T are incommensurable. The reason is as follows.

As soon as reference to certain observational material has been included in the characterization of what counts as a satisfactory explanation, in the very same moment the question arises as to how this observational material is to be presented. If it is correct, as has been argued all the way through the present paper, that the meanings of observational terms depend on the theory on behalf of which the observations have been made, then the observational material referred to in this modified sketch of explanation must be presented in terms of this theory also. Now incommensurable theories may not possess any comparable consequences, observational or otherwise. Hence, there may not exist any possibility of finding a characterization of the observations which are supposed to confirm two incommensurable theories. How, then, is the above account of explanation to be modified to cover the case of incommensurable theories also?[97]

It seems to me that the only possible way lies in closest adherence to the pragmatic theory of observation. According to this theory, it will be remembered, we must carefully distinguish between the *causes* of the production of a certain observational sentence, or the features of the process of production, on the one side, and the *meaning* of the sentence produced in this manner on the other. More especially, a sentient being must distinguish between the fact that he possesses a sensation, or disposition to verbal behaviour, and the interpretation of the sentence being uttered in the presence of this sensation, or terminating this verbal behaviour. Now our theories, apart from being pictures of the world, are also instruments of prediction. And they are good instruments if the information they provide, taken together with information about initial conditions characterizing a certain observational domain D_o would enable a robot without sense organs, but with this information built into it, to react in this domain in exactly the same manner as sentient beings who, without knowledge of the theory, have been trained to find their way about D_o and who are able to answer, 'on the basis of observation', many questions concerning their surroundings.[98] This is the criterion of predictive success, and it is seen not at all to involve reference to the *meanings* of the reactions carried out either by the robot or by the sentient beings (which latter need not be humans, but can also be other robots). All it involves is *agreement of behaviour*.

This criterion involves 'subjective' elements. Agreement is demanded between the behaviour of (non-sentient, but theory-fed) robots and that of sentient beings, and it is thereby assumed that the latter possesses a privileged position. Considering that perceptions are influenced by belief in theories, this criterion would seem to be somewhat arbitrary. It is easily seen, however, that it cannot be replaced by a less arbitrary and more 'objective' criterion. What would such an objective criterion be? It would be a criterion which is either based upon behaviour that is not connected with any theoretical element – and this is impossible (cf. my criticism of the theory of sense data above) – or it would be behaviour that is tied up with an

---

[97] As Herbert Feigl has pointed out to me, this difficulty also arises in the case of crucial experiments. See ch. 2.

[98] Of course, the *motivations* of the robot and of the sentient being must also be the same.

irrefutable and firmly established theory, which is equally impossible. We have to conclude, therefore, that a formal and 'objective' account of explanation cannot be given.

REFERENCES

1. Aristotle. *De Coelo.*
2. Aristotle. *De Generatione et Corruptione.*
3. Barker, S. 'The Role of Simplicity in Explanation', in *Current Issues in the Philosophy of Science*, ed. H. Feigl and G. Maxwell (New York, 1961), 265ff.
4. Bohm, D. *Causality and Chance in Modern Physics* (London, 1957).
5. Bohr, N. *Atomic Theory and the Description of Nature* (Cambridge, 1932).
6. Bohr, N. 'Discussions with Einstein', in *Albert Einstein, Philosopher–Scientist*, ed. P. A. Schilpp (Evanston, Ill., 1948), 201ff.
7. Born, M. *Natural Philosophy of Cause and Chance* (Oxford, 1948).
8. Brecht, B. *Schriften zum Theater* (Berlin and Frankfurt/Main, 1957).
9. Bunge, M. *Causality* (Cambridge, Mass., 1959).
10. Burnet, J. *Early Greek Philosophy* (London, 1930).
11. Carnap, R. 'Die Physikalische Sprache als Universalsprache der Wissenschaft', *Erkenntnis*, (1932), 432ff.
12. Carnap, R. 'Psychologie in Physikalischer Sprache', *Erkenntnis*, 3 (1933), 107ff.
13. Carnap, P. 'Über Protokollsätze', *Erkenntnis*, 3 (1933), 215ff.
14. Carnap, R. 'Testability and Meaning', *Phil. Sci.*, 3 (1936), 419ff, and (1937), 1ff.
15. Carnap, R. 'The Methodological Character of Theoretical Concepts', in *Minnesota Studies in the Philosophy of Science*, ed. H. Feigl and M. Scriven (Minneapolis, 1956), I, 38ff.
16. Clagett, M. *Greek Science in Antiquity* (London, 1957).
17. Clagett, M. *The Science of Mechanics in the Middle Ages* (Madison, 1959).
18. Conant, J. B. *Case Histories in the Experimental Sciences* (Cambridge, Mass., 1957), I.
19. Cohen, M. R. and Drabkin, I. E. eds. *A Source Book in Greek Science* (New York, 1948).
20. Danto, A. and Morgenbesser, S. eds. *Philosophy of Science* (New York, 1960).
21. Dewey, John. *The Quest for Certainty* (New York, 1960).
22. Fermi, E. *Thermodynamics* (New York, 1956).
23. Feigl, H. and Brodbeck, M. eds. *Readings in the Philosophy of Science* (New York, 1953).
24. Feigl, H. and Maxwell, G. eds. *Current Issues in the Philosophy of Science* (New York, 1961).
25. Feigl, H. and Scriven, M. eds. *Minnesota Studies in the Philosophy of Science* (Minneapolis, 1956), I.
26. Feigl, H., Scriven M., and Maxwell, G. eds. *Minnesota Studies in the Philosophy of Science* (Minneapolis, 1958), II.
27. Feyerabend, P. K. 'Carnap's Theorie der Interpretation Theoretischer Systeme', *Theoria*, 21 (1955), 21ff.

28. Feyerabend, P. K. 'Wittgenstein's *Philosophical Investigations*', vol. 2, ch. 7.
29. Feyerabend, P. K. 'Eine Bemerkung zum Neumannschen Beweis', *Phys.*, 14 (1956), 421ff.
30. Feyerabend, P. K. 'On the Quantum Theory of Measurement', this volume, ch. 13.
31. Feyerabend, P. K. 'An Attempt at a Realistic Interpretation of Experience', this volume, ch. 2.
32. Feyerabend, P. K. 'Complementarity', *2PROC. Arist. Soc.*, Suppl., 32 (1958), 75ff.
33. Feyerabend, P. K. 'Das Problem der Existenz Theoretischer Entitaeten', *Probleme der Wissenschaftstheorie* (Vienna, 1960), 35ff.
34. Feyerabend, P. K. 'O. Interpretacji Relacyj Nieokreslonosci', *Stud. Filoz.*, 19 (1960), 23ff.
35. Feyerabend, P. K. 'Patterns of Discovery', *Phil. Rev.*, 59 (1960), 247ff.
36. Feyerabend, P. K. 'Professor Bohm's Philosophy of Nature', this volume, ch. 14.
37. Feyerabend, P. K. 'Bohr's Interpretation of the Quantum Theory', in *Current Issues in the Philosophy of Science*, ed. H. Feigl and G. Maxwell (New York, 1961), 371ff.
38. Feyerabend, P. K. 'On the Interpretation of Microphysical Theories', unpublished paper.
39. Feyerabend, P. K. 'On the Interpretation of Scientific Theories', this volume, ch. 3.
40. Frank, P. *Relativity, a Richer Truth* (Boston, 1950).
41. Fuerth, R. 'Über einige Beziehungen Zwischen Klassischer Statistik und Quantenmechanik', *Z. Phys.*, 81 (1933), 143ff.
42. Goodman, N. *Fact, Fiction, and Forecast* (Cambridge, Mass., 1955).
43. Hanson, N. R. *Patterns of Discovery* (Cambridge, 1958).
44. Heisenberg, W. *Physics and Philosophy* (New York, 1958).
45. Hempel, C. G. 'Studies in the Logic of Confirmation', *Mind*, 54 (1945), 1ff, 97ff.
46. Hempel, C. G. 'A Logical Appraisal of Operationism', in *Validation of Scientific Theories*, ed. P. Frank (Boston, 1954), 52ff.
47. Hempel, C. G. and Oppenheim, P. 'Studies in the Logic of Explanation', *Phil. Sci.* 15 (1948), 135ff.
48. Jammer, M. *Concepts of Space* (Cambridge, Mass., 1957).
49. Körner, S., ed. *Observation and Interpretation* (London, 1957).
50. Körner, S. *Conceptual Thinking* (New York, 1960).
51. Kuhn, T. S. *The Copernican Revolution* (New York, 1959).
52. Landau, L. D., and E. M. Lifschitz. *Quantum Mechanics* (Reading, Mass., 1958).
53. Mach, E. *Waermelehre* (Leipzig, 1897).
54. Mach, E. *Zwei Aufsaetze* (Leipzig, 1912).
55. McLaurin, C. *An Account of Sir Isaak Newton's Philosophical Discoveries* (London, 1750).
56. Matson, W. I. 'The Naturalism of Anaximander', *Rev. Metaphys.*, 6 (1953), 387ff.

57. Matson, W. I. 'Cornford and the Birth of Metaphysics', *Rev. Metaphys.*, 8 (1955), 443ff.

58. Maier, A. *Die Vorlaeufer Galileis im 14, Jahrhundert* (Rome, 1949).

59. Morris, E. 'Foundation of the Theory of Signs', *International Encyclopaedia of Unified Science* (Chicago, 1942), section II/7.

60. Nagel, E. 'The Meaning of Reduction in the Natural Sciences', in *Science and Civilization*, ed. R. C. Stauffer (Madison, 1949), 99ff.

61. Nagel, E. *The Structure of Science* (New York, 1961).

62. Neumann, J. von *Mathematical Foundations of Quantum Mechanics* (Princeton, 1957).

63. Panofsky, E. 'Galileo as a Critic of the Arts', *Isis*, 47 (1956), 3ff.

64. Pliny, *Natural History*.

65. Popper, K. R. 'Naturgesetze und Theoretische Systeme', in *Gesetz und Wirklichkeit*, ed. S. Moser (Innsbruck, 1948), 65ff.

66. Popper, K. R. *The Open Society and Its Enemies* (Princeton, 1950).

67. Popper, K. R. 'The Aim of Science', *Ratio*, 1 (1957), 24ff.

68. Popper, K. R. 'Back to the Pre-Socratics', *Proc. Arist. Soc.*, New Series, 54 (1959), 1ff.

69. Popper, K. R. *The Logic of Scientific Discovery* (New York, 1959).

70. Reichenbach, H. *Experience and Prediction* (Chicago, 1948).

71. Sellars, W. 'The Language of Theories', in *Current Issues in the Philosophy of Science*, ed. H. Feigl and G. Maxwell (New York, 1961), 57ff.

72. Warnock, J. *British Philosophy in 1900* (Oxford, 1956).

73. Weizsaecker, C. F. von. *Zum Weltbild der Physik* (Leipzig, 1954).

74. Whorff, B. L. *Language, Thought, and Reality: Selected Writings*, ed. John B. Carroll, (Cambridge, Mass., 1956).

75. Wittgenstein, L. *Philosophical Investigations* (Oxford, 1953).

5
On the 'meaning' of scientific terms

1. In his criticism of Ryle, Hanson and me, Peter Achinstein[1] notices 'considerable oversimplification[s]' and discusses 'paradoxical consequences' arising therefrom. He points out that meanings do not always change with the theory to which they belong and suggests the existence of 'various kinds and degrees of dependence as well as independence' between terms and theories. He believes that awareness of these different kinds and degrees will eliminate the paradoxes and support two assumptions he finds plausible, namely '(A) . . . it is possible to understand at least some terms employed in a . . . theory before (and hence without) learning the principles of that theory'; and '(B) It must be possible for two theories employing many of the same terms to be incompatible . . . And this presupposes that at least some of the common terms have the same meaning in both theories'.

This belief of Achinstein's seems to be refuted by the existence of pairs of theories that may be regarded as competitors and yet do not share any element of meaning. Attention to 'various kinds and degrees of dependence' clearly cannot eliminate such cases – it will rather bring them to the fore. Two examples exhibiting the property just described will be discussed in the next two sections. It will then be argued that, from the point of view of scientific progress, the examples are to be welcomed. It will also be shown that (B), despite its *prima facie* plausibility, is of a very dubious nature. Finally, we shall arrive at the result that, in the decision between competing theories, 'meanings' play a negligible part and that attention to the 'variety of kinds and degrees of dependence', while certainly populating the semantic zoo, does not solve a single philosophical problem.

2. The first example is the pair $[T,T']$, where T is classical celestial mechanics and T' is the general theory of relativity. For the purpose of comparison I shall also consider a theory $\overline{T}$ which is the same as T except for a slight change in the strength of the gravitation potential.

Now T and $\overline{T}$ are certainly different theories – in our universe, where no region is free from gravitational influence, no two predictions of $\overline{T}$ and T will coincide. Yet it would be rash to say that the transition $T \to \overline{T}$ involves a change of meaning. For though the *quantitative values* of the forces differ

1 'On the Meaning of Scientific Terms', *J. Phil.*, 61 (1964), 497ff.

almost everywhere, there is no reason to assert that this is due to the action of different *kinds of entities*. After all, the existence of rubber bands of different strength does not indicate that there are various concepts of 'rubber band'. Nor does the existence of such a variety indicate that the notion of the surrounding spacetime continuum is not well defined. There is nothing mysterious about such stability. The concept 'rubber band' – or the more abstract concept 'force' – has been designed to cover a great variety of entities, among them also entities of different strength. Hence, they are not affected by any transition leading from one element of their extension to another.

This example shows that a diagnosis of stability of meaning involves two elements. First, reference is made to rules according to which objects or events are collected into classes. We may say that such rules determine concepts or kinds of objects. Secondly, it is found that the changes brought about by a new point of view occur *within* the extension of these classes and, therefore, leave the concepts unchanged. Conversely, we shall diagnose a change of meaning either if a new theory entails that all concepts of the preceding theory have zero extension or if it introduces rules which cannot be interpreted as attributing specific properties to objects within already existing classes, but which change the system of classes itself.

It is important to realize that these two criteria lead to unambiguous results only if some further decisions are first made. Theories can be subjected to a variety of interpretations, and the relation of concepts to practice can also be seen in many different ways. Not every interpretation leaves its mark on current procedures. Interesting ideas may therefore be invisible to those who are concerned with the relation between existing formalisms and 'experience' only.[2] It follows that we must (a) adopt a certain notion of 'interpretation'; and (b) choose from among the various kinds of interpretations consistent with this notion the particular one that we prefer. Questions concerning constancy or change of meaning have an unambiguous answer only *after* the decisions just described have been made. Otherwise we are dealing with pseudo-problems which, of course, we can 'solve' or 'refute' in any manner we please.

In what follows I shall decide (a) and (b) without giving detailed reasons for my decision. No such discussion is needed in the present brief note. All I intend to show is that a position which I hold can be presented coherently and that the alleged paradoxes it creates are harmless. I shall decide (a) by rejecting Platonism. This makes human practice the guide for conceptual considerations and the object of suggestions of conceptual reform. I shall

[2] Berkeley's notion of space as explained in *De Motu* was different from Newton's – but this difference appeared neither in experiment nor in the mathematical formalism accepted by either man. It consisted in an attitude influencing the *future development* of the theory of gravitation.

decide (b) by adopting an epistemological realism. This means regarding theoretical principles as fundamental and giving secondary place to the 'local grammar', i.e. to those peculiarities of the usage of our terms which come forth in their application to concrete and, possibly, observable situations. It is intended to subject the local grammar to the theories we possess rather than to interpret the theories in the light of the knowledge – or alleged knowledge – that is expressed in our everyday actions. Or, to put it differently, we want to analyse, to explain, to justify, and perhaps occasionally to correct the 'common knowledge' (which may also be the scientific knowledge of the preceding generation) by relating it to new theoretical ideas, rather than to interpret such ideas as new ways of talking about what is already well known. This is the way in which fundamental revolutions have taken place in the seventeenth century, and again in the twentieth century; and this is also the way that a reasonable theory of knowledge invites us to take.[3]

Let us now apply these decisions to the case at hand. Will our diagnosis be one of change or of stability? And if it is a diagnosis of change, then what kind of change can we expect? We see at a glance that the spatiotemporal frameworks of T and T' certainly have little in common (three-dimensional Euclidian continuum with absolute time constituting a four-dimensional continuum not permitting any non-singular metric in the case of T; four-dimensional Riemannian continuum with non-singular metric in the case of T': not even the overall topology of the two spaces is the same). Stability of meaning of the spatiotemporal terms can be diagnosed only if we can show that the transition $T \rightarrow T'$ occurs within the extension of a more general idea of space S that was established *before* the advent of T' (projecting more recent notions back into the past would mean revoking our decision on (a)). Now I think it will be agreed by everyone that an idea such as S cannot have been part of commonsense (it would be necessary to assume that commonsense is or was able to distinguish between topological, affine and metrical properties of space and that it is not committed to an unambiguous distinction between spatial and temporal properties). Nor is it possible to locate a suitable S in the empirical sciences. Riemann still retained an overall Euclidian topology. And the contribution of time to the metric did not occur before Einstein. It is of course quite conceivable that S may have been part of some metaphysical system, and I for one am prepared to accept a stability of meaning derived from such a background. But, first of all, reference to metaphysics is relevant only if the particular ideas needed are shared by the defenders of T and T'. This is not likely to be the case (absolutism on the part of Newton: relational theories on the part of the forerunners of relativity). Secondly, we may safely assume that

[3] See my 'Problems of Empiricism', in *Beyond the Edge of Certainty*, ed. Robert G. Colodny (Englewood Cliffs, N.J., 1965).

metaphysical Ss will be rejected by empiricists – and this includes my opponent. Result: the transition $T \rightarrow T'$ involves a change in the meaning of spatiotemporal notions.

This change is drastic enough to exclude the possibility of common elements of meaning between T and T'. To see this, consider the notion of the spatial distance between two simultaneous events, A and B. It may be readily admitted that the transition from T to T' will not lead to new methods for estimating the size of an egg at the grocery store or for measuring the distance between the points of support of a suspension bridge. But in considering (b) we have already decided not to pay attention to any *prima facie* similarities that might arise at the observational level, but to base our judgement on the principles of the theory only. It may also be admitted that distances that are not too large will still obey the law of Pythagoras. Again we must point out that we are not interested in the empirical regularities we might find in some domain with our imperfect measuring instruments, but in the laws imported into this domain by our theories. Now these laws are very different in T and T'. According to T, the distance (AB) is a property of the situation in which A and B occur; it is independent of signal velocities, gravitational fields, and the motion of the observer. An observer can influence (AB) only by actively interfering with either A or B. Any process on the part of the observer that does not reach either A or B leaves the distance unchanged. According to T', (AB) is a projection, into the spacetime frame of the observer, of the four-dimensional interval $[AB]$. $(AB)_{T'}$ will change even in those cases where a causal influence upon either A or B is excluded in principle. Now one might still wish to retain the idea of a common core of meaning by interpreting the difference between $(AB)_T$ and $(AB)_{T'}$ as being due to the different *assumptions* about space and time contained in T and T', respectively. And the locution 'space and time' would now refer to what can be characterized independently of either T or T', though in a manner that conflicts with neither (the last proviso is necessary in order to prevent a return to commonsense). Evidently it would correspond to the S mentioned above. We have already shown that no such idea can be assumed to exist. It follows, then, that the difference between $(AB)_T$ and $(AB)_{T'}$ is wholly due to the meanings of the notions used for explaining their properties. In traditional philosophical terminology: $(AB)_T$ and $(AB)_{T'}$ are *constituted* by the basic principles of T and T', *respectively*. These entities cannot be described, not even in part, by means that are independent of either theory at the time of the advent of T'. In chapter 4 I have expressed the fact by saying that '$(AB)_T$' and '$(AB)_{T'}$' are *incommensurable notions*.

3. The very same considerations apply if we consider the transition from (T) classical mechanics to (T'') quantum theory (I am now talking about

the elementary quantum theory in the form in which it has been developed by von Neumann, and not about the earlier and more intuitive ideas of Rutherford, Bohr and Sommerfeld). T'' introduces properties whose universalization is possessed by a physical system only if certain conditions are first satisfied.[4] This is true of all so-called 'dynamical' properties (angular momentum, which Achinstein discusses, is a dynamical property, and so are position, momentum, spin and so forth; electrical charge is not a dynamical property, at least not within the framework of the elementary theory). Now if we adopt an interpretation of the elementary theory that ascribes this feature to microproperties and macroproperties alike[5] and if we also decide to retain a two-valued logic,[6] we discover again that T and T''' are incommensurable. This feature of the pair has been discussed ever since Bohr introduced the principle of correspondence.

Some authors have commented on the difficulties connected with a principle that apparently ties together two theories whose concepts cannot be accommodated in a single point of view.[7] This is not the place to examine the matter in detail.[8] Let me only emphasize that the reappearance of conservation laws in the quantum theory cannot be regarded as an argument in favour of a common core of meaning (for such an argument, cf. Achinstein). For it is clear that the 'conservation laws' of the quantum theory share only the name with the corresponding laws of classical physics. They are expressed in terms of Hermitian operators, whereas the classical laws use ordinary functions that always have some value. They allow for 'virtual states' which are, strictly speaking, incompatible with conservation. No such states are possible in classical physics. They make use of properties that cannot be universalized simultaneously (position in the potential energy, momentum in the kinetic part) whereas classical properties can always be universalized. Only insufficient analysis could make one believe that the occurrence of so-called conservation laws in the two theories establishes a common core of meaning.

4. The above considerations, though not sufficient to settle the matter, still provide at least strong *prima facie* evidence for the existence of 'paradoxical' cases in Achinstein's sense. I do not see how the attention to details that Achinstein recommends is going to eliminate these cases. What we need are

[4] If P is a property, then $P \lor \sim P$ will be called its 'universalization'.
[5] This interpretation is suggested by Temple's proof of isomorphism (*Nature*, 135 (1935), 951; cf. also Groenewold, *Physica*, 12 (1946), 405ff). It agrees with our decision on (b). Cf. also ch. 2, n. 19.
[6] For reasons, see my paper 'Bemerkungen zur Verwendung nicht klassischer Logiken in der Quantentheorie', in *Publications of the Salzburg Institute for the Philosophy of Science* (Salzburg, 1965), 1.
[7] Cf. N. R. Hanson, *Patterns of Discovery* (Cambridge, 1958), as well as my comments in *Phil. Rev.* 69 (1960), 251.
[8] According to Bohr, the principle of correspondence is a theorem of the quantum theory rather than a 'bridge law' connecting quantum mechanics and classical physics.

not further details, but principles such as those involved in our decisions on (a) and (b), which teach us how to evaluate various proposals and which remove the ambiguities characteristic of all semantic information: we can relate the 'local grammar' of well-known expressions to different theories in different ways (realism, instrumentalism) and thereby give them different meanings. The adoption of such principles will be guided partly by scientific practice, partly by the demands of a reasonable methodology, such as maximum testability.[9] Now, according to both these guides, major revolutions are preferable to small adjustments, since they affect and, thereby, lay open to criticism even the most fundamental assumptions. Achinstein's rule (B) (which I would formulate as saying that competing theories must have common meanings) puts a restriction on the extent to which we are allowed to revise such assumptions. It does so in the belief that theories can compete only if they are incompatible and that they can be incompatible only if they have common meanings. Is this transcendental defence of an epistemological conservatism effective? In order to decide the question we return to our first example, the relation between the general theory of relativity and classical physics.

5. Combine T' with two assumptions which are contrary to fact, namely (i) the overall metric of space is almost Minkowskian; and (ii) the velocity of light is almost infinite. These two assumptions do not change the semantic properties of T'. Theories T and T' are still incommensurable. Yet it is possible, to a high degree of approximation, to establish an isomorphism between certain selected semantic properties of some (not all) descriptive statements of T' and of some (not all) descriptive statements of T (let the corresponding classes be C and C', respectively). This isomorphism will be valid for finite distances (AB), but not for distances approaching infinity. It will be valid for a finite number of parallel displacements of (AB) around a closed curve, but will cease to be so as this number approaches infinity. Considering that meanings are dependent on structure and not on the particular ways in which the structure is realized, we may say that, within the restrictions given, C and C' have a common core of meaning. We may even identify C and C'. (As $C \neq T$ and $C' \neq T'$, this does not affect the relation between T and T'.) However, C' is formulated in terms of T' and can, therefore, be examined in the manner preferred by Achinstein (meanings shared between the critic and the point of view criticized) within that theory. The examination will of course lead to the rejection of C' and, via the isomorphism, of C, and, as C is part of T, of T also. We see that the show can be rigged in such a manner that the demands for partial stability of meaning are satisfied. But the very method of rigging indicates that the demand is superfluous: when making a comparative evaluation of classical physics

[9] For details, cf. ch. 6.

and of general relativity we do not compare meanings; we investigate the conditions under which a structural similarity can be obtained. If these conditions are contrary to fact, then the theory that does not contain them supersedes the theory whose structure can be mimicked only if the conditions hold (it is now quite irrelevant in what theory and, therefore, in what terms the conditions are framed). It may well be that those champions of T or T' who see light only when they see – or believe they see – meanings 'are always at least slightly at cross purposes'.[10] The fact that argument proceeded even through the most fundamental upheavals, that it was understood, and that it led to results shows that meanings cannot be that essential.[11] I conclude then that principle B is neither necessary nor desirable.

6. These results can be immediately applied to such notorious museum pieces as the mind–body problem, the problem of the existence of the external world, the problem of free will, and to many other problems.[12] In all such cases 'new' points of view (which are actually as old as the hills) are criticized because they lead to drastic structural changes of our knowledge and are therefore inaccessible to those whose understanding is tied to certain principles. This conservatism may well have a physiological foundation. Education, as Professor Z. Young has put it so well,[13] consists in seriously damaging our central nervous system and in eliminating reactions of which it was initially capable. Admitting such damage and the consequent lack of imagination is one thing. However, one should never go so far as to try to inflict it upon others in the guise of a philosophical dogma.[14]

[10] T. S. Kuhn, *The Structure of Scientific Revolutions* (Chicago 1962), 147.
[11] The debates during the period of the older quantum theory are an excellent example of discussions of this kind.
[12] For details cf. again my 'Problems of Empiricism'.
[13] *Hitchcock Lectures* (University of California in Berkeley, 1964).
[14] For details, cf. ch. 6.

6

Reply to criticism

Comments on Smart, Sellars and Putnam

1. PROLIFERATION

In writing the papers which are discussed, and criticized by Smart, Sellars and Putnam, my aim was to present an abstract model for the acquisition of knowledge,[1] to develop its consequences, and to compare these consequences with science. With respect to the last point it may be expected that a comparison of historical phenomena with epistemological views or models will lead to new historical evidence, and to new ideas concerning the actual structure of science (or law, or philosophy, or commonsense). But it

[1] The papers by Smart, Sellars and Putnam appear in *Boston Studies in the Philosophy of Science* (New York, 1965), II, where this reply was also originally published. In his paper Smart concentrates on the views of Nagel and Sellars, and he also discusses some of my publications. Perhaps unintentionally he creates the impression that a new philosophical position is in the making, a kind of neorealism, and that I have contributed to its development. Such an impression would be both incorrect, and unfortunate. It would be incorrect as the present discussion does not lead beyond the arguments which have been used in the traditional issue between realism and instrumentalism. I for one am not aware of having produced a single idea that is not already contained in the realistic tradition. For example, most of the points which I have made in ch. 4, above appear in Boltzmann. But such an impression of novelty would also be very unfortunate. It would increase the tendency to disregard the connection between philosophy and the sciences which is so essential for the development of our problem, would further separate the most recent history of philosophy from its past and would thereby reinforce the provincialism and irrelevance that is a trademark of much of contemporary philosophy. Of course, many thinkers see the situation in a very different light. They believe that philosophy was born (by parthenogenesis?) in the twentieth century and that reference to the past can be only of the didactic kind, showing the mistakes committed by those not yet in possession of the One True Method. Of course, they never succeed in completely eliminating traditions either. The old problems have the habit of appearing again; but separated from the context in which they arose they now look very strange indeed. Compared with them, even the most pedestrian part of commonsense seems to be full of virtue and even the most ridiculous argument derived from it a source of sun-like clarity. There arises then an activity where the correction of what was once a sensible hypothesis and what has now become a silly mistake can pose as a major philosophical advance (Austin against Ayer on sense data); and where a bumbling and tedious recapitulation of the less interesting parts of the phylogeny of our thought can be hailed as a splendid achievement of a 'modern philosophy'. We need the historical background both again to give substance to our arguments, to show that they are relevant to a much wider class of problems than a specialist treatment would suggest, and for raising the level of discussion to the heights already achieved by Democritus, Galileo, Descartes, Faraday, Kant, Boltzmann, Maxwell, Einstein, Duhem. Considering such a wider background, Putnam's attempt 'to keep the literature straight', where by 'literature' he means Kemeny, Oppenheim and himself, is somewhat amusing.

is also important to realize that one must not allow this structure to interfere with the models. Such models tell us how to proceed if a certain aim is to be achieved. In this way they form a basis for the criticism as well as for the reform of what exists. It is clear that a theory of knowledge which is built up with such a purpose of criticism and reform in mind will be fairly different from an analytic theory that relies on the maxim that in the battle between the actual and the idea the latter must be regarded as a mere fancy, as a flight from reality, as a castle in the air[2] which shrinks into insignificance when confronted with the hard facts of (scientific, legalistic, commonsense) life.[3]

The model which underlies my own discussion has as its aim maximum testability of our knowledge. No argument will be given for this aim here.[4] All I intend to do is sketch some consequences and remove the air of paradox which, according to some critics, surrounds these consequences.

The main consequence is the *principle of proliferation: Invent, and elaborate theories which are inconsistent with the accepted point of view, even if the latter should happen to be highly confirmed and generally accepted.*[5] Any methodology which

---

[2] For a more detailed criticism of this view see my review of *Scientific Change*, ed. A. Crombie in *Br. J. Phil. Sci.*, 15 (1964), 244ff.

[3] Modern science and its methodology started as the rebellion of a few people against established modes of thought. These thinkers disagreed with the most fundamental beliefs of their time, which were held by scholars and by laymen alike and which also had impressive evidence in their favour. Yet it is only rarely that one finds in their work even a trace of hesitation and not once does the universal support for the established doctrine suggest to them that their own ideas might perhaps not be the whole truth. Commonsense, received opinion, the teachings of the schools count for nothing with Descartes. He is prepared to reject them all. In defending his own pet ideas Galileo was forced to contradict experience and reasonable conjecture. An immense optimism is at work here, an optimism which is prepared to abandon the products of the combined efforts of many generations and 'to begin the whole labour of the mind again' (Bacon). Ernst Mach, who initiated the movement of modern positivism, criticized the science of his time in the very same spirit. This is the attitude which has led to the sciences as we know them today. Every philosopher of science lives off the products of this great and distant past. But it is also clear that the contemporary tendency to dwell at length on *what is*, and either to oppose, or not to be concerned with, *what should be* would never have permitted such products to arise. It is time for the philosophers to recognize the calling of their profession, to free themselves from the exaggerated concern with the present (and the past) and to start again anticipating the future.

[4] For such arguments see K. R. Popper, *Conjectures and Refutations* (New York, 1962).

[5] When speaking of *theories* I shall include myths, political ideas, religious systems, and I shall demand that a point of view so named be applicable to at least some aspects of everything there is. The general theory of relativity is a theory in this sense; 'all ravens are black' is not. There are certain similarities between my use of 'theory' and Quine's 'ontology' ('On What There Is', *Proc. Arist. Soc.*, Suppl. vol. 25 (1951)), Carnap's 'linguistic framework' ('Empiricism, Semantics, and Ontology', *Revue Int. Phil.*, 4 (1950) 208ff), Wittgenstein's 'language game' (cf. my review 'Patterns of Discovery', *Phil. Rev.*, 59 (1960) 247ff), Pareto's 'theory' (*Treatise on General Sociology*, para. 7ff), Whorf's 'metaphysics' (*Language, Thought, and Reality: Selected Writings*, John B. Carroll, ed. (Cambridge, Mass., 1956), *passim*), Kuhn's 'paradigm' (*Structure of Scientific Revolutions* (Chicago, 1962), *passim*), etc. I prefer this 'accordion' use of the term because it provides a single name for problems which in my presentation are intimately related. However, the 'philosophical music' (Sellars, para. 4) I intend to make will

adopts the principle will be called a *pluralistic methodology*. The theories which the principle advises us to use in addition to the accepted point of view will be called the *alternatives* of this point of view.

Summarizing earlier arguments for proliferation[6] we may say (1) that no theory ever agrees (*outside* the domain of computed error) with the available evidence. Hence, if we do not want to live without any theory at all, we must have means of accentuating certain deviations, of lifting them out of the ocean of 'deviational noise' surrounding each theory. Alternatives provide such means.[7] (2) Theories agree with the *facts* (as opposed to *observations*) only to a certain extent. Indeed, it would be a complete surprise for everyone if a theory were found that represents all the facts perfectly. Some disagreement with facts is mirrored in the evidence. But there are other cases where physical laws prevent the discrepancy from ever appearing in the evidence.[8] Now if we find a theory which asserts the discrepancy, which is able to repeat the past successes of the accepted point of view, which has new and independent evidence in its favour, then we shall have good reason to abandon the accepted point of view despite its success. Alternatives are theories of the kind described. (3) One need not mention the *psychological* advantages flowing from the use of alternatives, i.e. the fact that a mind which is immersed in the contemplation of a single theory may not even notice its most striking weaknesses.[9]

not be extracted from the fact that the word is so used, but from the common features of the problems described. Cf. also section 9.

 Terms such as (1) 'consistent', (2) 'incompatible' (3) 'follow from' are applied to pairs of theories $[T, T']$, and they mean that (1) there do not, (2) there do, exist pairs of predictions, P and P', one following from T, one from T' (via corresponding initial conditions) which are incompatible. In a special case T' taken together with the conditions of validity of T will be (1) consistent with, (2) incompatible with, (3) sufficient to derive, T. For modifications of this explanation cf. section 5.

[6] Cf. e.g. ch. 4.

[7] Not a single planet moves in the orbit calculated in accordance with Newton's celestial mechanics (this has nothing to do with relativistic effects). There exist other as yet unexplained discrepancies exceeding the error of measurement by a factor of 10. These discrepancies provide a background of 'deviational noise' relative to which the 43″ of arc per century in the advance of the perihelion of Mercury lose much of their force. Moreover, there are many as yet unexplored possibilities of explaining this value within the framework of Newton's theory. An oblateness of the sun of magnitude 10^{-4} (corresponding to 0·09″ of arc) would increase the perihelion rotation by 7″. Owing to the turbulence of the atmosphere an oblateness of 0·05″ could be discovered only with great difficulty, so that at least 5″ of the famous 43″ depend on assumptions beyond the level of present observation. The as yet unexplained discrepancies mentioned above may be due to causes which are operative in the present case also. Is it not obvious that abandoning Newton's mechanics on the basis of so doubtful a number would be rather imprudent? And is it not also clear that only an account of the 43″ which explains them on the basis of new principles will lift them out of the background of deviational noise and turn them into an *effect* that is capable of *refuting* the Newtonian scheme? Naturally such an account will rest on a theory inconsistent with Newton – it will rest on an *alternative* to Newton's theory.   [8] Cf. Chs. 4 and 11.

[9] As an example one need only consider Galileo's intoxication with his preposterous theory of the tides. For an evaluation see A. Koestler, *The Sleepwalkers* (New York, 1959), 464ff.

The principle of proliferation not only recommends invention of new alternatives, it also prevents the elimination of older theories which have been refuted. The reason is that such theories contribute to the content of their victorious rivals. Thus certain contemporary attempts to give a classical account of typically relativistic effects provide further tests for relativity,[10] and their failure provides further corroborating evidence. The same is true of some well-known but unpopular approaches in the quantum theory.[11] Professor Naess has made it clear that such corrective use may well continue forever without our being able to indicate a point where it becomes preposterous.[12] Knowledge so conceived is not a process that converges towards an ideal view; it is an ever-increasing ocean of alternatives, each of them forcing the others into greater articulation, all of them contributing, via this process of competition, to the development of our mental faculties. All theories, even those which for the time being have receded into the background, may be said to possess a 'utopian' component in the sense that they provide lasting, and steadily improving, measuring sticks of adequacy for the ideas which happen to be in the centre of attention.[13] Hegel's assertion that theories emerge as aspects of the development of 'the idea' is therefore essentially correct, except that it confounds cause and effect and strives to retain a semblance of consistency by weakening the laws of logic. In our presentation, the competition of theories or, to speak in a less Platonist fashion, the discussion of alternatives by individual scientists, philosophers, politicians, etc. is the cause; the gradual improvement of all of them as well as of the minds of the participants, the effect. Hegel, on the other hand, regards the idea as the subject of development, and human thought as derivative.

There are many thinkers who find such considerations strange and not to their taste.[14] From what I have read I gather that Putnam is one of them.

---

[10] Cf. R. H. Dicke 'The Observational Basis of General Relativity' in *Gravitation and Relativity*, ed. Chiu H.-Y. and W. F. Hoffmann (New York, 1964), 1ff.

[11] Cf. vol. 2, ch. 7.

[12] In a manuscript which he kindly sent to me before publication.

[13] In this respect scientific theories are much more similar to works of literature than one would be inclined to believe. 'The domain of literature', writes Ingeborg Bachmann in her 'Frankfurter Vorlesungen' *Ingeborg Bachmann* (Munich, 1964), 298ff; esp. 33f), 'is an open domain – and this is true of ancient literature as well as of more recent works. It is less closed than any other domain – for example the sciences, where each new theory eliminates what has preceded [which is precisely the point at issue] – it is an open domain as its whole past intrudes into the present. With the force of all times it presses towards us, towards the threshold of time where we reside . . . and we are made to realize that none of its works is dated, or can be made ineffective. It contains all those presuppositions which resist final judgement and final categorization. These presuppositions which reside in the works themselves I would like to call "utopian" presuppositions.' A little less reliance on Wittgenstein whom Miss Bachmann seems to know quite well and a little more acquaintance with the history of science would have convinced her that the opposition between science and literature is much less pronounced than she seems to think.

[14] Proliferation of views has been regarded with suspicion in religion, in philosophy and,

Instead of further engaging in abstract argument, I would invite these opponents of pluralism to compare what I have just said with the actual history of thought. Are not the history of heliocentrism, and the history of atomism, splendid examples of the way in which antediluvian, preposterous ideas, ideas which are regarded with contempt and are ridiculed by the learned and by the common folk alike, may yet be turned against 'modern' views and *may even succeed in overthrowing them?* Are we not warned by the still quite recent attacks against Boltzmann and against his 'old-fashioned' atomic theory that things can never be regarded as settled,[15] and that the picture of a steady accumulative progress towards an aim which can be reached, at least approximately, either assumes too much knowledge on our part, or puts undue restrictions upon our instruments of criticism (in which latter case the possibility of a 'criticism from the past' will of course be

finally, in the sciences. This tendency towards monism seems to be a most persistent inheritance from our savage past. The *philosophical* debate between monism and pluralism has a long and interesting history. It started when Plato, heavily relying on common prejudice, ridiculed the Presocratics because of the variety of theories they produced (cf. n.13 above). Within empiricism the demand for monism was most clearly formulated by Newton (*Principia* book 3, rule 4, ed. F. Cajori, (Berkeley and Los Angeles, 1934), 400): 'In experimental philosophy we are to look upon propositions inferred by general induction from phenomena as accurately, or very nearly true, *notwithstanding any contrary hypothesis that may be imagined*, till such time as other phenomena occur by which they may either be made more accurate, or liable to exceptions' (my italics). Today Newton's rule is frequently invoked in the defence of the (slowly disintegrating) Copenhagen orthodoxy. Even in the fields which have traditionally exhibited a healthy disregard for parsimony of theory construction, for example in cosmology, some thinkers object to the proliferation of theories as being inconsistent both with the history of science, and with a sound methodology. Thus Heckmann, whose masterful analysis of Newton's cosmology and whose contributions to the relativistic theories have been a delight to all students of the subject, writes in an address delivered to the eleventh Solvay congress (E. Schuecking and O. Heckmann, 'World Models' in *Institut International de Physique Solvay, Onziéme Conseil de Physique* (Brussels, 1958), 1ff): 'The steady progress of science, in our opinion was possible only because scientists thought it not permissible to put forward new theories unless new data forced them to abandon the older concepts.' (This is historically incorrect. Galileo, Faraday, Boltzmann and Einstein brought about progress by the skilful application of alternatives. In many cases *no other way was possible* (cf. again my 'Problems of Empiricism', esp. section 6).) 'It would seem', so the authors continue, 'that some cosmologists have abandoned the aforementioned principle of methodology.' But, so the passage concludes, 'it is sound policy to refrain from theorizing along the lines of Bondi, Gold and Hoyle until there is strong empirical evidence for continuous creation of energy and momentum'. (Quite the contrary. Cosmology, more than any other discipline, offers such a close connection between fact and theory and such a vast amount of background noise that alternatives *will be quite indispensible* for the purpose of refutation.) T. S. Kuhn, to whom we owe a clear account of the function of alternatives in the development of science still shows some sympathy for monism (cf. vol. 2, ch. 8).

15 'It is not easy today to recapture the attitude that prevailed towards the end of the last century', writes von Smoluchowski ('Experimentell nachweisbare, der üblichen Thermodynamik widersprechende Molekularphänomene', *Phys. Z.* 13 (1912), 1070f). 'At that time the scientific leaders of Germany and France were convinced, with only very few exceptions, that the atomistic-kinetic theory had ceased to be of importance.' Boltzmann was then regarded as the 'last pillar' of that imposing edifice, as a somewhat backward defender of a lost cause.

excluded)? It is of course somewhat difficult to find events like those just described in the customary historical accounts which favour monism. But more recent research has gone a long way towards showing the decisive role that the principle of proliferation has played at almost all major stages of the history of thought.

2. STRONG ALTERNATIVES

Not all alternatives are equally suited for the purpose of criticism. Thus a theory which merely replaces a prediction P of T by another prediction, P', that is inconsistent with P and experimentally indistinguishable from it will not be satisfactory. It contradicts T, but it does not provide any reasons for removing T. Szilard's non-kinetic modification of the second law of phenomenological thermodynamics is of this kind.[16] It is assumed that deviations from the second law will occur which average out over many cycles. Of course, it is quite possible that such deviations do actually exist. But they will then refute the law directly, and without help from Szilard's account. A similar remark applies also, to a certain extent, to the older quantum theory which was largely a 'transcription of the Balmer formulae'.[17] The first condition to be satisfied by a theory that is fit for criticism, going beyond the criticism provided by the observations themselves and strengthening their refuting power, is, therefore that it must contain assertions over and above the assertion (the prediction) which leads to the contradiction. Secondly, the additional assertions must be connected with the contradicting assertion more intimately than by mere conjunction. Thirdly, we shall not be inclined to eliminate T simply because there exists a different story, T', which contradicts it, even if this second story should possess considerable interest. There ought to be some independent reason to accept T', and one of the best reasons is, of course, evidence in favour of it. Such evidence need not be there from the beginning and an alternative T' should not be eliminated simply because it does not yet provide it. But (a) it must be *possible* to produce it; and (b) such evidence must at some time be *actually* produced. Finally, the critic should also be able to account for the earlier success of the criticized theory for it is only in this case that a clear judgement of superiority can be made and that correspondingly clear reasons for temporary removal can be given. The second, third and fourth conditions show that a 'strong' connection is needed between the refuting prediction and the remainder of the theory (which includes the independent evidence, and the evidence supporting the theory to be criticized).

Now it would certainly be desirable to possess a more detailed account of

[16] Z. Phys., 32 (1925), 753ff.
[17] B. Hoffmann, *The Strange Story of the Quantum* (New York, 1959), 58.

the 'strength' of the connection between different parts of a theory, or of its *coherence*. There are some mathematical functions for which such an account is already available. Thus complex functions of a single variable which are defined in a domain D and differentiable everywhere in the domain, so-called *regular functions*, coincide everywhere in D if they coincide along a path of arbitrary length, and even if they coincide in a denumerable manifold of points, provided the manifold converges (identity theorem for regular functions). Conversely two regular functions which differ along a path of arbitrary length in D will differ everywhere else in D. We see here how the high degree of coherence of the regular functions is directly connected with their critical ability. This leads to a more general conjecture, that two coherent theories ('coherent' in the as yet unspecified sense introduced at the beginning of the present paragraph) which approximately agree in a certain domain (and are confirmed here by the same evidence) will differ noticeably almost everywhere else if they differ in a finite, or in a denumerably infinite, number of points. Theories of this kind (*strong alternatives* as I shall call them) would be especially well suited for the purpose of criticism (as their postulates make them differ everywhere, they are not *ad hoc* anywhere).

It is interesting to note that the four conditions and the demand for coherence which characterize severe critics coincide with the conditions characterizing acceptable explanatory systems. This means that a strong alternative T' of T which has received independent support will not only remove T but will also be a satisfactory successor of T: *the best criticism is provided by those theories which can replace the rivals they have removed.*

3. A MODEL FOR PROGRESS

Let us now consider the features of an activity which consists in the invention of theories for the purpose of explanation, in the criticism of these theories with the help of strong alternatives, and in the replacement of refuted theories by the rivals which brought about their downfall. Such an activity will lead to a succession of theories $T, T', T'', T''', \ldots$ We want to investigate the relation between these theories. This investigation clearly has nothing whatever to do with historical fact. It starts from a certain aim – to obtain testable knowledge – it indicates how the aim can be reached, and it then goes on to describe what will happen if people proceed accordingly. It provides a model for possible action without claiming that actual science conforms to the model.

It is not unimportant to discuss the features of models of this kind. First, the model has great advantages and one may hope that scientists will try to act in accordance with it. Secondly, using the model in the course of historical research we may discover features unnoticed by those who favour monism and we may come to realize that actual science is much closer to pluralism than one would expect when consulting the (usually monistic) historians. Thirdly, discussion of the model and removal of apparent 'paradoxes' will lend support to those who intuitively favour pluralism and strong alternatives, but who are prevented by the general philosophical climate from fully relying on their intuitions. I really hope that the occasional disparity between the model and actual scientific practice will be regarded as a criticism of the latter, not of the former. In the struggle between an ideal and actual reality the ideal must always be given the upper hand. After all, we do not want to leave the historical development of a discipline to chance (or to Oxford commonsense); we want to shape it, and improve it in accordance with ideas we find reasonable. This was the way in which modern science and its philosophy started (cf. n.3 above). And this is also the way in which we may hope further to improve them both.

4. CONSISTENCY

What is the relation between two theories, T and T' (where T' succeeds T as the result of criticism)? We see at once that

  (i) T cannot be subsumed under T' (T' is the successor of T)

 (ii) T cannot be explained on the basis of T'

(iii) T cannot be reduced to T'

if we retain the demand that subsumption, explanation, reduction must be by logical derivation. The reason is simply that T', being a critic of T, is also inconsistent with T. It is gratifying to see that in actual science, too, the transition to a new theory is often the transition to a theory which contradicts what went on before. The so-called 'layer model' which has been elaborated by Nagel therefore agrees neither with actual science, nor with the arguments supporting proliferation.

Of course, it is always possible to preserve the layer model and the philosophy on which it rests. This can be done either by replacing strict rules of derivation with approximate rules,[18] or by replacing T with another theory, $\overline{T}$, which does follow from T'.[19]

The first suggestion is unacceptable; it eliminates the very same procedure of criticism by theories which has brought about the transition $T \rightarrow T'$.[20] But assume that there exists strong refuting evidence of T that is independent of T'. Then the removal of the inconsistency between T and T' will still lower the empirical content of T, will make T more 'metaphysical',

---

[18] Smart (161).

[19] Putnam (206ff).

[20] The same applies to Sellars' suggestion in paragraphs 62–4 of his essay insofar as we are here advised to weaken the relation between what Sellars calls 'the theory' and the 'antecedent generalizations'.

and why? – just in order to preserve a philosophical model that would otherwise have to go. Is it really necessary to point to the similarity between this move and the manoeuvres of the much maligned metaphysicians of the past?

The second suggestion is nothing but a concealed admission of defeat. For what is so attractive about a philosophical point of view concerning the relation between theories, which can be upheld only if *real* theories, theories which have been discussed in the scientific literature, are replaced by emasculated caricatures of them? Consider Putnam's own example. He admits, and says Nagel admits, that geometrical optics (O) is inconsistent with wave optics (W).[21] Now he looks for a theory, $\bar{O}$, which is vaguely similar to $\bar{O}$ and can therefore be called a geometrical optics, but which follows from W. I am quite prepared to admit the existence of such an $\bar{O}$. It will not be the theory Putnam suggests (in his comments, p. 207): light rays are normal to the wave front only in isotropic media; and what about those cases where no wave front exists, such as inside a container with perfectly reflecting walls? Nor will it be possible to write the desired $\bar{O}$ down explicitly ('rays' must be replaced by sausage-like structures etc.). But let us disregard these difficulties and let us be content with the fact that an $\bar{O}$ of the desired kind can be imagined abstractly. What has been shown? It has been shown that while almost all that was actually said about light before the advent of W contradicted W and could therefore be criticized by W (and criticize W in turn) there are some logical creatures, such as $\bar{O}$, which do not contradict W.[22] This is a great discovery indeed, especially in view of the trivial fact that all the logical consequences of W follow from it and provide a reservoir for the choice of some suitable (though perhaps not analytically presentable) $\bar{O}$. But surely it was not the intention of those who formulated the layer model to assert that the *logical consequences* of the theories in the top level follow from them, can be reduced to them, or can be explained on the basis of them. What they asserted was that certain theories entail other theories *which have been formulated independently*. And *this* assertion is refuted by my argument and by the historical examples provided.[23] Result: re-

---

[21] This inconsistency was an important factor in the refutation of the particle theory of light which may be regarded as a wider theory, embracing O. Newton was acquainted with diffraction phenomena and he treated them extensively in his *Opticks* (book 3, part 1). He suggested an explanation in terms of an interaction between the particles and the bending edge. He was therefore able to accommodate these phenomena. It was only through the successful elaboration of the wave theory (which also gave an account of the rectilinear propagation of light – Newton's main objection against waves) that the inadequacy of the particle theory, and of its kinematics, viz. geometrical optics, was finally demonstrated. The inconsistency between geometrical optics and wave optics had an important function in this demonstration.

[22] Newton's objection against the wave theory (see the last footnote) can be construed as an indirect criticism of a primitive form of W via O.

[23] This remark seems also to apply to Achinstein's criticism 'On the Meaning of Scientific Terms', *J. Phil.*, 61 (1964), 497ff.

course to 'approximate rules of correspondence' (Smart) and to 'approximate theories'[24] *is out*.[25]

5. MEANING INVARIANCE

In his paper Putnam creates the impression that I am mainly interested in meanings and that I am eager to find change where others see stability. This is not so. As far as I am concerned, even the most detailed conversations about meanings belong in the gossip columns and have no place in the theory of knowledge. This is true even in those cases where meanings are invoked to force a decision about some different matter. For even here their only function is to conceal some dogmatic statement which would not be accepted, if presented by itself, and without the chatter of semantic discussion.[26]

Still, the fact that meanings occur so often in philosophical debate forces

[24] The term 'approximate theories' is due to Putnam (206); see also Sellars, para. 65.

[25] A word about *thermodynamics*. It is still far from clear how the relation between the *existing* phenomenological theories (*P*) and the *existing* statistical theories (*S*) is to be conceived. From the point of view of our model which *prescribes* how to construct theories a contradiction between *S* and *P* would be desirable. Historically, one of the most decisive arguments against *P*, Einstein's theory of Brownian motion, did use an *S* that was incompatible with the second law of the phenomenological theory (for details see again my 'Problems of Empiricism'). But apart from this one case there reigns confusion. It seems to me that neither Nagel, nor Putnam, are aware of these facts. Nagel only discusses the schoolbook relation $T \sim E_{kin}$, and Putnam speaks (219) of the 'well known reduction' of *P* to *S*. What is this 'well known reduction' he refers to? Where in the literature is it to be found? As far as I can see there exists a great variety of different approaches, ranging from Boltzmann's original ideas, and investigations into the ergodic theorem arising therefrom, via Tolman's equal-probability approach, Khinchin's work (where we also find pertinent criticism of Nagel's 'formula') to the very confused considerations which are based upon the papers of Szilard and upon von Neumann's book. There are of course also more popular accounts and I cannot entirely get rid of the suspicion that Putnam, when speaking of the 'well known reduction', is speaking mainly of them. But this is much too slender and too inarticulate material for philosophical generalizations. Does this mean that the case *S* versus *P* is not yet ripe for discussion? Not at all. One of the basic results of a statistical physics that has not been emasculated, and turned into a Putnamian 'approximate theory', is the existence of *fluctuations*, even between different levels of temperature. This phenomenon is in flat contradiction with the second law of *P*. The attempt, due to Szilard and others, to preserve the second law of *P* by corrective terms expressing information on part of the observer fails (a) because the fluctuations occur independently of any observer (sudden freezing of undercooled fluids); and (b) because we can easily imagine quite simple mechanisms which violate the second law (see my note 'On the Possibility of a *Perpetuum Mobile* of the Second Kind' in *Mind and Matter, Essays in Honor of Herbert Feigl*, ed. P. Feyerabend and G. Maxwell, (Minneapolis, 1965). Such a reinterpretation of *S* is also undesirable for methodological reasons for it robs the theory of its ability to criticize *P* and thereby reduces the empirical content of the latter. I agree not all is well with thermodynamics. However, it would be very unwise to defend an unreasonable method (layer-monism) by reference to a mess in physics. And whatever the judgement on these matters is going to be, a third-hand or fourth-hand account of science is too little to be content with in basic discussions, even if such an account, because of its simplicity, should have managed to become 'well known' in some circles.

[26] For an analysis of the arguments against a mind–body materialism see ch. 10.

us to deal with them too, at least occasionally, and some terminological hints are therefore indicated.[27] On the basis of these hints we now put the following question: assume that T and T' are two strong alternatives. What relation may we expect to find between their primitive descriptive terms?

In order to answer the question we examine the following three theories: T = classical celestial mechanics, including the spacetime framework; T' = general relativity; $\bar{T} = T$, except for some slight modification in the strength of the gravitational interaction. $\bar{T}$ is inconsistent with T. In our universe, which does not contain regions free from gravitational influence, $\bar{T}$ and T will nowhere lead to the same predictions. Yet $\bar{T}$ cannot aspire to be a strong alternative of T. The difference between the predictions of the two theories can be accounted for by changing part of T. No modification is needed of other principles (the law of inertia, the spatiotemporal framework, can both be retained in their original form). A strong confirmation of $\bar{T}$ endangers only a small part of T, it does not necessitate a reconsideration of the 'nature' of space, time, force, mass. Using the terminology of n. 27 we may express this by saying that the transition $T \rightarrow \bar{T}$ is not of the fundamental kind and that the 'meanings' of the primitive descriptive terms of T and $\bar{T}$ will largely coincide.

The transition $T \rightarrow T'$ is an altogether different matter. The fact that within T' all events are embedded in a four-dimensional Riemannian continuum which acts not merely as a somewhat unusual background of coordinates, but is supposed to describe *intrinsic properties* of physical processes now necessitates the reformulation of all the basic laws of T and the redefinition of all its basic concepts (spatial distance, temporal distance, mass, force, etc.). The new laws will not only read differently, they will also conflict in content with the preceding classical laws. And this is not just a

---

[27] In what follows I shall assume that the rules (assumptions, postulates) constituting a language (a 'theory' in our terminology) form a *hierarchy* in the sense that some rules presuppose others without being presupposed by them. A rule R' will be regarded as being more fundamental than another rule R'', if it presupposed by more rules of the theory, R'' included, each of them being at least as fundamental as the rules presupposing R''. It is clear that a change of fundamental rules will entail a major change of the theory, or of the language in which they occur. Thus a change in the spatiotemporal ideas of Newton's celestial mechanics makes it necessary to redefine almost every term, and to reformulate every law of the theory, whereas a change of the law of gravitation leaves the concepts, and all the remaining laws, unaltered. The former ideas therefore are more fundamental than the law of gravitation.

Now it seems reasonable to assume that the customary concept of meaning is closely connected, not with *definitions* which after all work only when a large part of a conceptual system is already available, but with the idea of a *fundamental rule*, or a *fundamental law*. Changes of fundamental laws are regarded as affecting meanings while changes in the upper layers of our theories are regarded as affecting beliefs only. There exists therefore a rather close connection between meanings and certain parts of theories. Meanings in the sense just explained are also independent of the analytic–synthetic issue. Moreover, it is now quite plausible to assume that meaning-talk can be replaced, without residue, by theory-talk. For a more detailed account see ch. 5.

matter of words. The classical, or absolute idea of mass, or of distance, cannot be defined within T'. Any such definition must assume the absence of an upper limit for signal velocities and cannot therefore be given within T'. *Not a single primitive descriptive term of T can be incorporated into T'.* Using the terminology of n. 27 we may express this by saying that the change of rules accompanying the transition $T \rightarrow T'$ is a fundamental change, and that the meanings of all descriptive terms of the two theories, primitive as well as defined terms, will be different: T and T' are *incommensurable theories* (other incommensurable theories have been discussed in earlier papers).

Now one of the most frequent objections against admitting incommensurable theories into the class of the strong alternatives of a given theory T is that they cannot contradict T and cannot, therefore, be used to criticize T. Professor Shapere expresses most clearly what has been said by many critics:

> For in order for two sentences to contradict one another one must be the denial of the other; and this is to say that what is denied by the one must be what the other asserts; and this in turn is to say that the theories must have some common meaning. On the other hand, two sentences which do not have any common meaning can neither contradict, nor not contradict, one another.[28]

This remark is entirely correct and it indicates that the use of incommensurable theories for the purpose of criticism must be based on methods which do not depend on the comparison of statements with identical constituents.[29] Such methods are readily available.

One of the main properties of strong alternatives is that they disagree everywhere if they disagree in a finite number of points (see section 2). Moreover, the disagreement is due to basic rules, and not to *ad hoc* adaptations from point to point (strong alternatives are *coherent* theories – see again section 2). Now one may be interested either in the relation between different theories, or in the features of the world as seen from the point of view of a single theory. Questions of the first kind may be called, by analogy to terms which Carnap has used on a similar occasion,[30] *external questions*; questions of the second kind may be called *internal questions*. External questions can be decided without bringing in meanings at all. We simply compare two infinite sets of elements with respect to certain structural properties and inquire whether or not an isomorphism can be established. Considering a Euclidian plane and an arbitrarily bent surface with a Riemannian metric imposed upon it, we realize that there is nothing corresponding to a large-scale Pythagorean triangle on the latter which

---

[28] Letter of August 18, 1964. I am grateful to Professor Shapere for having provided me with some very detailed and clear criticisms of earlier papers.

[29] A 'constituent' of a statement containing logical constants is any statement that is part of it and does no longer contain any logical constants.

[30] Cf. the reference in n.5 above.

establishes that, in two dimensions at least, Newtonian mechanics and general relativity are different theories. The actual situation (three-dimensional Euclidian space and four-dimensional Riemannian manifolds) is of course much more drastic and shows even more clearly that appeal to meanings is not needed to prove difference. Adding the usual physical interpretation, we may conclude that the validity of the one is incompatible with the validity of the other (I omit the detailed argument for lack of space). Of course, there is still the question as to which theory is to be preferred. Usually, one tries to solve the question on the basis of *crucial experiments*. This is quite proper, but may again lead into trouble. For if one assumes that a crucial experiment consists in selecting, on the basis of observation, one of two contradictory statements, then we seem again to be forced to use only such theories as have a certain minimum of similarity in meanings.[31] However, there may be empirical evidence against one, and for another theory without any need for similarity of meanings. This can occur in various ways. First, we may introduce incompatibility of theories as above (lack of isomorphism with respect to certain basic relations plus consequences following therefrom) and then confirm one of two incompatible theories. If the confirmed theory and its rival are strong alternatives, then the rival will have to go. Secondly, both theories may be able to reproduce the 'local grammar' of sentences which are directly connected with observational procedures.[32] In this case the utterance of one of the sentences in question in accordance with the rules of the local grammar, or the utterance of a local statement, as we shall say, can be connected with *two* 'theoretical' statements, one of T, and one T' respectively (theoretical statements, corresponding to a given local statement S will be called statements associated with S within the theory in question). We may now say that the empirical content of T' is greater than the empirical content of T, if for every associated statement of T there is an associated statement provided by T', but not vice versa. And we may also say that T' has been confirmed by the very same evidence that refutes T if there is a local statement S whose associated statement in T' confirms T' while its associated statement in T refutes T. Similar remarks apply to the question of independent evidence. Thirdly, we may construct a model of T within T' and consider its fate. The proof of the irrationality of $\sqrt{2}$ may be seen in this way. To start with, rational and irrational numbers form two distinct domains, R and I, so that it does not seem to be possible to relate a single entity, $\sqrt{2}$, to both of them. However, we may select a subdomain, R', of I which is isomorphic to R with respect to all the properties we think essential

---

[31] It was Herbert Feigl whose incessant arguments on this point woke me from my dogmatic slumber and made me realize the need for a more detailed theory of test. Cf. n.21 of ch. 2.

[32] The local grammar of a statement is that part of its rules of usage which is connected with such direct operations as looking, uttering a sentence in accordance with ostensively taught (*not* defined) rules, etc.

of the latter, and interpret the 'experiment' as being one between I and R'.[33]
Within physics the construction, in T', of a model $\bar{T}$ of T usually is not
possible without violation of highly confirmed laws. This, then, is sufficient
evidence for the rejection of $\bar{T}$ and, via the isomorphism, of T.

To sum up: it *is* possible to use incommensurable theories for the purpose
of mutual criticism. This removes, I think, one of the main 'paradoxes' of
the approach suggested by me.[34]

Now for the internal questions. The disagreement between strong
alternatives is due to basic rules, and not to *ad hoc* adaptation from point to
point. According to n. 27 above, we must conclude that the difference will
turn up in the descriptive concepts, and that the theories will be incommensurable. We see here quite distinctly the function of the descriptive concepts
of a coherent theory. *They reflect, within each individual description and prediction
the peculiarities of the fundamental principles of the theory.* That is, they will be
applicable whenever the principles are adopted, and inapplicable otherwise. Hence, the concepts of theories which have been shown to be incompatible by the first method above cannot be applied simultaneously, provided, of course, that the basic principles *are* allowed to contribute to their
content. This proviso leads at once to the next three questions:

 (i) the historical question: *do* we allow our basic beliefs, principles, etc., to
      influence our concepts?

 (ii) the methodological question: *should* we allow our basic principles to
      influence our concepts?

(iii) the physiological question: *can* all our basic beliefs, principles, etc.,
      influence our concepts?

It is in the consideration of these three questions that the greatest
confusions arise. Usually (ii) is rejected by appeal to the fact that crucial
experiments would then become impossible. This point has already been
dealt with. Or it is rejected because empirical statements are supposed to
retain an invariable meaning. Why? In order to provide a neutral basis for
our decision between different theories. This point can be reduced to the
previous one and is removed with it. But this last point, which is a logical
point (and an invalid one) is then often interpreted as establishing limits to
our interpretation of concepts, and especially of observational concepts.
Now it is one thing to say that some interpretations of observational
statements make some theories irrefutable. But it is quite a different thing

---

[33] For details of the relation between I, R and R' see e.g. F. Waismann, *Introduction to
Mathematical Thinking* (New York, 1951).

[34] For a discussion of such paradoxes see also P. Achinstein, 'On the Meaning of Scientific
Terms', *J. Phil.*, (1964), 497ff, as well as ch. 5. The argument in the text also invalidates one
of my main criticisms of Professor N. R. Hanson's *Concept of the Positron* (the criticism was
published in a rather unkind review, *Phil. Rev.*, (1964), 266. Looking back upon this criticism
I find it somewhat amusing that I should have used the very same argument against Hanson
which Achinstein, Shapere, Putnam and others are now using against me. It is sometimes
difficult to recognize one's friends.

to say that such undesirable interpretations *can never be carried out*. We meet here Kant's familiar transcendental argument, only it is now presented upside-down, not as an argument showing which theories are certain, but as an argument showing how certainty can be avoided at one place and how it therefore turns up at another. I am sure that Strawson's assumption that there are elements of meaning common to all languages has been arrived at in some such manner – if it is based upon argument, that is. But it is equally evident that while he and others move here in a logical fairyland the problem is a factual one, to wit (question (iii) above) whether all principles, including the most esoteric ones, can be incorporated, *physically*, into our reaction patterns (with possible effect upon our perceptions), or whether some such principles are not just 'too much' for our organism. It is the great merit of Smart that he has drawn attention to this side of the problem, and of Sellars that he has arrived at his conclusion with this feature of the problem in sight (cf. his reference to neurological facts in para. 58 (b)). This is an important step beyond the pseudological rigmaroles we are usually invited to consider. I hope that I shall be able at some time to give a more detailed account of the matter. At the present, with but two days to go for completing the paper (including, of all days, Christmas Eve!), removed from anyone who might provide inspiration, either by discussion, or by beauty, I can only offer the following brief considerations.

6. THE HISTORICAL QUESTION

The answer to the historical question is that people very often *did* allow fundamental principles to influence the most concrete parts of their every-day lives (or the 'local grammar' of the words used in this context). As proof I invite the reader to consult historical accounts of the way in which religious beliefs have influenced all aspects of human activity, the most pedestrian parts included, so that even the distinction between waking and dreaming changed its position and its nature. This applies also to such apparently 'rational' people as the Greeks.[35] All this refutes the idea, held by Austin[36] and others, that such beliefs were metaphysical superstructures erected on a single unchangeable basis – Oxford commonsense. In chs. 4, 9 and 10, I have given very few examples of this all-pervasive character of basic theory indicating that the historical and sociological literature is full of cases of this kind. Let me briefly repeat one of these examples and deal with Putnam's objections to it. The example is the way the pair 'up-down' was used in antiquity (and in the early Middle Ages). Now first of all this example deals with concepts, and not with words, so that Putnam's two-word excursion into Greek scholarship is irrelevant. (Besides, his interest in

[35] Cf. E. R. Dodds, *The Greeks and the Irrational* (Berkeley, 1956).
[36] Private communication.

the Greek *words* seems to indicate that it is he, and not I, who deals with occurrences, rather than with usages.) Secondly the grammar of the pair is not at all dependent on physical assumptions concerning *forces*.[37] 'Aut est quisquam tam ineptus qui credat esse homines quorum vestigia sint superiora quam capita. Aut ibi quae apud nos jacet inversa pendere? Fruges et arbores deorsum versus crescere?' (Lactantius, *Divinae Institutiones*, III, *de falsa sapientia*; apologies for the Latin – however, in arguments with linguistic philosophers, it is better to use the original *words* rather than a translation, which would preserve the *concepts* only). Here reference is made to unusual *positions*, and not to unusual *motions* (of course, if we start from unusual positions, then even the usual forces will bring about unusual motions!). From considerations like these it is then inferred that the heavens cannot be spherical, which is indeed evidence that 'down is always the same direction' was linguistically required (Putnam, 211). The fairy-tales which Putnam mentions only apply to unusual *forces*: they are still compatible with the absolute sense of 'up–down'. This pair was therefore used much in the same way in which Max Black thinks the pair 'sooner–later' is used today.[38] And *this* use has disappeared today, at least in some circles (we no longer talk of the *lower* side of the earth, but of the *other* side). The change admittedly is not a universal one. This now leads at once to the following question: should we not educate people in a manner that makes everyday behaviour agree with basic science as it agreed with basic theology before? This is question (ii).

7. THE METHODOLOGICAL QUESTION

The power of a theory can be fully utilized only if it is not treated as an instrument for prediction, so that the local grammar is allowed completely to determine the 'nature of things'. It is doubtful whether Newton would ever have identified the Galilean orbit of a projectile with the Keplerian orbit of the moon[39] had he restricted himself to data of observation (cf. fig. 1 after his very instructive drawing in the *System of the World*). No data were available for C, D, E; the orbits of projectiles on the one side, and the orbit of the moon on the other were thereby shown to belong to different domains. However, quite apart from facts like these, it has never been explained why a new and satisfactory theory should be interpreted as an instrument, while the principles behind the established usage, including the local grammar, which are shown to be inadequate by the new theory,

[37] Talking of forces, by the way, is not as anachronistic as Putnam makes it appear. There were of course no Newtonian forces as they are understood now. There were forces of a different kind, vitalistic entities, quite similar to some of the entities considered by Newton's contemporary, Henry More (who had a considerable influence upon Newton's thought).

[38] 'The Direction of Time' in *Models and Metaphors* (Ithaca, 1962).

[39] For this formulation see A. Koestler, *The Sleepwalkers*, 504.

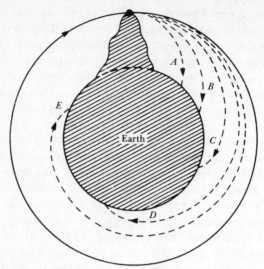

Fig. 1 Sketch from Newton's *System of the World* showing the relation between terrestrial and celestial orbits

are not so interpreted. After all, the only advantage of the latter is that they are *familiar*, an advantage which is a psychological and historical accident and which should therefore have no influence upon questions of interpretation and of reality. One may try to answer this criticism by ascribing an instrumental function to *all* principles, old and new, and not only to those contained in the most recent theory. Such a procedure means acceptance of a sense-datum account of knowledge. Having shown elsewhere that such an account is impossible,[40] I can now say that this consequence of a *universal* instrumentalism is tantamount to its refutation. Result: neither a restricted, nor a universal instrumentalism can be carried through in a satisfactory manner. This disposes of the instrumentalist move.

It does not yet dispose of the arguments against the realistic interpretation of new, and successful theories, or against the *complete* replacement, in accordance with the principle enunciated at the end of section 2, of older theories by their victorious rivals. Of these arguments there are four:
Argument 1: T, the older theory, or the corresponding principles of common-sense, express what a person *means* who describes events within the domain of overlap D, between T and the theory T'. Thus a layman talking about the temperature of his milk 'surely' does not consider 'the kinetic energy of the molecular constituents of the liquid' (an argument against the universal use of statistical physics, discussed by Nagel).[41] And 'one can talk of trees, chimneypots, and telescopes in a way which is independent of the choice

[40] 'Das Problem der Existenz Theoretischer Entitaeten' in *Probleme der Wissenschaftstheorie*, ed. E. Topitsch (Vienna, 1960), 35ff.   [41] Cf. *The Structure of Science* (New York, 1961).

between Newtonian and Einsteinian physics' (an argument against the universal use of relativity, discussed in Smart's paper). Or, 'I *know* [pronounced indignantly, and emphatically] what pains are, and I also *know* [same as above] that a reference to material constituents is not only not implied, but is even nonsensical' (familiar argument against the universal use of materialism).

Argument 2: It would be practically impossible, too complicated, to use T' inside D. 'For example', writes Smart, 'it would be impossible in practice to use general relativity in celestial mechanics (though it may be needed to explain isolated anomalies, as with the perihelion of Mercury).' This point is sometimes given great weight in arguments against the feasibility of a purely physiological account of the mind.

Argument 3: Scientists (or the common folk) *in fact* use T inside D.

Argument 4: T is our *observation language*. Eliminating *it* means eliminating the observational basis on which science rests, and thereby also T' (this argument has been used by Feigl, Nagel and others).

All these arguments have one thing in common: they would have provided excellent weapons for the Aristotelians against Galileo; for the defenders of witchcraft against their more enlightened opponents; and for the backward side of many other debates. As a matter of fact some of these arguments have been used for the purpose of defending an older doctrine which was in danger of being overthrown, and especially in the disputes which accompanied the emergence of modern science. It is therefore a little late to try to enforce them now, after the rash opponents of Aristotelianism (to mention the most popular example) have succeeded in eliminating it. Besides, if it is indeed so important to stick to what is commonly accepted, be it now part of commonsense,[42] or of some more esoteric doctrine, then consistency would seem to force us to retreat even further, at least to the conceptual system of *Pithecanthropus Oxoniensis*, and to start from there. On the other hand, if such a procedure is deemed too radical, if we are allowed to use the ideas of classical physics (for example) because they are accepted now, then what prevents us from voting now for the future, and from bringing about *its* general acceptance by proper methods of education (such as the use of educated mothers or nurses)? Moreover, the fact that comprehensive conceptual schemes have been replaced *in toto* (women having no longer the opportunity of betraying their ineffective husbands with the devil) would seem to indicate that the difficulties mentioned in the arguments are largely imaginary. A more detailed examination fully confirms this guess.

The first argument is incomplete. Different people, members of different language groups will 'mean' different things when proceeding descriptively

---

[42] Sellars' objection against equating commonsense with theories such as the theory of relativity will be dealt with in section 9.

within D.[43] Whom shall we choose as a point of reference? Moreover, it cannot be denied at this stage of the inquiry that a practitioner of T' may become accustomed to his theory to such an extent that in the end he will see D in its terms. The argument even admits this possibility and tries to prevent it from becoming an actuality (cf. the above quotation from Nagel with its *two* alternative descriptions of the same situation – the caloric properties of milk). It admits the possibility of language groups whose members 'mean' T' when proceeding inside D. Why not take their meanings as a measure of adequacy? Why not try even to form such groups through education, especially in view of the fact that T' has advantages over and above T? (After all, we have assumed that T was removed by T', and T' is a strong alternative of T. According to the statement at the end of section 2 this means that T' is superior to T.) The argument in the form in which we have presented it so far cannot answer this question.[44]

We may now strengthen the argument by adding the assertion (which I shall call the *universality thesis*) that everybody, and not only a few laymen, thinks in terms of T. The universality thesis is silently assumed to be correct by many linguistic philosophers. Ryle's *Concept of Mind*, for example, is not intended as a field study of the conceptual system of *homo Oxoniensis*; it is supposed to possess universal validity. In section 6 we have referred to material indicating that the universality thesis is *false*.[45] It remains to point out that it is also irrelevant.[46] A language (a theory) is not only a vehicle for meanings, it is also supposed to inform us about what exists. The universally accepted idiom (if there is such a thing) may not satisfy the second demand. It may be a universal mistake. Universality of T therefore cannot convince us that T' must not be universalized. Quite the contrary, it is only through the attempt to universalize a new theory that we can discover the mistakes of the accepted ideas (for this see again sections 1 and 2 above). The universality thesis therefore is not only incorrect, and irrelevant. It is also desirable that it be false.

However – and this at last brings us up to a real difficulty – is it not

[43] Cf. section 6 and ch. 10.

[44] Reference to the empirical success of T, or of some common idiom mimicking T, is clearly irrelevant, for we have admitted that T is successful in D. But so is T' which also has additional confirmations elsewhere (see section 2).

[45] Cf. also ch. 10. The popular argument that a layman can function as an observer for the theoretician and that the latter is therefore committed to the categories of the layman's report, and perhaps even to his whole language, is easily refuted. The scientist cannot use the layman's report without some interpretation of his own (consider what would happen were he to take over the interpretation of a primitive Zulu's account of a comet). He very often uses not the *content* of this report, *but the fact that the sentence expressing it has been uttered*. After all, he might have employed a well conditioned dog without being thereby forced to include barks into his theory, and to 'mean' with them that the dog 'means'.

[46] This was of course well known to all those philosphers of the past who objected to regarding common consent as a basis for truth. For a more recent discussion with ample material see para. 591ff of V. Pareto, *Treatise on General Sociology*.

plausible to assume that the (physiological) organization of the human mind puts limits to the replacement of theories and to scientific progress? Are there not very general theories, which cannot be replaced except by altogether destroying the functioning of the mind? This consideration, if I understand him correctly, is the basis of Sellars' objections against me. It will be dealt with in the next section.

The second argument mentions formal difficulties: it may be impossible to develop the formalism of T' in such a manner that everyday occurrences can be described in its terms. Usually a theory deals with fairly idealized elementary situations which are not accessible to direct observation. Concrete events, such as the processes occurring at the level of everyday observation, involve the interaction of many elementary processes and it is not always possible to calculate their interaction in terms of the most recent theory. This applies even to theoretical cases. For example, we do not yet see any way of completely replacing Newton by relativity in planetary astronomy (this has been observed by Smart). However, the universal application of a theory, or of some other general point of view, does not require the universal use of its *syntactical apparatus*; all it requires is that the *categories* of the theory be regarded as basic and that these categories replace the categories of the preceding point of view.[47] Thus the adoption of a materialistic position with respect to the mind–body problem does not entail that from now on we shall have to employ density functions in phase space for the description of mental events, and it is not exposed as impossible by the fact that the explicit form of these functions is not available, and will perhaps never be available. Nor does the universal application of general relativity entail that the *formalism* of classical mechanics must go and must be replaced by the formalism of general relativity (not even classical mechanics itself has as yet achieved coherence of expression in all domains of application, and it will perhaps never achieve it). In the first case all that is needed is the realization that we are dealing with complex patterns and sequences of physical events and that the appropriate categories will therefore be those of physical processes (localizability is one of them; mass is another). In the second case we may retain all the formulae of classical physics, but we must now interpret their key terms after the fashion of general relativity, either as four-dimensional invariants, or as projections into the space – or into the time – part of the four-dimensional Riemannian continuum. It would of course be desirable to add coherence of formal expression to coherence of content. The main thing is however the latter. Without it there would not even exist a motive for the attempt to achieve coherence of formal expression. Extension of a new theory to the domain in which an older point of view is still doing well is therefore much

---

[47] This, I hope, resolves the difficulties that Hanson has felt in connection with the principle of correspondence. Cf. also my review in *Phil. Rev.*, 69 (1960), 251.

less difficult than might appear at first sight. We may retain the familiar words (such as 'pain', 'thought', etc.) or the familiar formulae; we may even retain the local grammar (see n.32) of these words which is that part of their grammar that is restricted to the domain D of the overlap of T and T'. We may also continue to use all those reflexes which guide their automatic application in observational situations. Only, this local grammar cannot any longer determine *all* our linguistic reactions. It must be incorporated into a new framework (which corresponds to the takeover of a new level of neurological organization). Such integration may then retroactively lead to a partial reorganization even of the local grammar itself.[48] It may lead to 'seeing things differently' and it may in this way precipitate the discovery of quite obvious but still unnoticed weaknesses of T *inside D* (the realization that we can see the fixed stars only on a photon account of light is an example). All this must of course be examined in greater detail, especially via a more thorough examination of physiological principles. Still, the difficulties insinuated (but never stated) in arguments 1, 2, and 3 seem to be largely imaginary. They seem to be due not to the existence of objective barriers, but rather to the non-existence of that amount of boldness that is necessary for overcoming the problems which arise in the course of our attempt to comprehend the world.

8. OBSERVATION

Argument 4 is not easy to evaluate. It may be given a physiological twist, but it also contains an element of logic. This dual character of the argument considerably improves the position of its defenders. They start with the stronger version and create the impression that their argument has *substance*. When attacked, they withdraw to the weaker version and can so pretend that their argument is *invulnerable*. A shifty procedure like this should be easy to look through – but this is not what happens. Quite the contrary, the procedure seems to possess some kind of intuitive appeal. Exhibition of the dubious character of this appeal is overdue.

The *logical* version of the argument assumes that there are statements, so-called basic statements, or observation statements, which can be absolutely confirmed by observational evidence. This property is thought to be connected with the 'meaning' of their main descriptive terms (the observation terms). If this is true, then a radical enough change of the meanings of all observation terms will eliminate the property, and destroy the observa-

---

[48] Physiologically speaking this reorganization may be quite minute (a point already made by F. A. Lange). This removes the objection (which has some real force in other contexts) that the necessity to stay alive and to function as an organism puts limits upon our ability to 'see things in a new light'. We are here confronted with some very interesting connections between physiology on the one side, and the realism–instrumentalism on the other. For a more detailed discussion cf. section 9.

tional basis of our knowledge. One may therefore formulate the logical version by saying that a logical connection is assumed to exist between the meaning of a term (apologies for again bringing in semantic matters) and the fact that it is, or is not, an observational term. I shall call a theory of observation which is based upon such a connection a *semantic theory of observation*.

A semantic theory of observation is usually supported by an analysis of such verbs as 'to see', or 'to observe', or 'to hear', or else by an analysis of the nouns which form their objects. If such an analysis is correct (a matter that is open to considerable doubt) then the semantic theory forms an essential part of commonsense. There are also other, and less linguistic ways, of supporting it. Whatever the means of support, the semantic theory is unacceptable to anyone who rejects the synthetic *a priori*. For consider the principles P which guarantee that observational terms can be applied. These principles describe some general features of the world (or of the mind) and are therefore synthetic. On the other hand a theory implying their denial would eliminate its own observational basis, and would therefore be unacceptable. We can say in advance that such a theory must not be formulated – which guarantees the perennial correctness of P. P is synthetic *a priori*.

The undesirable consequences can be avoided by redefining observational terms so that the logical connection between meaning and observability is eliminated (as it is in any case within commonsense). Only if we admit that meaning and observability are two different and independent (though empirically related) dimensions of a term, only then shall we be sure that *this* source of a synthetic *a priori* has been destroyed. The new theory of observability which results from the described procedure (and which was formulated very clearly in the early thirties by Popper, Carnap and Neurath)[49] may be called the *pragmatic theory of observation*. The choice between the pragmatic theory and the semantic theory is of course purely a matter of convention. If we want synthetic *a priori* statements, if we want to be able to derive eternal laws from facts of observation (as was done by Aristotle, and as is done again by the school of 'pure phenomenology'), then the semantic theory will most certainly be a Splendid Thing. However, if it is our intention not to except any part of our knowledge from revision, then we shall have to choose the pragmatic theory. This choice eliminates argument (4).

For it is clear that the question whether or not a certain term, such as the term 'electron' is, or can be, an observational term, is now a question of empirical fact. We have already dealt with this question in section 6 when asking to what extent a theory can be integrated into our behaviour so that

---

[49] R. Carnap, 'Ueber Protokollsaetze', *Erkenntnis* 3, (1933), 215ff expresses the theory thus: 'It stipulates that . . . *any* concrete statement may be regarded as a protocol statement.'

we can 'see' the world in its terms. We have pointed out that so far no limit of integration has been found (for this see also the next section). However, physics provides also the opportunity of an *external integration*, as one might call it,[50] i.e. of an integration which makes use of devices bridging the gap between the entities considered, and some of the more 'basic' parts of human behaviour. In this case the observation sentence is produced by the combined system [apparatus+observer]. There are no limits to a procedure of this kind. This refutes the 'physiological' version of argument (4).

A variant of argument (4) consists in pointing out that without some restriction in the use of new theories which allows for the definition of a neutral 'experience', it may become impossible to obtain *impartial* evidence for a theory, or to stage a single *crucial experiment*. This has already been discussed in section 5. However, I would like now to pursue the matter a little further as it is important to be quite clear about the situation in which we are here involved. As we have indicated above, it *is* possible to retain older ideas, and to allow modifications only at the outskirts of our knowledge. Putnam's 'approximate theories' would be one way of doing this. It is possible to obtain what many empiricists seem to regard as such an important feature of knowledge, viz. a relatively neutral 'experience' that can decide between two universal theories without being influenced by either. But such a procedure introduces the synthetic *a priori* and may prevent discovery of weaknesses of the accepted point of view right in the centre of its most successful domain of application. The empiricists therefore cannot have their cake and eat it too. *I* can (for this see again section 5 above).

It remains to make a few comments on argument (3). This argument, or rather an attitude corresponding to it, is rather widespread among scientists. The attitude invites concentration on successful theories and deplores consideration of alternatives. It is the attitude of monism with which we are already familiar. Again, there are two kinds of argument which may be used, and have been used to support the attitude. The first argument, which goes back to Newton (and was refuted by Goethe, among others), assumes that it is possible to prove a theory 'directly, and positively' (Newton) and *not* only via the elimination of alternatives. This argument excludes alternatives as not being empirically adequate. The second argument admits that any theory may sooner or later break down, but adds that one should make the best of it while it is still successful. This latter argument (for which see n.14 above) excludes alternatives as bringing about an unwarranted dissipation of the valuable energy of the scientists. In practice one usually meets a combination of argument one and argument two, argument one providing the punch, argument two a safe second line of retreat (this fluctuation between a more logical and a more practical position is quite characteristic of the 'classical' empiricism of Newton and of his followers; it still appears in Niels Bohr). Neither the combination nor the single arguments are acceptable, for which see above.

We may now summarize by presenting a scoreboard of arguments:

Argument 1: incomplete without the universality thesis, irrelevant and leading to undesirable conclusions with it

Argument 2: irrelevant

Argument 3: invalid

Argument 4: invalid

As far as I can see these are the only arguments available against the idea that a good theory should replace a bad theory completely, and not only in isolated cases (the only argument not yet dealt with will be discussed in the next section). The arguments have also been presented here in a rather strong fashion. What one usually finds in the literature on this point is an attitude of conservatism struggling to become an argument. None of the arguments is satisfactory. This is the one side of the issue. On the other side, on our side, there exist very good *abstract reasons* why a new theory should be used everywhere: only this procedure will lead to the strongest possible criticism of the received point of view. These abstract reasons are fortified by *concrete examples* of extremely successful scientific research making use of the very procedure that arguments (1–4) want to forbid. Finally it is merely *plain commonsense* (though perhaps not Oxford commonsense) to prefer total improvement to piecemeal tinkering and to accept such total improvement wherever it can be had. The case for a complete takeover therefore seems to be well established – but for one point to which we now at last turn.

9. THE PHYSIOLOGICAL QUESTION

It is reported that Pavlov once criticized the early Bolshevists' theory of education as being incompatible with the laws of behaviour and thinking which he himself had discovered. This criticism may be removed by the following consideration. Assume that the educational *aims* of the Bolshev-ists are attractive, and that their realization is therefore desirable. Then the attempt to realize them may either lead to success which is a good thing, and thereby to the refutation of Pavlov's theory which, being a step for-ward, is also a good thing. Or this attempt may fail and thereby confirm Pavlov's view, which is again a good thing. Hence, both scientific method and proper education make it desirable to attempt to carry out the Bolshevists' aim.

The very same argument can be applied to those who resist the complete replacement of commonsense by some scientific ideology on the grounds that the attempt is utopian. For if the attempt succeeds, then man will have succeeded in ridding himself of prejudices which would have prevented him from ever seeing the world as it is and from figuring out the details of its real structure, which is a good thing; at the same time the laws of the mind on which the opponents base their charge of utopianism will be refuted, which is also a good thing. If the attempt does not succeed, then this provides confirming evidence for these alleged laws, which is again a good thing.

These two arguments are precisely analogous except for the fact that Pavlov *could point to the laws* he thought would prevent certain attempts from succeeding whereas no one has as yet formulated the laws of the mind which allegedly prevent the universalization of scientific ideologies. Even worse, most arguments are nowadays based not on physiology, but on meanings, and are therefore entirely irrelevant. Among past arguments we may mention Kant's proposal to derive the structure of the mind from the structure of a theory then widely held. This proposal rested on the features of a *single* theory. The possibility that other theories might bend the forms of perception and the categories to accommodate *their* structure was either not considered, or was rejected outright, and without any experimental investigation. The opposite view, the view that we see things as we believe them to be – 'you see a bird, I see an antelope; the physicist sees an X-ray tube, the child a complicated light-bulb; the microscopist sees coelenterate mesoglea, his new student sees only a gooey, formless stuff; Tycho and Simplicius see a mobile sun, Kepler and Galileo see a static sun'[51] – has again been held without experimental support. I strongly sympathize with this latter view, but I must now regretfully admit that it is incorrect. Experiments have shown that not every belief leaves its trace in the perceptual world and that some fundamental ideas may be held without any effect upon perception.[52] The laws of perceptual organization are then not wholly dependent on the information available to us even in those cases where the information has already become part of our automatic behaviour. All this clearly cries for further research. Is there the slightest suspicion that such research may support the Strawsonian thesis of a 'massive central core of human thinking that has no history' and is never going to have a history?[53]

Before turning to this question, let me first briefly deal with some points which were raised by Sellars. His paper is so rich in content and so full of interesting conjectures that it is perhaps somewhat impudent to deal with it

[51] N. R. Hanson, *Patterns of Discovery* (Cambridge, 1958), 17.

[52] Thus Professor Michotte in his epoch-making investigations of phenomenological causality has produced certain cases where 'causal impressions appear in conditions which, from the point of view of commonsense mechanics, are paradoxical', *The Perception of Causality* (New York, 1963), 87. These cases refute the idea that the causal aspects of certain perceived processes are due to an 'act of interpretation' on our part or that 'under the influence of past experience or in some other way, we ourselves invest certain basic impressions of movement with a "meaning"' (p. 87). Conversely, he has shown that 'it seems impossible to produce a visual causal impression of one object being attracted by another' (p. 125).

[53] *Individuals* (London, 1959), xiv.

in such a summary fashion. Yet, the editor beckons urgently! Here, then, the comments.

I start with a few remarks on Sellars' attempt to distinguish between theories which express beliefs, and other principles which 'are constitutive of the very concepts in terms of which we experience the world' (para. 5). It seems to me that this distinction does not separate science from commonsense. The spacetime framework of classical physics and of the theory of relativity, to mention but one example, have quite a lot in common with frameworks of commonsense as described by Sellars. They have 'no external subject matter' (para. 6); their concepts help 'to project, or transpose the attributes of sense impressions into the categorial framework of physical things and processes' (para. 57) so that they are (repeat) 'constitutive of the very concepts in terms of which we experience the world' (para. 5). Now this similarity is strong enough to give the same treatment to both and to express this fact by a unified terminology ('theories' – see n.5 above). For example, we shall have to admit that a relativistic account of the spatiotemporal aspects of commonsense is possible *here and now* (paras. 48, 74) – after all, the theory of relativity is 'already at hand', it is not 'a "regulative ideal"' (para. 50). Sellars' reluctance to admit this, his emphasis on the difference between science and commonsense, is due to his preoccupation with *microtheories* (paras. 47f, 67, 9. Putnam, too, seems only to consider microtheories and not framework theories such as the theory of relativity. Cf. his comment on the chimneypots on p. 215.). Microtheories indeed do not seem to contain any large-scale principles. They seem to permit a safe distinction between the inevitable external subject matter (gases, large-scale objects) and some internal subject matter. One sees that both are necessary and one can at least envisage the possibility of *retaining* the customary account of perceived objects, and of interpreting the micro-account as *additional information*, filling in holes, or empty spots, not dealt with by the macroconcepts. True, Sellars does not himself subscribe to this 'hole theory' or 'Swiss cheese theory' of commonsense (which Smart has conjured up on another occasion in order to make a scientific materialism more palatable to the pundits of ordinary speech). Still, by likening scientific theories to microtheories, and by relating the latter to an external subject matter, independent of it, indispensable, and rooted in commonsense, he has created the impression that a particular shape of commonsense is 'methodologically indispensable' (para. 59), too. The fact that there are scientific theories without any external subject matter refutes this impression and imposes the task of *modifying* those aspects of commonsense which are incompatible with them. Now according to Sellars the framework of commonsense 'is a framework of (among other things) coloured physical objects extended in space and enduring through time' (para. 53). Question: in Euclidian space, and with time running down separately? If the answer is yes, i.e., if the shape of the objects is fixed independently of their motion, then a revision is clearly needed. The intuitive content of this revision can be shown by large-scale models (such as the behaviour of an image in a spherical mirror, or the behaviour of the normal world when seen through distorting glasses) and may at some time become visible even in the real world. Four-dimensional invariants are not entirely beyond our faculty of perception (just consider our ability to perceive the *changing* character of a person behind the quite indescribable changes of his face, behaviour, tone of voice). The question that remains is, of course: is it possible *always* to carry out such a modification even of the most primitive perceptual elements of commonsense?

The answer to this last question must of course remain hypothetical. No stronger claim can be made at the present time. There are some arguments in favour of an affirmative answer. To start with, we should consider that until now only two or three per cent of the inbuilt circuits of the brain have been utilized. A large variety of further integrations is therefore possible. There may of course be difficulties: present education after all consists, according to an excellent remark of Professor Z. Young, in badly damaging our learning apparatus.[54] 'The moulding of the baby into a transmission receiver', writes Professor Waddington in the same vein, 'seems a difficult, and complicated, and even slapdash process, and not at all what one might have thought out if one had set out to design this job. A frequent result of the process seems to be that people believe too much, and believe it much too strongly';[55] or, to use Gantt's even more drastic characterization, they increasingly become 'a museum of antiquities'.[56] (Note, that the main intention of our linguistic friends is to find 'logical' reasons for continued infliction of such damage, and for leaving the museum unaired). It seems reasonable, then, to assume that a new mode of education whose principles have yet to be found will make it easier to change points of view, to integrate theories into behaviour, and in this way to see the world in terms of such theories. The education will have to be careful not to let the world of everyday objects come forth too impressively, and it will also have to keep the division between dream and reality somewhat open. The Greeks, whose perceptual world seems to have been much less settled than ours,[57] also

---

[54] *The Hitchcock Lectures* (Berkeley, 1964).
[55] 'Issues in Evolution', *The University of Chicago Darwin Centennial Discussions*, (Chicago, 1960), III, 172ff.   [56] *Darwin Centennial*, III, 199; II, 235.
[57] 'We have to take into consideration that man awake comes to know that he is awake only through the rigid and lawful net of the concepts he uses; this is why he sometimes believes that he is dreaming when this net of concepts has been torn to pieces . . . Pascal is quite correct when saying that a recurring dream would occupy us just as much as things we see every day . . . The waking day of a mythically excited people, such as the older Greeks is indeed closer to dreaming than the day of a sober thinker. The constantly acting miracle, as it is assumed by the myth, is responsible for this. If each tree can now talk as a nymph, if a god can rob a virgin under the cover of a bull, if Athene herself is suddenly seen, passing in a

produced thinkers with a range of imagination quite inconceivable today. For those already grown up, and messed up, a psychological procedure like the procedure of alienation suggested by Brecht may do the trick.[58] All these are problems which need investigating, especially through a more detailed study of the central nervous system. The mentalistic bias of much of contemporary philosophy is one of the most widely used means to prevent such investigation from becoming effective; and this is, and always has been, its ideological function.

So far we have been dealing with the higher levels of mental organization only. Now the structure of the central nervous system has been described by various authors as consisting of layers of organization, ranging from highly organized, involuntary levels of low complexity to lowly organized, voluntary and very complex higher levels. The higher levels control and overlay the activity of the lower levels. If they are put out of action, then the primitive levels break through and impose their own laws. This creates the 'positive' components of 'dissolution', first described by Hughlings Jackson. The whole system is somewhat precarious and always threatened by a 'revolution from below'. Now one might be inclined – though this is of course entirely hypothetical – to correlate certain basic features ('archetypes'?) of common languages with lower levels and the abstract notions of theoretical science with higher levels. If such a correlation does in fact exist, then the life of a theoretician who wants to 'see' the world in terms of the most recent scientific ideology will not be an easy one. But neither was the transition from animal life to civilization with all its additional repressions and its precarious balance between sophistication and savagery. The argument from common usage, amplified by these physiological considerations, is therefore correct insofar as it maintains that certain modes of thinking, being physiologically more basic, are also more firmly integrated into human life, behaviour, perception. The argument certainly goes too far when asserting that these modes of thinking are the only ones which can be integrated. After all, we do overcome our lower nature and nobody would be inclined to attack civilized morality by pointing out that we are apes with inbuilt constraints.

These are only some of the problems arising in connection with what we have above called the physiological question. It is evident that they are not even touched upon in the grammar-school philosophy of our friends, the contemporary school-philosophers of grammar.

beautiful carriage on the side of Peisistratus through the marketplaces of Athens – and this every honest Athenian believed to be the case – then like in a dream everything is possible at every moment and nature surrounds man like a masquerade of the gods.' F. Nietzsche, 'Ueber Wahrheit und Luege im aussermoralischen Sinn', *Werke*, ed. Schlechta, (Munich, 1969), III, 319f.) Cf. also Dodds, *The Greeks and the Irrational*, and the literature quoted there.
[58] Cf. the excellent analyses in B. Brecht, *Versuche 25/26/35* (Berlin, 1958), 102ff, concerning the role of what he calls the V-effect in everyday life.

7

Science without experience

One of the most important properties of modern science, at least according to some of its admirers, is its *universality: any* question can be attacked in a scientific way, leading either to an unambiguous answer or else to an explanation of why an answer cannot be had. In the present note I shall ask whether the *empirical hypothesis* is correct, i.e. whether experience can be regarded as a true source and foundation (testing ground) of knowledge.

Asking this question and expecting a scientific answer assumes that a science *without* experience is a possibility; i.e., it assumes that the idea is neither absurd nor self-contradictory. It must be possible to imagine a natural science without sensory elements, and it should perhaps also be possible to indicate how such a science is going to work.

Experience is said to enter science at three points: testing; assimilation of the results of test; understanding of theories.

A test may involve complex machinery and highly abstract auxiliary assumptions. But its final outcome has to be recognized by a human observer who *looks* at some piece of apparatus and *notices* some observable change. Communicating the results of a test also involves the senses: we *hear* what somebody says to us; we *read* what somebody has written down. Finally, the abstract principles of a theory are just strings of signs, without relation to the external world, unless we know how to connect them with experiment, and that means, according to the first item on the list, with experience, involving simple and readily identifiable sensations.

It is easily seen that experience is needed at none of the three points just mentioned.

To start with, it does not need to enter the process of *test*: we can put a theory into a computer, provide the computer with suitable instruments directed by him (her, it) so that relevant measurements are made which return to the computer, leading there to an evaluation of the theory. The computer can give a simple yes–no response from which a scientist may learn whether or not a theory has been confirmed without having in any way participated in the test (i.e. without having been subjected to some relevant experience).

Learning what a computer says means being informed about some simple occurrence in the macroscopic world. Usually such information travels via the senses, giving rise to distinct sensations. But this is not always the case. Subliminal perception leads to reactions directly, and without sensory data. Latent learning leads to memory traces directly, and without sensory data. Post-hypnotic suggestion leads to (belated) reactions directly, and without sensory data. In addition there is the whole unexplored field of telepathic phenomena. I am not asserting that the natural sciences as we know them today could be built on these phenomena alone and could be freed from sensations entirely. Considering the peripheral nature of the phenomena and considering also how little attention is given to them in our education (we are not trained to use effectively our ability for latent learning) this would be both unwise and impractical. But the point is made that sensations are not *necessary* for the business of science and that they occur for practical reasons only.

Considering now the objection that we understand our theories, that we can apply them, only because we have been told how they are connected with experience, one must point out that experience arises *together with* theoretical assumptions, *not* before them, and that an experience without theories is just as incomprehensible as is (allegedly) a theory without experience: eliminate part of the theoretical knowledge of a sensing subject and you have a person who is completely disoriented, incapable of carrying out the simplest action. Eliminate further knowledge and his sensory world (his 'observation language') will start disintegrating; even colours and other simple sensations will disappear until he is in a stage even more primitive than a small child. A small child on the other hand, does not possess a stable perceptual world which he uses for making sense of the theories put before him. Quite the contrary, he passes through various perceptual stages which are only loosely connected with each other (earlier stages disappear when new stages take over) and which embody all the theoretical knowledge achieved at the time. Moreover, the whole process (including the very complex process of learning up to three or four languages) gets started only because the child reacts correctly to signals, interprets them correctly, because he possesses means of interpretation even before he has experienced his first sensation. Again we can imagine that this interpretative apparatus acts without being accompanied by sensations (as do all the reflexes and all well-learned movements such as typing). The theoretical knowledge it contains certainly can be applied correctly, though it is perhaps not understood. But what do sensations contribute to our understanding? Taken by themselves, as they would appear to a completely disoriented person, they are of no use, either for understanding, or for action. Nor is it sufficient just to link them to the existing theories. This would mean extending the theories by further elements so that we obtain longer expressions, not the understanding of the shorter expressions that we wanted. No – the sensations must be incorporated into our behaviour in a manner that allows us to pass smoothly from them into action. But this returns us to the earlier situation where the theory was applied, but allegedly not yet understood. Understanding in the sense demanded here thus turns out to be ineffective and superfluous. Result: sensations can be eliminated from the process of understanding also (though they may of course continue to accompany it, just as a headache may accompany deep thought).

I conclude with a few remarks on the observational–theoretical dichotomy.

Most of the time the debates about this dichotomy concentrate on the question of its existence, *not* on the question of its purpose. We may readily admit the existence of statements that are examined by looking and of other statements that are examined with the help of complicated calculations, involving highly abstract theoretical assumptions. There are observational statements and theoretical statements in that sense. But there are also statements expressed by long sentences and statements expressed by short sentences, intuitively plausible statements and statements that either sound absurd or leave our intuitions unmoved, and so on and so forth. Why is it preferable to interpret theories on the basis of an *observation* language rather than on the basis of a language of intuitively evident statements (as was done only a few centuries ago and as must be done anyway, for observation does not help a disoriented person), or on the basis of a language containing short sentences (as is done in every elementary physics course)? Because observation is supposed to be a source (a testing ground) of knowledge. Is this supposition correct? And does it justify the use of observational languages for the explanation of theories?

It justifies such use only if observation can be shown to be the *only* or the *only trustworthy* source of knowledge. We have just seen that the first part is far from true. Knowledge can *enter* our brain without touching our senses. And some knowledge *resides* in the individual brain without ever having entered it. Nor is observational knowledge the most reliable knowledge we possess. Science took a big step forward when the Aristotelian idea of the reliability of our everyday experience was given up and was replaced by an empiricism of a more subtle kind. Later on progress was often made by following theory, not observation, and by rearranging our observational world in conformance with theoretical assumptions. In the struggle for better knowledge, theory and observation enter on an equal footing, as do intuitive plausibility and intuitive absurdity: the absurd theory may win the day and the plausible theory may have to be given up just as the refuted theory may win the day, pushing aside, and making irrelevant, the refuting observations (this is what happened, for example, at the time of Galileo). Empiricism, insofar as it goes beyond the invitation not to forget considering observations, is therefore an unreasonable doctrine, not in agreement with scientific practice.

To sum up: a natural science without experience in *conceivable*. Conceiving a science without experience is an effective way of examining the empirical hypothesis that underlies much of science and is the *conditio sine qua non* of empiricism. Proceeding in this way, we may find methods that are more effective than plain and simple observation (just as Galileo found certain illusory phenomena to be more effective sources of astronomical knowledge than plain, direct, undiluted observation). Proceeding in this way of course means leaving the confines of empiricism and moving on to a more comprehensive and more satisfactory kind of philosophy.

Part 2

Applications and criticisms

8

Introduction: proliferation and realism as methodological principles

Proliferation of views and forms of life was recommended by John Stuart Mill 'on four different grounds'.[1] First, because a view one may have reason to reject may still be true. 'To deny this is to assume our own infallibility.' Secondly, because a problematic view 'may and very commonly does, contain a portion of truth; and since the general or prevailing opinion on any subject is rarely or never the whole truth, it is only by the collision of adverse opinions that the remainder of the truth has any chance of being supplied'. Thirdly, even a point of view that is wholly true but not contested 'will . . . be held in the manner of a prejudice, with little comprehension or feeling of its rational grounds'. And, fourthly, one will not understand its meaning, subscribing to it will become 'a mere formal confession' unless a contrast with other opinions shows wherein this meaning consists.

The first two reasons are amply supported by the history of science. We find that ideas are often rejected before they can show their strength. Even in a fair competition one ideology, partly through accident, partly because greater attention is devoted to it, may assemble successes and overtake its rivals. This does not mean that the beaten rivals are without merit and have ceased to be capable of making a contribution to knowledge. It only means that they have temporarily run out of steam. They may return and defeat their defeaters. The philosophy of atomism is an excellent example. It was introduced (in the West) in antiquity with the purpose of 'saving' macro-phenomena such as motion. It was overtaken by the dynamically more sophisticated philosophy of the Aristotelians, returned with the scientific revolution, was pushed back with the development of continuity theories, returned again late in the nineteenth century and was again restricted by complementarity. Or take the idea of the motion of the earth. It arose in antiquity, was defeated by the powerful arguments of the Aristotelians, regarded as an 'incredibly ridiculous' view by Ptolemy, and yet staged a triumphant comeback in the seventeenth century. What is true of theories is true of methods: knowledge was founded on speculation and logic; then Aristotle introduced a more empirical procedure which was replaced by the

[1] 'On Liberty', quoted from *The Philosophy of John Stuart Mill*, ed. M. Cohen (New York, 1961), 245f.

mathematical methods of Descartes and Galileo, which in turn were combined with a fairly radical empiricism by the members of the Copenhagen school. The lesson to be drawn from his historical sketch is that a temporary setback for a theory, a point of view, an ideology must not be taken as a reason for eliminating it. A science interested in finding truth must retain all the ideas of mankind for possible use or, to put it differently: *the history of ideas is an essential part of scientific method*.

The third and fourth reasons receive support from some very interesting and depressing phenomena that occur wherever an idea manages to become the centre of attention. The rise, success and triumph of a new theory, point of view, or philosophy almost always leads to a considerable decrease of rationality and understanding. When the view is first proposed it faces a hostile audience; excellent reasons must be produced to gain for it an even moderately fair hearing. The reasons are often disregarded, or laughed out of court, and unhappiness is the fate of the bold inventors. But if the reasons are understood and accepted, if there are some temporary successes, then interest may be aroused and people may devote themselves to the study of the theory. Professional groups form and make the theory sufficiently respectable to be represented at conferences and meetings. Even the diehards of the status quo now feel an obligation to study the one or the other paper and to make a few grumbling comments. There comes then a moment when the theory is no longer an esoteric discussion topic for advanced seminars and conferences, but enters the process of education itself. Introductory texts are written, popularizations appear, examination questions contain problems to be solved in its terms. Scientists and philosophers from distant fields, trying to show their erudition, drop a hint here and there and this often quite uninformed desire to be on the right side is in turn taken as a sign of the great importance of the theory. But this increase in importance unfortunately is not accompanied by better understanding – quite the contrary. Problematic aspects which were originally introduced with the help of arguments now become basic principles; doubtful points, having been generally accepted turn into slogans; debates with opponents became standardized and also more and more unrealistic, for the opponents, having to express themselves in terms of the new views, now seem to raise only quibbles, and besides, their complaints, involving a misuse of well-established terms, often sound outlandish. Alternatives are still employed but they do not contain realistic counter-proposals; they only serve as a background for the splendour of the new theory (example: the role of 'inductivism' in Popperian arguments). So finally we have success – but it is the success of a manoeuvre carried out in a void, not the success of a reasoned view overcoming difficulties which were not set up for easy solution in the first place. An empirical theory (as opposed to a philosophical theory) such as quantum mechanics, or an empirical practice such as modern scientific medicine with its materialistic background, can of course point to numerous positive results, but note that *any view* and *any procedure* that is developed and applied by intelligent beings has results; the question is whose results are better and more important for those who receive them. The question cannot be answered, for there are no alternatives to provide a point of comparison.

Mill gives a clear and compelling description of these phenomena. Debates and reasoning, he writes, are features

> belonging to periods of transition, when old notions and feelings have been unsettled and no new doctrines have yet succeeded to their ascendancy. At such times people of any mental activity, having given up their old beliefs, and not feeling quite sure that those they still retain can stand unmodified listen eagerly to new opinions. But this state of things is necessarily transitory: some particular body of doctrine in time rallies the majority round it, organizes social institutions and modes of action conformably to itself, education impresses this new creed upon the new generation *without the mental processes that have led to it* and by degrees it acquires the very same power of compression, so long exercised by the creeds of which it had taken place.[2]

An account of the alternatives replaced, of the process of replacement, of the arguments used in its course, of the strength of the old views and the weaknesses of the new, not a 'systematic' account but *a historical account of each stage of knowledge,* can alleviate these drawbacks and increase the rationality of one's allegiance to the views one regards as the most satisfactory.

Proliferation, for Mill, is therefore not only an expression of his liberal attitude, of his realization that people have a right to live as they see fit, and of his faith that such a plurality of life styles and modes of thought will be of advantage to all of us, *it is also an essential part of any rational inquiry concerning the nature of things.* 'Outmoded' views are kept alive because they please some people and because the most advanced theories cannot be understood and examined without their help.

It is depressing and a sign of the historical illiteracy of most modern philosophers (including modern philosophers of science) to see that they are either unaware of the importance of proliferation or ascribe desiccated versions of it to authors far removed from Mill's humanity, simplicity and perceptiveness. Popperians even confront us with the amusing spectacle of fights for a priority none of the participants deserves. Thus Imre Lakatos grants Popper that he has introduced proliferation as an 'external catalyst' of progress but reserves for himself the discovery that 'alternatives are not

---

[2] 'Autobiography' quoted from *Essential Works of John Stuart Mill*, ed. M. Lerner (New York, 1965), 149, my italics. Note how clearly (and simply) this describes the transition from a 'revolutionary' stage (in Kuhn's sense) to a stage of 'normalcy'.

merely catalysts, which can later be removed from the rational reconstruction' but '*necessary parts*' of the falsifying process;[3] while some Germans hail Helmut Spinner as the real inventor of proliferation. The true story is of course that the only contribution these gentlemen made was to bowdlerize Mill's great principle.[4]

Interestingly enough, elements of the principle are found even in Mill's *Logic*.[5] According to Mill, hypotheses, i.e suppositions 'we make (either without actual evidence, or on evidence avowedly insufficient)'[6] and for which 'there are no other limits . . . than those of the human imagination; we may, if we please, imagine, by way of accounting for an effect, some cause of a kind utterly unknown, and acting in accordance to a law altogether fictitious'[7] – such hypotheses 'are absolutely indispensable in science'.[8] Using them,

> we begin by making any supposition, even a false one, to see what consequences will follow from it; and by observing how these differ from the real phenomena, we learn what corrections to make in our assumption. The simplest supposition which accords with the more obvious facts, is the best to begin with; because its consequences are the most easily traced. This rude hypothesis is then rudely corrected, and the operation repeated; and the comparison of the consequences deducible from the corrected hypothesis with the observed facts, suggests still further correction. . .[9]

These 'rules of hypothesis'[10] contain all the elements of the method of conjectures and refutations. Mill even makes it clear that a hypothesis in agreement with all known phenomena is not thereby shown to be true 'since this [i.e. agreement with all known phenomena] is a condition sometimes fulfilled tolerably well by two conflicting hypotheses; while there are probably many others which are equally possible but which, for want of anything analogous in our experience, our minds are unfit to conceive'.[11] A

[3] *Philosophical Papers* (Cambridge, 1978), I, 109, n.4 and text. For the correctness of the claim cf. ch. 6.1 above and *A M*, 48, n.2.

[4] Does it matter who the inventor was? Well, it seems to matter to Popperians. Besides, different authors have different background philosophies which put a different complexion on their (alleged) discoveries. For Mill proliferation is an instrument of understanding and social reform. For the Popperians it is a clever trick within a narrowly technical philosophy.

[5] *System of Logic* (London, 1879), II, ch. 14.

[6] *Ibid.*, 9.   [7] *Ibid.*, 10.   [8] *Ibid.*, 16.

[9] *Ibid.*, 17.   [10] *Ibid.*, 19n.

[11] Whewell, in his review of Mill's *Logic* (quoted from R. E. Butts, ed., *William Whewell's Theory of Scientific Method* (Pittsburgh, 1968), 292) objects: 'I know of no such case in the history of science when the phenomena are at all numerous and complicated', which attests to his historical knowledge but not to his imagination (note that in his reply to Huyghens, Newton takes the devising of numerous alternative hypotheses 'to be no difficult matter': I. B. Cohen, ed., *Isaac Newton's Papers and Letters on Natural Philosophy* (Cambridge, 1958), 144). Historically it was only the general theory of relativity 'which showed . . . that we can point to two essentially different principles, both of which correspond to experience to a large extent' (Albert Einstein 'On the Method of Theoretical Physics', quoted from *Ideas and Opinions* (New York, 1954), 273f).

further function of alternatives is therefore to show that a hypothesis, while successful, is still not the one and only possible account of the facts.[12]

In section 6 of ch. 4 I added a further and somewhat more technical argument for proliferation. The argument was this: there exist cases where a well-known and easily accessible fact F conflicts with a widely held theory T but laws of nature (and not only observational inaccuracies) prevent us from ever discovering this conflict and, therefore, from using F to refute T.[13] An example is the Brownian particle.[14] Looked at from the point of view of

[12] According to Mill, hypotheses are not the only ingredients of scientific knowledge. They are necessary, *there is a rational way of using them*, they produce discoveries, they may open 'paths of inquiry full of promise, the results of which none can foresee' (p. 19n – comment on Darwin's 'hypothesis') but *they are not proven* and therefore not 'inductive truths' (p. 12). Hypotheses can be turned into inductive truths by a uniqueness proof that shows 'that no other hypothesis would accord with the facts' (p. 12). Newton, according to Mill, gave such a uniqueness proof for his law of gravitation. He 'began by an assumption that the force which at each instant deflects a planet from its rectilinear course, and makes it describe a curve around the sun, is a force tending directly towards the sun. He then proved that if this be so, the planet will describe, as we know by Kepler's first law that it does describe, equal areas in equal times; and, lastly, he proved that if the force acted in any other direction whatever, the planet would not describe equal areas in equal times. It being thus shown that no other hypothesis would accord with the facts, the assumption was proved; the hypothesis became an inductive truth. Not only did Newton ascertain by this hypothetical process the direction of the deflecting force; he proceeded in exactly the same manner to ascertain the law of variation of the quantity of that force. He assumed that the force varied inversely as the square of the distance; showed that from this assumption the remaining two of Kepler's laws might be deduced; and finally, that any other law of variation would give results inconsistent with those laws and inconsistent, therefore, with the real motions of the planets, of which Kepler's laws were known to be a correct expression' (12f).

No reference to uniqueness occurs in *On Liberty* and the reference to human fallibility in the first reason (above p. 139) makes us suspect that Mill regarded a proof of this kind not as the last word in the matter and as no hindrance to proliferation: *the idea that a certain point of view represents an inductive truth is for Mill a fallible idea and capable of being corrected by further research.* The difference between him and Popper lies in the fact that Popper rejects inductive truths in Mill's sense, declares that all statements of science are and must be hypothetical and so accepts and repeats only *part* of Mill's methodology. This is connected with his fondness for another well-known feature of science – its *occasional* aversion to ad hoc hypotheses. The aversion makes good sense in a (qualitatively and quantitatively) infinite universe for it instructs us not to be content with the land already conquered but to proceed further into the unknown ('Go West, young Man!'). It started when the new geographical discoveries of Henry the Seafarer, Columbus and Magellan suggested that there were new continents of knowledge as well, that there was an 'America of Knowledge' just as there had been a geographical continent, and when it was believed that the infinity of the world and the infinity of man's resources would never put an end to the search for more and more knowledge. The search ceases to make sense in a finite world explored with finite means. Mill's methodological tool-box contains means for dealing with both kinds of world and is therefore richer than Popper's.

[13] Note the restriction. I do not say as some Popperian critics have suggested, that *any two* conflicting alternatives are related in this way but that *there exist alternatives* which have this property.

[14] In an unpublished MS 'Conflict and Order in Science and Methodology' Elie Zahar discusses another example, asserting that I myself 'adduce' it in *A M*. This I do – but in order to make a point entirely different from the one at issue here. *Added 1980.* John Worrall has criticized my treatment in a different but equally incompetent manner. See his contribution to H. P. Durr, ed., *Versuchungen* (Frankfurt, 1981), II, as well as my reply in the same volume.

the statistical theory, it is a perpetual-motion machine of the second kind and therefore a refuting instance of the second law of the strict (phenomenological) theory of thermodynamics. Is it possible to discover this feature of Brownian motion without help from an alternative to the phenomenological theory? It is not. To show that Brownian motion is due to a conversion of the heat of the fluid into the motion of the particle one would first have to measure the velocity changes of the particle and therefore its velocity, which is impossible,[15] and then compare it with the changes of the heat content of the surrounding medium, which is equally impossible:[16] we cannot use the Brownian particle to refute the phenomenological theory in a direct way. But using the kinetic theory we can connect the deviations with F as well as with novel facts, thus providing indirect evidence for them and against the phenomenological theory. Schematically (fig. 1) (the phenomenological theory) predicts p while actually $p' \neq p$. Natural laws forbid a direct separation of p and p' (for example, by direct measurements). p' but not p triggers F so that F refutes P. The refutation remains unknown unless it is asserted by the kinetic theory K which also leads to novel predictions N. Note how this example differs from Duhem's discussion of the relation between Kepler's laws and Newton's theory of gravitation, and from Popper's repetition of it: according to Duhem (and Popper) a new theory may conflict with an established law and thereby inspire new tests of the law *which could have been carried out without it*:[17] the facts revealed by these tests existed, and were observable independently of our having introduced the

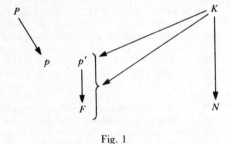

Fig. 1

[15] For details cf. Einstein, Z. *Elektrochemie*, 13 (1907), 41ff, translated in Albert Einstein, *Investigations on the Theory of Brownian Movement* (New York, 1956), 63ff, esp. 67: 'since an observer operating with definite means of observation in a definite manner can never perceive the actual path transversed in an arbitrarily small time, a certain mean velocity will always appear to him as an instantaneous velocity. But it is clear that the velocity ascertained thus corresponds to no objective property of the motion under investigation.' Actual measurements such as Exner's (*Ann. Phys.* 2 (1900), 843ff) therefore always led to results in conflict with the kinetic theory. Cf. also Mary Jo Nye, *Molecular Reality* (London, 1972), 98ff.
[16] The impossibility is here due to the fact that all thermometers have fluctuations of exactly the size we want to measure.
[17] This feature of the case was already noticed by Newton, *Principia*, ed. F. Cajori (Berkeley, 1960), 421. Cf. also the argument in ch.12.

new theory. In the Brownian motion example, however, there is no way of identifying such facts independently of the kinetic theory.

With this we have also a reply to Elie Zahar's attempt to assimilate Brownian motion to the Duhem–Quine case and so to make it accessible to neo-Popperians. In the Duhem–Quine case we have a theory P, a set of conditions A, initial conditions I and $PA \rightarrow (I \rightarrow Q)$, while actually $I\bar{Q}$, and we decide to *save* A by replacing P by K such that $KA \rightarrow (I \rightarrow \bar{Q})$ (alternatively we could have saved P by using a different A).

Now let us assume P to be phenomenological thermodynamics, Q the conditions in the close surroundings of a Brownian particle according to phenomenological thermodynamics, and A the statement that the system containing the Brownian particle is a closed system in thermodynamic equilibrium. Then we know at once that P forbids any fluctuations: $PA \rightarrow Q$. How do we establish that *in the conditions already set up*, i.e. in A, there actually occur fluctuations sufficiently large to make the Brownian particle move? Or, formally, how do we establish that what happens around the particle is not Q but Q^* where $Q^* \rightarrow \bar{Q}$? That is, how do we establish AQ^*? Only by finding a theory that

explains the successes of P, makes novel predictions N not obtainable from P (C)

and is such that $KA \rightarrow Q^*$.

Condition C is precisely the condition on which Popperians would prefer K to P. K *is* preferred to P, it produces Q^*, thereby $\bar{Q}$, and so refutes P. The formal presentation of the case resembles the formal presentation of the Duhem–Quine case (just as the formal presentation of a refutation may be identical with that of a proof), but the interpretation is entirely different: in Duhem we replace P by K *because we have found $\bar{Q}$ instead of Q*. Here we replace P by K *in order to show* that $\bar{Q}$ and not Q.

In chs. 2 to 7 and 10 to 15 proliferation is combined with the 'realism' of thesis I (ch. 2.6) to criticize philosophical ideas and certain developments in the sciences. This assumes (A) that we *can* be scientific realists in deeds as well as in words, that the world is built in such a way that the demand for maximum testability helps us to explore it and to discover its features; and (B) that the *epistemological* command: Increase testability by pushing your views to the limit! *does not conflict with more important commands elsewhere*, such as *moral* commands. These assumptions will be criticized in the second volume, which also contains a brief history of conflict with scientific practice and a humanitarian attitude (see vol. 2, ch. 1). A brief account of the problems is given in the Introduction to vols. 1 and 2.

9

Linguistic arguments and scientific method

1. Linguistic arguments play a very important role in contemporary philosophical discussion. It is believed that they are capable of refuting age-old philosophical theses and providing a short cut to the solution of philosophical problems. If this belief should turn out to be correct, then we would possess here a very powerful eliminative instrument which could be applied even *before* a theory has been developed to such an extent that empirical examination becomes possible. However, such an instrument is a two-edged sword. Curtailing philosophical speculation, it may curtail scientific investigation also. It may even arrest scientific progress. It is this conjecture which has prompted me to examine the matter in some detail.

This examination will be carried out by keeping close to a single paper of the linguistic point of view: Norman Malcolm's 'Moore and Ordinary Language'.[1] I have found it advantageous to proceed in this fashion rather than to introduce a variety of points of view. Malcolm's paper is clear and straight-forward. It is free from that kind of sophistication which hides difficulties beneath brilliance. And yet it is typical of the attitude I want to examine.

2. Moore has asserted that 'the "commonsense view of the world" is in certain features wholly true'.[2] Moore has never given reasons for this

---

[1] 'Moore and Ordinary Language' in *The Philosophy of G. E. Moore*, ed. P. A. Schilpp (Evanston; Ill., 1956).
Added 1980. More recent writings in linguistic philosophy have moved away from simple forms of the paradigm case argument but they still maintain that 'successful communication proves the existence of a shared, and largely true, view of the world' (D. Davidson, 'The Method of Truth in Metaphysics', *Midwest Studies in Philosophy* (Minneapolis, 1977, II, 245). One reason given by Davidson, that 'much of what is agreed on must be true if some of what is agreed is false', is a wider version of a paradigm-case argument, using as a paradigm not particular objects or events, but the activity of criticizing and ascribing error. The argument in the paper that follows can easily be extended to this version. On the other hand, it must be admitted that the overthrow of an entire world view, including the most familiar assumptions, can be stopped by the *decision* to make commonsense (and the views of man it contains) an essential part of any form of knowledge. Such a decision was made by Aristotle and, much later, by Niels Bohr, in his interpretation of the quantum theory. Cf. the Introduction to vols. 1 and 2, as well as vol. 2, chs. 1 and 5.
[2] 'A Defence of Commonsense', in *Philosophical Papers* (London, 1959), 44.

particular belief of his.[3] This is why Malcolm, who subscribes to Moore's belief, regards his method of philosophizing – his way of confronting a doctrine which on general grounds denies the truth of such statements as 'I am absolutely certain that I am now sitting in my office and thinking of Spund', with these statements themselves – as incomplete, or as a 'first step'.[4] Malcolm himself proceeds to provide *arguments* in favour of the truth of the statements Moore uses. To be more specific, he provides arguments, among other things, against such assertions as 'There are no material things'; 'No material thing exists unperceived'; 'No empirical statement can be known to be true'; 'No empirical statement can ever be certain'. In the present section I shall concern myself with the first two statements only as they are statements of fact. Malcolm's attack against the last two statements will be briefly discussed in section 10. The present discussion is of importance for the sciences. For example, it is important for physiology: the statement 'There are only material things in this world and material processes' might be repudiated in a very similar fashion.

3. If I understand Malcolm correctly, then his argument against 'There are no material things' and his reason for calling this statement the expression of a 'philosophical paradox' would be as follows: the statement does not commit any mistake of fact. It is erroneous because it uses 'improper language'.[5] If a child were to say, on the occasion of entering a room full of furniture 'There is not a single material object here', 'we should smile and correct his language'.[6] To the objection that somebody believing in ghosts might use a similar argument in order to refute 'Ghosts don't exist', he replies that the word 'ghost', although having a descriptive use (it could be used in a descriptive sentence if there were ghosts), can be explained without production 'of an instance of the true application of the word' whereas 'material object' (as well as 'behind', 'below', 'earlier', 'later') cannot be so explained.[7] No one 'could . . . have learned the difference between "seeing a material thing" and "seeing an after-image" . . . unless [he] had been actually acquainted with cases of seeing material things'.[8] Result:

> In the case of all expressions the meaning of which must be *shown* and cannot be explained, as can the meaning of 'ghost', it follows from the fact that they are ordinary expressions [which fact implies that they are being used and that

---

[3] At least I am not aware that he has. Nor are those writers on Moore whom I have consulted. Warnock's insistence (*British Philosophy in 1900* (Oxford, 1956), 22f), that 'Moore was quite exceptionally careful always to say exactly what he meant' and that his insistence on the *truth* of certain statements should therefore not be misconstrued as being thought by him to be capable of further analysis points in the same direction.

[4] 'Moore and Ordinary Language', 367.

[5] *Ibid.*, 365.   [6] *Ibid.*, 353.

[7] *Ibid.*, 360f.   [8] *Ibid.*, 361.

people know how to use them] that there have been many situations of the kind they describe, otherwise so many would not have learned the correct use of those expressions. Wherever a philosophical paradox asserts, therefore, with regard to such an expression, that always when the expression is used the use of it produces a false statement, then to prove that the expression is an *ordinary expression* is completely to refute the paradox.[9]

4. Starting the criticism of this argument, I first mention a minor point: the argument assumes that there is a clear and unambiguous distinction between expressions, the meaning of which must be *shown* and cannot be explained, and other expressions, the meaning of which can be *explained* and need not therefore be shown. In order to be able to apply Malcolm's method of refuting 'philosophical paradoxes' one would have to be given a clear criterion as to which meanings can be explained and which cannot. No such criterion appears anywhere in the paper; it is just asserted, without further argument, that 'material thing' (and I presume, also 'mental process') belongs to the latter class. This means that the problem that has been raised in connection with Moore's philosophy, its dogmatism, has been put into a different place; it has not been solved. And as far as I am concerned there is not much to choose between a philosophy which says 'so-and-so is *true*' and makes deductions from this; and a philosophy which says 'so-and-so cannot be explained, only shown' and starts its deductions there.

5. But – and with this a much more serious objection commences – how can the meaning of 'material object' ever be 'shown' to anybody? According to Malcolm (and according to Moore, whom he defends here), the phrase contains the idea of unperceived existence. It also contains the idea that the only influence an observer is capable of exerting upon a material object is a *causal* influence. The behaviour of an observer who is not causally interacting with a material object leaves unchanged all the important properties of the latter such as its size, its mass, its colour (*objective* colour, that is, and not *perceived* colour) and the rate of change of any periodic process that might be going on in its inside. We shall call these two ideas the idea of *unperceived existence* and the idea of *observer independence* respectively. And our problem now is: how is it possible to 'show' these two ideas rather than to explain them? And what procedure counts as 'showing', say, observer independence?

6. It is clear that observational results will not be sufficient for this purpose. Observational results admit of many different interpretations.[10] They are incapable of leading to unique meanings. It is also evident that there cannot exist any observational result *directly* corresponding to the idea of unper-

---

[9] *Ibid.*, 361.   [10] For this point, see ch. 2.4, 2.5.

ceived existence, or to the idea of observer independence. The process of 'showing' the idea of a material object must therefore contain a further element which is not observational, although it is used in connection with observational results. This further element is instruction in the proper use of material-object words.

As has been made clear by Wittgenstein (and by some of his nominalistic predecessors such as Berkeley) such instruction is not verbal explanation (presentation of definitions). Verbal explanations and definitions *presuppose* that a great deal of the language spoken is already understood. There must therefore exist a kind of instruction in the proper use of terms which, although employing words, does not *presuppose* understanding of these words, and which nevertheless *terminates* in understanding. It is such instruction to which Malcolm seems to refer when asserting that the meaning of some terms must be shown and cannot be explained.

7. Now it can be easily shown that the success of this non-explanatory method of teaching (which is the one method used by parents and nurses) *does not at all presuppose the existence of material objects*, and success of the method (i.e. existence of material-object words, properly used, in the ordinary language) is therefore no indication whatever of the existence of material objects. For how does the method proceed? The teacher shows the pupil various objects and instructs him by gesture, by talk in the presence and in the absence of the objects, to make the appropriate responses. What influences the pupil are the *perceptions* he has (of the objects shown and of the instructions of the teacher) *and not the objects themselves*. Blind pupils will have to be taught in a very different fashion and so will pupils exhibiting certain perceptual irregularities. Any object, process, relation, or manipulation leading to the perceptions necessary for success of the method used for teaching material-object talk will therefore do. It can be shown that the manifold of the entities satisfying this condition is at least Aleph One. Objects differing imperceptibly from material objects in the sense analysed by Malcolm and Moore, for example, objects which are observer-dependent, but so weakly that no perceptual difference results, belong to the manifold and so do objects which possess all the perceptual manifestations of material objects but are not material objects. What follows from the existence of material-object talk in the ordinary language (provided, which is very doubtful, that the notion of a material object can only be shown and not explained) is therefore not the existence of *material* objects, but the existence of *some* objects (relations, processes, etc.) *from a much wider class of entities* whose extension is not even well defined. This finishes our criticism of the way in which Malcolm resolves the 'paradox', 'There are no material objects.' The application to the resolution of 'There are no mental processes' should be obvious.

8. Malcolm's argument is an instance of what has been called the 'paradigm case argument', or the 'argument from standard examples'. We may state the argument briefly by saying that the extension of concepts which have been, and must be, *taught* ostensively (note that I do not assume here the usual cluster of ideas connected with the notion of ostensive *definition*), is not the zero class. The argument is invalid, as the meaning of most concepts which are taught ostensively is much richer than are the perceptions employed in the process of ostension. Thus, for example, the concept of a material object contains the idea of unperceived existence and also the idea of observer independence, whereas all that can be shown perceptually are things which are *more or less* stable and *more or less* independent of the position of the onlooker.[11] These perceptions being *sufficient* (together with the instruction received) for establishing the concept, the existence of this concept (or the existence of the corresponding use) does not guarantee the existence of the objects in its extension. To infer existence of an object from the existence of a use that has been taught ostensively therefore involves an invalid transition from perceptions to objects bringing about these perceptions.[12]

9. The fact that such a transition is nevertheless made exhibits another feature of paradigm-case arguments which is rather surprising: the existence of material objects is inferred from the fact that material-object

[11] The discussion is based on a very simplified and to that extent incorrect account of the process of ostensive teaching (and learning). It is assumed that at the beginning, both the pupil and the teacher possess exactly the same perceptions and that all the former lacks is a language. This condition is almost never satisfied in reality. Teaching a language with well-defined categories and concepts very often leads to a decisive change in the perceptual processes of a pupil. What is first perceived only vaguely and indistinctly becomes definite; initially unrelated impressions are combined into wholes and expectation leads to perception of such wholes even when many of the constituting expressions are missing. A good example is the improvement of hearing that goes along with musical instruction. In this way instruction in a language can increase the number of perceptions which count as verifying the ideological background of the language spoken – a further reason why the paradigm-case argument cannot be successful.

[12] E. Gellner, *Words and Things* (London, 1959), 34 argues that reference to paradigm cases involves a confusion between connotation and denotation, or a confusion between use and legitimate use. This is a somewhat unfair criticism. If we do not want to assume that all our ideas are inborn, then we must admit the existence of methods of teaching certain ideas, and if the latter ideas are to be factually relevant the teaching procedure must somehow make use of the objects the ideas are supposed to refer to. This is the reasonable core of the assumptions centred around the paradigm-case argument. However, the argument fails as it assumes *too simple a relation* between our (not inborn) ideas and the objects they are supposed to refer to. It assumes that the object itself is needed for teaching the proper use of factually relevant terms, whereas all that is needed are certain perceptions together with certain beliefs. The perceptions can be illusory and the beliefs can be false. A direct transition from use to existence is therefore impossible even in those (perhaps non-existent) cases where explanation is impossible and where the meaning must be shown. (For a criticism of this distinction cf. also J. W. N. Watkins, 'Farewell to the Paradigm-Case Argument', *Analysis*, 18 (1957), 25f as well as Flew's reply, *Ibid.* 34ff.)

words are being used in the ordinary language. The procedures leading to such use consist in the exhibition of ordinary situations and corresponding verbal instruction. The idea that existence can be derived from the success of this rather pedestrian method assumes that the concept of a material object is, as it were, a repetition of the situations employed in the process of teaching: tables are things on which one can sit, under which one can hide, which do not move by themselves but need effort in order to be moved and so on. This is a very poor account even of the concepts of the ordinary language. *Ordinary language philosophers underestimate the idiom on which they base their most decisive arguments.*[13]

To take another example: the notions 'up' and 'down' are taught in such a fashion that one would be inclined to infer that they can be meaningfully used only in such contexts as 'Come down and eat your breakfast' or 'Why don't you come up and have a drink with us?' Yet from the very beginning even ordinary men have applied these notions to astronomical domains. The popular argument that the earth cannot be spherical as the antipodes would otherwise fall 'down' is one of the most conspicuous examples of such application, an example moreover which quite clearly shows the need for revising the ordinary notions.[14] Or take the notion of solidity. In a well-publicized passage,[15] Miss Stebbing criticizes 'the nonsensical denial of [the] solidity of' material objects, which some authors inferred from the kinetic theory of matter and the electron theory.[16] She defends her charge of nonsense by pointing out that the kinetic theory does not make tables collapse when we sit upon them, that it does not make them swallow up objects which have been put on top. Using this kind of argument she insinuates that 'solidity' as used in ordinary language covers just these everyday occurrences and nothing else. This, I submit, is far from true. The notion of solidity does not only make assertions about macro-occurrences; it also contains a micro-account of the objects to which it is being applied. That is, it not only asserts that they resist penetration, but it also asserts or implies that they resist penetration *because of their being full of compact stuff.*[17]

The examples which I have mentioned so far, and other examples, which

---

[13] Their belief in the correctness of the paradigm-case argument is a direct consequence of this.

[14] For further discussion and literature see ch. 4.7.

[15] The argument used in this passage is mentioned with approval by J. O. Urmson in his essay 'Some Questions Concerning Validity' in *Essays in Conceptual Analysis* (Oxford, 1956), 120ff. 'Miss Stebbing', it reads here 'shows conclusively that the novelty of scientific theory does not consist, as has been unfortunately suggested [why *unfortunately*?], in showing the inappropriateness of ordinary descriptive language'. Except for such 'trivial' uses (122) Urmson is, however, critical of the argument.

[16] *Philosophy and the Physicists* (London, 1937), 53.

[17] According to the *Oxford English Dictionary* 'solid' means, among other things, 'free from empty spaces, cavities, interstices, etc.; having the interior completely filled up or in . . . of material substance; of a dense and massive consistency; composed of particles which are firmly and continuously coherent'.

could be provided in great number, support the hypothesis, which I have held for some time, that many philosophers referring to the 'ordinary language' refer to an *ideal language* of their own making, to a language that is fairly cagey and non-committal and therefore rather safe from revision (in this respect the language is close to the sense-datum language of the positivist, or to the language envisaged by operationalistic philosophers) they do not refer to any actually spoken idiom.[18]

This, of course, completely undermines an argument to which appeal is frequently made, the argument from empirical success. The core of this argument consists in the assumption that an idiom that is being used and has been used for a long time has thereby proved its mettle: it may be regarded as a true mirror of reality. It now turns out that the idiom upon which linguistic philosophers *do* base their investigations (with certain exceptions, that is, including Austin, at least in his more linguistic moods) is not spoken by anybody and cannot therefore profit from the alleged practical success of ordinary English. The situation is therefore as follows: the argument from practical success does not work. Even if it did work, linguistic philosophers could not use it, for they base their work not upon the common idiom but upon some artificial idiom (which, however, is becoming more and more common in certain circles).

10. All the arguments dealt with so far concern cosmological assumptions. As has been pointed out in the last paragraph of section 2, linguistic arguments concerning matters of *logic, conduct,* and *method,* need a different treatment. The discussion of their weaknesses is not directly connected with the topic of the present paper. However, the points of interest can be made very briefly and are worth making.

Consider first the argument against 'No empirical statement is ever, and under any circumstances, certain'. According to Malcolm this statement involves a misuse of language: 'if a child who was learning the language were to say, in a situation where we are sitting in a room with chairs about, that it was "highly probable" that there were chairs, then we should smile *and correct his language'.*[19] This argument asserts that there are situations when refusal to apply the word 'certain' is in conflict with the rules of ordinary English. Such refusal therefore cannot claim the praise of laudable philosophical acumen, being possible only for a person who has not yet successfully passed all his English tests.

I do not want at this juncture to contest the point made about the proper use of 'certain'. It may well be (although I doubt it) that Wittgenstein and

---

[18] This may be the reason why some empiricists, including Herbert Feigl, regard ordinary language as an ideal observation language: using this language, they think they are not committed to any far-reaching assumptions.

[19] 'Moore and Ordinary Language', 345f, original italics.

Moore are right and that the English language *does* contain empirical statements which closely resemble mathematical statements 'in the respect that experience could not refute them'.[20] The question is: is this a satisfactory state of affairs, or should we not rather change the meanings of those statements in such a fashion that they *do* become refutable? The content of a statement increases with the number of statements which could count as refuting evidence. If the statement does not not exclude anything, if it is valid in all possible worlds, then it is incapable of selecting situations of the real world in which we live, and is therefore dead weight. Languages containing such dead weight ought to be rebuilt to become more effective and informative means of communication. Some statement of the language which results after the improvement has been carried out, will, of course, violate the rules of the old and unsatisfactory idiom. These violations are due to an improvement in the light of insight rather than childish mistakes, and they are therefore to be welcomed. Excluding them on the basis of factual arguments concerning linguistic usage means rejecting suggestions for improvement by reference to the old, and undesirable, facts – an excellent example of what has been called a naturalistic fallacy. Now assume that a language has been reformed in such a fashion that it no longer contains irrefutable statements of apparently empirical character. Then 'X is certain', with 'certain' equivalent to 'irrefutable' or 'incorrigible', will be false in this language for all empirical sentences and so will 'X is certain in conditions C' (this takes care of Austin's analysis according to which certainty is not a property belonging to a select class of statements but rather a property which a statement may acquire under some circumstances and may not possess under other circumstances).[21] Of course, 'certain' may not always mean 'irrefutable' (it does not seem to mean this in ordinary English), but whatever it means, 'X is certain' cannot now make us complacent with respect to X, it cannot make us believe that the question concerning the truth of X 'is settled',[22] and this even if we had the 'best possible reasons' for such a belief.[23]

11. I now return to Malcolm's paper and to the arguments he uses in order to refute statements such as 'There are no material objects'. There is a passage in Malcolm's essay where he seems to consider the question of empirical evidence in addition to the question of paradigmatic use.

---

[20] For this cf. Malcolm's notes on Wittenstein's lectures of 1949, published in *Ludwig Wittengenstein, A Memoir* (Oxford, 1958), 87ff. The quotation is from p. 89.

[21] See *Sense and Sensibilia* (New York, 1964), ch. 10.

[22] *Ibid.*, 115.

[23] The phrase after the comma refutes the theory of induction as expounded e.g. by P. Edwards, 'Russell's Doubts about Induction', *Mind* 58 (1949), 230ff and as insinuated in Wittgenstein's *Philosophical Investigations* (Oxford, 1953), 116, 477–85; cf. also the discussion in A. Pap, *Analytische Erkenntnislehre* (Vienna, 1955), 104ff.

Counting the ways in which some of Moore's true sentences might turn out to be false, he mentions two: the sentences might be self-contradictory; and they might be empirically false. I shall not discuss the first possibility. Concerning the second possibility, Malcolm points out that

> the only ground for maintaining that when people use the expression then what they say is always false, will have to be the claim that *on the basis of empirical evidence* it is known that the sort of situation described by the expression has never occurred and will never occur. But it is abundantly clear that the philosopher offers no empirical evidence for his paradox.[24]

Two comments are in order, one general, one more specific.

12. The more general comment is this: Malcolm and many philosophers who think along similar lines assume that philosophical positions are only *verbally* different from the beliefs held by commonsense, and therefore cannot be regarded as starting-points of progress. This assumption implies that neither independent empirical evidence, nor independent arguments (independent, that is, from what is believed by commonsense) are relevant for their defence. There is no such evidence, and there are no such arguments, and this not only accidentally, but because of the very nature of philosophical positions. They can stand on their own feet, they create their own evidence, and they do not need extraneous support.

The interesting thing is that there *are* philosophical positions which have exactly these properties. A dogmatic system which is introduced out of the blue, which refuses to give arguments but at the same time claims to possess absolute truth, which presents us with its *dictum*, which we may then either take or leave alone, a system which is explained only by reference to itself, such a system comes very close indeed to the ideas which some linguistic philosophers seem to have about philosophy.

If such a system is general enough, then it may be able to accompany even the most ordinary situations with its rather unusual enunciations. Not being given independent information, a nominalist, or contextualist (in matters of meaning), will be able to understand the system only from what it says in *such* situations. There being no excess of evidence, or of argument, he will have to conclude that all the system does is describe ordinary situations in a rather extraordinary fashion; i.e. he will have to conclude that all that has been achieved was a pointless change of meaning. Dogmatic philosophies therefore do not advance our knowledge, all they do is describe ordinary and well-known things in an extraordinary and not so well-known fashion. With this judgement I am inclined to agree.[25] (The judgement is by no means particularly new or revolutionary – see Galileo's arguments against the verbalism of some of his opponents.)

---

[24] 'Moore and Ordinary Language', 358.
[25] *Added 1980.* I no longer agree. Cf. the end of n.1 above.

However, it is one thing to criticize a certain kind of philosophy on the grounds that it does not improve knowledge and introduces purely verbal changes; and it is quite a different thing to apply the criticism to *any* philosophical position, including those that might lead to a drastic reformation both of belief and of language. It may well be true that at some time some philosopher 'offers no empirical evidence for his paradox' (that he offers no empirical evidence for the statement that there are no material objects, or for the statement that all thoughts are material processes). Empirical evidence is difficult to come by. Philosophers discussing the atomic theory often gave little consideration to questions of evidence. But this does not turn atomism into a paradoxical theory in Malcolm's sense; nor does it mean that the atomists regarded empirical evidence as irrelevant to the truth of their theory. The same may be said of any of the paradoxical statements on Malcolm's list. Only if it could be shown that the philosophers holding such views regard them as *absolutely true* and reject as irrelevant both questions of independent argument and questions of test, only then would Malcolm's position possess a semblance of plausibility. (Moreover it could then also be defended in a much simpler and straightforward fashion.)

13. I come now to my second, and more specific point. Malcolm says that a philosopher denying the existence of material objects has no empirical evidence showing that 'the sort of situation described by the expression [material object] has never occurred and will never occur'. In short, he possesses no falsifying evidence for 'There are material objects' (we shall call this sentence *S*) *and he has therefore no right even to consider statements which are inconsistent with S* (except, perhaps, as examples of false or absurd statements). The *principle* (I shall call it principle *P*) behind this move is: don't introduce a theory contradicting a well-confirmed and commonly accepted theory unless the latter theory has been refuted. This is a very plausible principle and it is held by a good many scientists and philosophers. A scientist may try to support it in the following fashion. He may admit that experience does not allow us to select a single theory to the exclusion of all others. Being limited and indefinite, experience supports theories which are mutually inconsistent.[26] Assuming one of these theories to be the generally accepted one, we have to admit, then, that its alternatives cannot be *eliminated* by factual reasoning. Still, there are very forceful reasons against introducing them. It is bad enough, so a defender of the above principle might point out, that the accepted point of view, and the accepted language, do not possess full empirical support and are not uniquely selected by facts. Adding a new theory of an equally unsatisfactory character will not improve the situation; nor is there much sense in

[26] For details see ch. 4.6 and ch. 6.

trying to replace the accepted theories or the accepted language by some of their possible alternatives. Such replacement will not be an easy matter. New grammatical rules, a new formalism, will have to be learned, and familiar problems will have to be calculated in a new way. Textbooks must be rewritten, university curricula readjusted, observational facts reinterpreted. And what will be the result of all the effort? Another theory, or another language which, from an empirical point of view, has no advantage whatever over the theory and the language it replaces. The only real improvement, so a defender of principle *P* will continue, derives from the *addition of new facts*. Such facts will either support the current theories and the common language used for their expression; or they will force us to modify them by indicating precisely where they go wrong. In both cases, they will precipitate real progress and not just arbitrary change. The proper procedure must therefore consist in the confrontation of the accepted point of view and the accepted language with as many relevant facts as possible. The exclusion of alternatives is then required for reasons of expedience: their invention not only does not help, but it even hinders progress by absorbing time and manpower that could be devoted to better things. And the function of principle *P* lies precisely in this: it eliminates such fruitless discussion and it forces the scientist to concentrate on the facts which, after all, are the only acceptable judges of a theory. This is how the practising scientist will defend his concentration on a single theory to the exclusion of all empirically possible alternatives.

It is worthwhile repeating the reasonable core of this argument: theories and languages should not be changed unless there are pressing reasons for doing so. The only pressing reason for changing a theory is disagreement with facts. Discussion of incompatible facts will therefore lead to progress. Discussion of incompatible alternatives will not. Hence, it is sound procedure to increase the number of relevant facts. It is not sound procedure to increase the number of factually adequate, but incompatible, alternatives. One might wish to add that formal improvements such as increase of elegance, simplicity, generality, and coherence should not be exluded. But once improvements have been carried out, the collection of facts for the purpose of test seems indeed to be the only thing left to the scientist.

These are the reasons which could be adduced in favour of the principle I abstracted from Malcolm's remark concerning the lack of evidence, refuting the existence of material objects. These reasons go far beyond anything to be found in the linguistic camp, and especially in Malcolm's paper. Still, I am prepared to give Malcolm the benefit of the doubt and I have therefore tried to make his case as convincing and as interesting as possible. It will soon emerge that even these arguments are untenable. What I would like the reader to remember is that the linguistic case is much weaker that has been presented here and much less articulate.

14. The argument in the last section is perfectly all right *provided the facts to be collected for the purpose of test exist, and are available independently of whether or not one considers alternatives to the theory to be tested, or the language under consideration.* This assumption, on which the validity of the argument in the last section depends in a most decisive manner, I shall call the assumption of the relative autonomy of facts, or the *autonomy principle*. It is not asserted by this principle that the discovery and description of facts is independent of *all* theorizing and language-building. But it *is* asserted that the facts which belong to the empirical content of some theory and which are described with the corresponding language are available whether or not one considers alternatives to *this* theory. I am not aware that this very important assumption has ever been explicitly formulated as a separate postulate of the empirical method. However, it is clearly implied in almost all investigations which deal with questions of confirmation and test. All these investigations use a model in which a single theory or a single language is compared with a class of facts (or observation statements) which are assumed to be 'given' somehow. I submit that this is much too simple a picture of the actual situation.

Facts and theories, facts and languages are much more intimately connected than is implied by the autonomy principle. Not only is the description of every single fact dependent on *some* theory (which may, of course, be very different from the theory to be tested). As is shown by special examples, such as the example of Brownian motion,[27] there exist also facts which cannot be unearthed except with the help of alternatives to the theory to be tested, and which become unavailable as soon as such alternatives are excluded. This suggests that the methodological unit to which we must refer when discussing questions of test and empirical content is constituted by a *whole set of partly overlapping, factually adequate, but mutually inconsistent, theories*. Both the relevance and the refuting character of many very decisive facts can be established only with the help of other theories which, although factually adequate, are not in agreement with the view to be tested. This being the case, production of such refuting facts may have to be preceded by the invention and articulation of alternatives to this view, i.e. it may have to be preceded by the invention of theories whose main principles contradict at least some of the principles of the accepted point of view, and of languages whose grammar is different from the grammar of the accepted idiom (which *may* be some commonly spoken idiom). Demanding that a language be left unchanged unless one discovers facts inconsistent with the principles implicit in its grammar, therefore, means putting the cart before the horse. The invention of uncommon theories and uncommon languages *comes first* and it is *with their help* that inadequacies of the common idiom can be discovered. The fate of the idea of observer independence is an excellent example of what has just been said.

[27] See ch. 4.6, ch. 8 and ch. 12.

15. The idea of observer independence which was briefly mentioned in section 9, is part of Newtonian physics as well as of the common point of view concerning material objects. Its incorrectness was realized in the following fashion. First, there appeared certain difficulties in electro-dynamics which demanded a solution (absence of second-order effects of motion). It was assumed, for some time, that a solution could be found within the framework of classical physics; i.e. it was assumed that the difficulties were difficulties of the *application* of the classical point of view rather than of some basic assumptions of this point of view itself. Lorentz provided what seemed to be a satisfactory account, in these terms. Einstein, on the other hand, suspected that the classical point of view itself was incorrect and constructed a theory which was inconsistent with the idea of observer independence. In comparing this theory with the classical solution (and in carrying out this comparison, the general theory of relativity must not be forgotten; it could be developed on the basis of the special theory but not at all on the basis of the ideas of Lorentz), it was found to possess tremendous advantages, and it was therefore adopted. This adoption was, of course, also due to the empirical success of the ideas of special and general relativity. But one must note that this empirical success was not *preceded* by the complete failure, on empirical grounds, of the classical point of view *taken by itself*. This failure was rather a *consequence* of the success of the new theory and could not have been demonstrated without this theory. Of course, the old theory was in difficulty. Every theory is always in some difficulty or other. The question is whether the difficulty of classical physics was a *fundamental* difficulty, or whether it was accessible to treatment with the help of additional *classical* hypotheses. It is clear that the answer to this question cannot consist in the collection of further *facts*. No amount of facts, however carefully collected, can show that there exist no classical hypotheses explaining a certain difficult situation. The only way such an idea can be supported consists in the development of an alternative *theory* (an alternative to any classical point of view, that is) which 'turns [the difficulty] into a principle',[28] elaborates its consequences, and compares the resulting point of view with the classical ideas. And this is how Einstein actually proceeded.

Observer dependence was therefore not refuted by the production of observations inconsistent with it, or by the realization that a language using it was impractical. It was refuted by the construction, and confirmation, of a theory which worked with observer dependence, and it could not have been refuted in any other way. The consequence of all this is, of course, that material objects in the sense explained above (i.e. objects which are obser-ver-independent and which exist unperceived) *do not exist*. And as tables,

---

[28] Einstein, 'Zur Electrodynamik bewegter Korper', *Annalen der Physik*, 12 (1905), reprinted in Lorentz–Einstein–Minkowski, *Das Relativitaetsprinzip* (Leipzig, 1923), 26.

chairs, bookcases are supposed to be material objects in Malcolm's (and other peoples') arguments, it follows that they do not exist either. No doubt this assertion sounds strange to those unfamiliar with conceptual revolutions, but it is not a bit more strange than the denial of angels, demons, or the devil seemed to the faithful who had been brought up in the corresponding beliefs and who in addition had experiences such as hearing voices, partially split personality, fear of corruption, and the like. And we should realize that for them the spiritual world – even if it was populated by evil spirits – was much more important and much more secure than the transitory material world of tables, chairs and philosophy books.[29]

[29] Wittgenstein seems to admit that factual discoveries may force one to revise the conceptual system one possesses. 'If anyone believes that certain concepts are absolutely the correct ones, and that having different ones would mean not realizing something that we realize – then let him imagine certain very general facts of nature to be different from what we are used to, and the formation of concepts different from the usual ones will become intelligible to him.' (*Philosophical Investigations*, 230.) However, I do not think he makes it clear that these 'very general facts', if they exist, could not make themselves noticed *except with the help of a conceptual system which takes them into account* and is therefore very different from whatever language is the commonly accepted one. Such systems cannot be built up in a second. They appear in the form of non-testable, i.e. *philosophical* positions which conflict with the common views of the time and whose grammar differs from the grammar of the accepted language. But their use, which is the first step towards a reform of this language is excluded by Wittgenstein's demand that 'philosophy may in no way interfere with the actual use of language'. (*Philosophical Investigations*, para. 124.)
There are some philosphers, among them my former colleague Stanley Cavell, who admit that ordinary language may be in need of change but who deny at the same time that such a change can be brought about by philosophical reasoning. Discussing the above quotation from Wittgenstein, Cavell points out 'that, though of course there are any number of ways of changing ordinary language, *philosophizing* does not change it'. (I am quoting from a MS which Cavell let me have prior to its publication.) More specifically he repeats the charge that a philosopher who defends an apparently absurd thesis such as 'There are no material objects', or 'No statement is certain', does not assert anything contrary to what is believed by the ordinary man. All he does is present this belief in misleading language. 'The assumption, shared by [the] *ordinary language critic and [the] defender of the tradition*, that [philosophical] words are not meant in their ordinary senses destroys the point . . . of such statements' (original italics). This is quite correct provided the philosophical statement is dogmatic and *ad hoc*. i.e. provided it is not intended to imply anything beyond the things which were known before. But it may not be meant in this sense. It may be the first step in the development of a new conceptual scheme which, when elaborated in sufficient detail, makes assertions which conflict with assertions implicit in the use of language ('ontological consequences' – cf. ch. 2). What he seems to overlook (and what many contemporary physicists seem to overlook) is that developing a system in this fashion is not an easy matter, and that a thinker (or a tradition) intending to develop such a new way of speaking (and thinking and seeing) may be stuck for a long time with the mere idea of the non-existence of material objects without as yet being able to give it concrete content (a very good example is the early history of atomism). Now if he is forbidden to make the first step on the grounds that it is pointless, then he will not be able to make the second step and a change of the kind admitted by Cavell (and, as he says, by Wittgenstein) *will never occur*. Cavell's criticism does eliminate a *dogmatic* philosophy for which the inarticulate and untestable first step is also the last step and the end of the matter. But he seems to assume that all philosophy must be dogmatic in this sense (and here he of course follows in the footsteps of his master, Wittgenstein – cf. vol. 2, ch. 7.9), which is far from true. Nor does one need to be a Wittgensteinian in order to see how barren dogmatic thinking is.

16. We are now in a position to give substance to the conjecture that was formulated in the first section. The result of our examination is as follows: first it has been shown that idioms which are employed by many and serve their practical purposes cannot on that account alone be used as a basis for cosmological argument. Secondly it has been pointed out that the adequacy of widely spoken idioms must be investigated with the help of alternative idioms and alternative beliefs. A direct comparison with 'the facts' is not sufficient. Such a wider investigation will be the more thorough, the greater the difference between the investigated idiom (belief) and the alternative idiom(s) (beliefs(s)) used in the investigation. However closely a language seems to reflect the facts, however universal its use and however necessary its existence may seem to those speaking it, its factual relevance can be asserted only *after* it has been confronted with such alternatives *whose invention and detailed development must therefore precede any assertion of practical success or factual relevance.* Conversely, a method which frowns upon the development of new idioms or regards assertions such as 'There are no material objects' as *a priori* nonsensical will make it impossible to investigate the validity of the beliefs implicit in the common idiom, and it will thereby turn these beliefs into metaphysical dogmas. The progress of science, or of any rational inquiry that wants to examine prejudice, will be severely interfered with by such a method. The question as to what should be retained – the linguistic method of embalming unexamined prejudice, or the rational method of trying to criticize and thereby to progress – must therefore be answered in favour of the latter.

10

Materialism and the mind–body problem

1. This paper has a twofold purpose. First, it defends materialism against a certain type of attack which seems to be based upon a truism but which is nevertheless completely off the mark. And secondly, it intends to put philosophy in its proper place. It occurs only too often that attempts to arrive at a coherent picture of the world are held up by philosophical bickering and are perhaps even given up before they can show their merits. It seems to me that those who originate such attempts ought to be a little less afraid of difficulties; that they ought to look through the arguments which are presented against them; and that they ought to recognize their irrelevance. Having disregarded irrelevant objections they ought then to proceed to the much more rewarding task of developing their point of view *in detail*, to examine its fruitfulness and thereby to get fresh insight, not only into some generalities, but into very concrete and detailed processes. To encourage such development from the abstract to the concrete, to contribute to the invention of further ideas, this is the proper task of a philosophy which aspires to be more than a hindrance to progress.

2. The crudest form of materialism will be taken as the basis of argument. If *it* can successfully evade the objections of some philosophers, then a more refined doctrine will be even less troubled.

Materialism, as it will be discussed here, assumes that the only entities existing in the world are atoms, aggregates of atoms and that the only properties and relations are the properties of, and the relations between, such aggregates. A simple atomism such as the theory of Democritos will be sufficient for our purpose. The refinements of the kinetic theory, or of the quantum theory, are outside the domain of discussion. And the question is: Will such cosmology give a correct account of human beings?

3. The following reason is put forth why this question must be answered in the negative: human beings, apart from being material, have *experiences*; they *think*; they *feel* pain; etc. These processes cannot be analysed in a materialistic fashion. Hence, a materialistic psychology is bound to fail.

The most decisive part of this argument consists in the assertion that experiences, thoughts and so on, are not material processes. It is customary to support this assertion in the following manner.

4. There are statements which can be made about pains, thoughts, etc., which cannot be made about material processes; and there are other statements which can be made about material processes but which cannot be made about pains, thoughts, etc. This impossibility exists because the attempt to form such statements would either lead to results which are *false*, or else to results which are *meaningless*.

Let us consider meaninglessness first. Whether or not a statement is meaningful depends on the grammatical rules guiding the corresponding sentence. The argument appeals to such rules. It points out that the materialist, in stating his thesis, is violating them. Note that the particular words he uses are of no relevance here – whatever the *words* employed by him, the resulting *system of rules* would have a structure incompatible with the structure of the idiom in which we usually describe pains and thoughts. This incompatibility is taken to refute the materialist.

It is evident that this argument is incomplete. An incompatibility between the materialistic language and the rules implicit in some other idiom will criticize the former only if the latter can be shown to possess certain advantages. Nor is it sufficient to point out that the idiom on which the comparison is based is in common use. This is an irrelevant historical accident. Is it really believed that a vigorous propaganda campaign which makes everyone speak materialese will turn materialism into a correct doctrine? The choice of the language that is supposed to be the basis of criticism must be supported by better reasons.[1]

---

[1] *Added 1980.* I no longer agree with the assumption, implicit in the argument from the propaganda campaign, that the 'correctness' of an idiom, or of the statements that can be formulated in its terms, empirical statements included, is independent of the (linguistic) practice to which the statements belong: the truth, even of 'empirical' statements, may be *constituted* by the fact that they are part of a certain form of life which assembles evidence in a certain way. 'Athene has provided me with new strength' can be a true observation statement for a Homeric warrior; it is false, or simply nonsense for Xenophanes, and our modern materialists (cf. vol. 2, ch. 1 as well as *AM*, ch. 17). There exist numerous experimental statements about the properties of the ether in nineteenth-century electromagnetic theory – but none of these statements makes any sense today. Phlogiston was weighed and its effects were demonstrated (as oxygen is weighed and its effects are demonstrated today) – and yet phlogiston is now believed to be a mere fiction. There are philosophers such as Ian Hacking (cf. his review of the work of Imre Lakatos in *Br. J. Phil. Sci.* 1979) who call such assertions 'implausible' but this only means that they are unaware of my reasons for making them. In ch. 17 of *AM* I give strict criteria for recognizing truth and observability and I then show that, and why, statements about gods satisfied these criteria. Research and not vague impressions should decide the matter.
Secondly, the choice of an idiom for the description of mental events cannot be decided by considerations of testability and 'cognitive content' alone. It may well be that a materialistic language (if it ever gets off the ground) is richer in cognitive content than commonsense and contains physiological knowledge that did not exist when the common idioms arose and were shaped by the demands of a complex and demanding life (cf. the Austin quotation in the next

5. As far as I am aware there is only one further reason that has been offered: it is the *practical success* of ordinary English which makes it a safe basis for argument.

> Our common stock of words [writes Austin] embodies all the distinctions men have found worth drawing, and the connexions they have found worth marking, in the lifetime of many generations: these surely are likely to be more numerous, more sound, since they have stood up to the long test of the survival of the fittest, and more subtle . . . than any that you or I are likely to think up . . .[2]

This reason is very similar to, and almost identical with, a certain point of view in the philosophy of science. Ever since Newton it has been assumed that a theory which is confirmed to a very high degree is to be preferred to more tentative general ideas and it has been, and still is, believed that such general ideas must be removed in order not to hinder the course of factual discovery. 'For if the possibility of hypotheses', writes Newton (in a reply to a letter by P. Pardies), 'is to be the test of truth and reality of things, I see not how certainty can be obtained in any science; since numerous hypotheses may be devised, which shall seem to overcome new difficulties.'[3] I mention this parallel in order to show that philosophical points of view which *prima facie* seem to bear the stamp of revolutionary discoveries, especially to those who are not too well acquainted with the history of ideas, may in the end turn out to be nothing but uncritical repetitions of age-old prejudices. However, it must also be emphasized, in all fairness to the scientists, that the parallel does not go very far. Scientific theories are constructed in such a way that they can be *tested*. Every application of the theory is at the same time a most sensitive investigation of its validity. This being the case there is indeed some reason to trust a theory that has been in use for a considerable time and to look with suspicion at new and vague ideas. The suspicion is mistaken, of course, as I shall try to point out presently. Still, it is not completely foolish to have such an attitude. At least *prima facie* there seems to be a grain of reason in it.

The situation is very different with 'common idioms'. First of all, such idioms are adapted not to *facts*, but to *beliefs*. If these beliefs are widely accepted; if they are intimately connected with the fears and the hopes of the community in which they occur; if they are defended, and reinforced

section). But it will be much poorer in other respects. For example, it will lack the associations which now connect mental events with emotions, our relations to others, and which are the basis of the arts and the humanities. We therefore have to make a choice: do we want scientific efficiency, or do we want a rich human life of the kind now known to us and described by our artists? The choice concerns the *quality of our lives* – it is a *moral* choice. Only a few modern philosophers have recognized this feature of the strife about materialism.

[2] 'A Plea for Excuses' in *Philosophical Papers*, ed. J. O. Urmson and G. J. Warnock (Oxford, 1961), 130.

[3] *Isaac Newton's Papers and Letters on Natural Philosophy*, ed. I. B. Cohen (Cambridge, 1958), 106. See also vol. 2, ch. 2.

with the help of powerful institutions; if one's whole life is somehow carried out in accordance with them – then the language representing them will be regarded as most successful. At the same time it is clear that the question of the truth of the beliefs has not been touched.

The second reason why the success of a 'common' idiom is not at all on the same level as is the success of a scientific theory lies in the fact that the use of such an idiom, *even in concrete observational situations*, can hardly ever be regarded as a *test*. There is no attempt, as there is in the sciences, to conquer new fields and to try the theory in them. And even on familiar ground one can never be sure whether certain features of the descriptive statements used are confronted with facts, and are thereby examined; or whether they do not simply function as accompanying noises. Some more recent analyses concerning the nature of facts seem to show that the latter is the case. It is clear that the argument from success is then inapplicable.

Assume thirdly – and now I am well aware that I am arguing contrary to fact – that the idiom to which reference is made in the above argument *is* used in a testable fashion and that the parallel, alluded to above, with scientific method is a legitimate one. Is it *then* possible to reject materialism by reference to the success of a non-materialistic language?

The answer is 'No' and the reason is as follows: in order to discuss the weaknesses of an all-pervasive system of thought such as is expressed by the 'common' idiom, it is not sufficient to compare it with 'the facts'.[4] Many such facts are formulated in terms of the idom and therefore already prejudiced in its favour. Also there are many facts which are inaccessible, *for empirical reasons*, to a person speaking a certain idiom and which become accessible only if a different idiom is introduced. This being the case, the construction of alternative points of view and of alternative languages which radically differ from the established usage, far from precipitating confusion, *is a necessary part of the examination of this usage* and must be carried out *before* a final judgement can be made. More concretely: if you want to find out whether there *are* pains, thoughts, feelings in the sense indicated by the common usage of these words, then you must become (among other things) a materialist. Trying to eliminate materialism by reference to a common idiom, therefore, means putting the cart before the horse.

6. The argument presented so far has some further features which are in need of criticism. Let us take it for granted that incompatibility with ordinary (or other) usage and the meaninglessness arising from it are sufficient reasons for eliminating a point of view. Then it must still be made clear that while the grammar of the *primitive terms* of the point of view may be incompatible with accepted usage, the grammar of the *defined terms* need not

[4] For an explanation see chs. 4, 8 and 11.13ff.

be so incompatible. The same applies to the 'grain' of both:[5] it has some-times been objected that a sensation is a very simple thing, whereas a collection of atoms has a much more complex structure (it is 'spotty'). This is correct. But there are still *properties* of such collections which do not participate in their 'grain'. The density of a fluid is an example. The fluid itself has the same 'grain' as a heap of atoms. The density has not. It ceases to be applicable in domains where the fine structure of the fluid becomes apparent. There are infinitely many other properties of this kind. The defender of the customary point of view has therefore much too simple an idea of the capabilities of materialism. He overlooks that materialism might even be able to provide him with the synonyms he wants; he overlooks that the materialistic doctrine might be able to satisfy his irrelevant demand for at least partial agreement of grammar.

7. While the argument from meaninglessness is wholly based upon language, the argument from falsity is not. That a thought cannot be a material process is, so it is believed, established *by observation*. It is by observation that we discover the difference between the one and the other and refute materialism. We now turn to an examination of this argument.

8. To start with we must admit that the difference does exist. Introspection does indicate, in a most decisive fashion, that my present thought of Aldebaran is not localized whereas Aldebaran is localized; that this thought has no colour whereas Aldebaran has a very definite colour; that this thought has no parts whereas Aldebaran consists of many parts exhibiting different physical properties. Is this character of the introspective result proof to the effect that thoughts cannot be material?

The answer is 'No' and the argument is the truism that what appears to be different does not need to be different. Is not the seen table very different from the felt table? Is not the heard sound very different from its mechanical manifestations (Chladni's figures; Kundt's tube; etc.)? And if despite this difference of appearance we are allowed to make an identification, postulat-ing an object in the outer world (the physical table, the physical sound), then why should the observed difference between a thought and the impres-sion of a brain process prevent us from making another identification, postulating this time an object in the inner (material) world, viz. a brain process? It is of course quite possible that such a postulate will run into trouble and that it will be refuted by independent tests (just as the earlier identification of comets with atmospheric phenomena was refuted by inde-pendent tests). The point is that the *prima facie* observed difference between thoughts and the appearance of brain processes does *not* constitute such trouble. It is also correct that a language which is based upon the assumption

[5] The 'grain' objection is due to Wilfrid Sellars.

that the identification has already been carried out would differ significantly from ordinary English. But this fact can be used as an argument against the identification only *after* it has been shown that the new language is inferior to ordinary English. And such disproof should be based upon the fully developed materialistic idiom and not on the bits and pieces of materialese which are available to the philosophers of today. It took a considerable time for ordinary English to reach its present stage of complexity and sophistication. The materialistic philosopher must be given at least as much time. As a matter of fact he will need more time as he intends to develop a language which is fully testable, which gives a coherent account of the most familiar facts about human beings as well as of thousands more recondite facts which have been unearthed by the physiologists. I also admit that there are people for whom even the reality of the external world and the identifications leading to it constitute a grave problem. My answer is that I do not address them, but that I presuppose a minimum of reason in my readers; I assume they are realists. And assuming this I try to point out that their realism need not be restricted to processes outside their skin – unless of course one already presupposes what is to be established by the argument, that things inside the skin are very different from what goes on outside. Considering all this I conclude that the argument from observation is invalid.

9. It is quite entertaining to speculate about some results of an identification of what is observed by introspection with brain processes. Observation of microprocesses in the brain is a notoriously difficult affair. Only very rarely is it possible to investigate them in the living organism. Observation of dead tissue, on the other hand, is applied to a structure that may differ significantly from the living brain. To solve the problems arising from this apparent inaccessibility of processes in the living brain we need only realize that the living brain is already connected with a most sensitive instrument – the living human organism. Observation of the reactions of this organism, introspection included, may therefore be much more reliable sources of information concerning the living brain than any other 'more direct' method. Using a suitable identification-hypothesis one might even be able to say that introspection leads to a *direct observation* of an otherwise quite inaccessible and very complex process *in the brain*.[6]

10. Against what has been said above it might be, and has been, objected that in the case of thoughts, sensations, feelings, the distinction between what they are and what they appear to be does not apply. Mental processes are things with which we are directly acquainted. Unlike physical objects

---

[6] This seems to have been one of the reasons why Ernst Mach suggested combining introspection with physiology. See vol. 2, chs. 5 and 6.

whose structure must be unveiled by experimental research and about whose nature we can make only more or less plausible conjectures, they can be known completely, and with certainty. Essence and appearance coincide here, and we are therefore entitled to take what they seem to be as a direct indication of what they are. This objection must now be investigated in some detail.

11. In order to deal with all the prejudices operating in the present case, let us approach the matter at a snail's pace. What are the reasons for defending a doctrine like the one we have just outlined? If the materialist is correct, then the doctrine is false. It *is* then possible to test statements of introspection by physiological examination of the brain, and reject them as being based upon an introspective mistake. Is such a possibility to be denied? The doctrine we are discussing at the present moment thinks it is. And the argument is somewhat as follows.

When I am in pain, then there is no doubt, no possibility of a mistake. This certainty is not simply a psychological affair, it is not due to the fact that I am too strongly convinced to be persuaded of the opposite. It is much more related to a logical certainty: there is no possibility whatever of criticizing the statement. I might not show any physiological symptoms – but I never meant to include them in my assertion. I might not even show pain behaviour – but this is not part of the content of my statement either. Now if the difference between essence and appearance were applicable in the case of pains, then such certainty could not be obtained. It *can* be obtained as has just been demonstrated. Hence, the difference does not apply and the stipulation of a common object for mental processes and impressions of physiological processes cannot be carried out.

12. The first question which arises in connection with this argument concerns the *source* of this certainty of statements concerning mental processes. The answer is very simple: it is their *lack of content* which is the source of their certainty. Statements about physical objects possess a very rich content. They are vulnerable because of the existence of this content. Thus, the statement 'There is a table in front of me' leads to predictions concerning my tactual sensations; the behaviour of other material objects (a glass of brandy put in a certain position will remain in this position and will not fall to the ground; a ball thrown in a certain direction will be deflected); the behaviour of other people (they will walk around the table; point out objects on its surface); etc. Failure of any one of these predictions may force me to withdraw the statement. This is not the case with statements concerning thoughts, sensations, feelings; or at least there is the impression that the same kind of vulnerability does not obtain here. The reason is that their content is so much poorer. No prediction, no

retrodiction can be inferred from them, and the need to withdraw them can therefore not arise. (Of course, lack of content is only a *necessary* condition of their empirical certainty; in order to have the character they possess, statements about mental events must also be such that in the appropriate circumstances their production can be achieved with complete ease; they must be *observational* statements. *This* characteristic they share with many statements concerning physical objects.)

13. The second question is how statements about physical objects *obtain* their rich content and how it is that the content of mental statements as represented in the current argument is so much poorer.

One fairly popular answer is by reference to the 'grammar' of mental statements and of physical statements respectively. We mean by pains, thoughts, etc., processes which are accessible only to one individual and which have nothing to do with the state of his body. The content of 'pain' or of 'thinking of Vienna' is low because 'pain' and 'thought' are mental terms. If the content of these terms were enriched, and thereby made similar to the content of 'table', they would cease to function in the peculiar way in which mental terms do as a matter of fact function, and 'pain', for example, would then cease to mean what is meant by an ordinary individual who, in the face of the absence of physiological symptoms, of behavioural expression, of suppressed conflicts, still maintains that he is in pain. This answer may be correct, and it will be taken to be correct for the sake of argument. However, in order to defeat the materialist it must also be shown that a language structured in this way will describe the world more correctly, and more efficiently than any language the materialist could develop. No such proof is available. The argument from 'common' usage and, for that matter, from any established usage is therefore irrelevant.

14. There is only one point on which this argument may possess some force, and this point concerns the use of *words*: having shown that a materialistic pain and an 'ordinary' pain would be two very different things indeed, the defender of the established usage may forbid the materialist to employ the word 'pain' which for him rightfully belongs to the ordinary idiom. Now, quite apart from the fact that this would mean being very squeamish indeed, and unbearably 'proper' in linguistic matters, the desired procedure *cannot be carried out*. The reason is that changes of meaning occur too frequently, and that they cannot be localized in time. Every interesting discussion, i.e. every discussion which leads to an advance of knowledge, terminates in a situation where some decisive change of meaning has occurred. Yet it is not possible, or it is only very rarely possible, to say *when* the change took place. Moreover a distinction must be drawn between the *psychological circumstances* of the production of a sentence, and the *meaning* of

the statement that is connected with that sentence. A new theory of pains will not change the pains; nor will it change the causal connection between the occurrence of pains and the production of 'I am in pain', except perhaps very slightly. It *will* change the *meaning* of 'I am in pain'. Now it seems to me that observational terms should be correlated with causal antecedents and *not* with meanings. The causal connection between the production of a 'mental' sentence and its 'mental' antecedent is very strong. It is learned very early in life. It is the basis of all observations concerning the mind. To sever this connection is a much more laborious affair than a change of connections with meaning. The latter connections change all the time anyway.[7] It is therefore much more sensible to establish a one to one connection between observational terms and their causal antecedents, than between such terms and the always variable *meanings*. This procedure has great advantages and can do no harm. An astronomer who wishes to determine the rough shape of the energy output (dependence on frequency) of a star by looking at it will hardly be seduced into thinking that the word 'red', which he uses for announcing his results, refers to sensations. Linguistic sensitivity may be of some value. But it should not be used to turn intelligent people into nervous wrecks.

15. Another reply to the question in section 13 which is *prima facie* satisfactory is that we know quite a lot about physical objects and that we know much less about mental events. We use this knowledge not only on the relatively rare occasions when we answer questions involving it, but we infuse it also into the notions with which we describe material objects: a table *is* an object which deflects a ball thrown at it; which supports other objects; which is seen by other people; and so on. We let this knowledge become part of the language we speak by allowing the laws and theories it contains to become the grammatical rules of this language. This reply would seem to be supported by the fact that objects of a relatively unknown kind always give rise to fewer predictions and that the statements concerning them are therefore relatively safe. In many such cases the only tests available are the reports of others which means that mass hallucinations can still count as confirming evidence.

Now this reply, however plausible, does not take into account that a considerable amount is known about mental processes also, and this not only by the psychologist, or the physiologist, but even by the common man, whether British, or a native of Ancient Greece, or of Ancient Egypt. Why has *this* knowledge not been incorporated into the mental notions? Why are these notions still so poor in content?

16. Before answering the question we must first qualify it. It is quite

[7] See ch. 2.5.

incorrect to assume that the relative poverty of mental notions is a common property of all languages. Quite the contrary, we find that people have at all times objectivized mental notions in a manner very similar to the manner in which we today objectivize materialistic notions. They did this mostly (but not always – the witchcraft theory of the Azande constitutes a most interesting *materialistic* exception) in an objective-*idealistic* fashion and can therefore be easily criticized, or smiled about, by some 'progressive' thinkers of today. In our present discussion such criticism is off the mark. We have *admitted*, in section 8, that the materialistic type of objectification may at some future time run into trouble. What we wanted to defend was the *initial* right to carry it out and it was this *initial* right that was attacked by reference to 'common usage'. Considering this context, it is important to point out that there is hardly any interesting language, used by a historical culture, which is built in accordance with the idea of acquaintance. This idea is nothing but a philosophical invention. It is now time to reveal the motives for such an invention.

17. We start the discussion with one further argument intending to show that, and why, the knowledge we may possess about mental events must not be incorporated into the mental terms and why their content must be kept low. This argument is apparently factual and it consists in pointing out *that there is knowledge by acquaintance*, or, alternatively, that there are things which can be known by acquaintance; we *do* possess direct and full knowledge of our pains, of our thoughts, of our feelings – at least of those which are immediately present and not suppressed.

This argument is circular. If we possess knowledge by acquaintance with respect to mental states of affairs, if there seems to be something 'immediately given', then this is the *result* of the low content of the statements used for expressing this knowledge. Had we enriched the notions employed in these statements in a materialistic (or an objective-idealistic) fashion *as we might well have done*, then we would no longer be able to say that we know mental processes by acquaintance. Just as with material objects we would then be obliged to distinguish between their nature and their appearance, and each judgement concerning a mental process would be open to revision by further physiological (or behavioural) inquiry. The reference to acquaintance cannot therefore justify our reluctance to use the knowledge we possess concerning mental events, their causes, their physiological concomitants (as their physiological content will be called *before* the materialistic move) for enriching the mental notions.

18. What has just been said deserves repetition. The argument which we attacked was as follows: there is the *fact* of knowledge by acquaintance. This fact refutes materialism which would exclude such a fact. The attack

consisted in pointing out that although knowledge by acquaintance may be a fact (which was, however, doubted in section 16), this fact is the result of certain peculiarities of the language spoken *and therefore alterable*. Materialism (and, for that matter, also an objective spiritualism like the Egyptian theory of the *ba*, or Hegel's spiritualism) recognizes the fact and suggests that it be altered. It therefore clearly cannot be refuted by a repetition of the fact. What must be shown is that the *suggestion* is undesirable, and that acquaintance is desirable.

19. We have here discovered a rather interesting feature of philosophical arguments. The argument from acquaintance presents what seems to be fact of nature, viz. our ability to acquire secure knowledge of our own states of mind. We have tried to show that this alleged fact of nature is the result of the way in which any kind of knowledge (or opinion) concerning the mind has been incorporated, or is being incorporated into the language used for describing facts: the knowledge (or opinion) is not used for *enriching* the mental concepts; it is rather used for making predictions in terms of the still unchanged, and poor concepts. Or, to use terms from technical philosophy, this knowledge is interpreted instrumentalistically, and not in a realistic fashion. The alleged fact referred to above is therefore a projection, into the world, of certain peculiarities of our way of building up knowledge. Why do we (or why do philosophers who use the language described) proceed in this fashion?

20. They proceed in this fashion because they hold a certain philosophical theory. According to this theory, which has a very long history and which influences even the most sophisticated and the most 'progressive' contemporary philosophers (with the possible exception of Wittgenstein), the world consists of two domains, the domain of the outer, physical world, and the domain of the inner, or mental world. The outer world can be experienced, but only indirectly. Our knowledge of the outer world will therefore forever remain hypothetical. The inner world, the mental world, on the other hand, can be directly experienced. The knowledge gained in this fashion is complete, and absolutely certain. This, I think, is the philosophical theory behind the method we described in the last section.

I am not concerned here with the question of whether this theory is correct or not. It is quite possible that it is true (though I am inclined to doubt this, especially in view of the fact that it presents what should be the result of a decision, viz. the richness or the poverty of the content of a statement and its corresponding property of being either hypothetical or certain, *as a fact of nature*, and thereby confounds the basic distinction between the *ought* and the *is*). What I *am* interested in here is the way in which the theory is *presented*. It is not presented as a hypothesis which is open to criticism and which can be rationally discussed. In a certain sense it is not even presented. It is rather incorporated into the language spoken in a fashion which makes it inaccessible to any debate – whatever the empirical results, they are not used for enriching the mental concepts which will therefore forever refer to entities knowable by acquaintance.

This procedure has two results. It hides the theory and thereby removes it from criticism. And it creates what looks like a very powerful fact supporting the theory. As the theory is hidden, the philosopher can even *start* with this fact and reason from it, thereby providing a kind of inductive argument for the theory. It is only when we examine what independent support there exists for this alleged fact that we discover that it is not a fact at all but rather a reflection of the way in which empirical results are handled. We discover that 'we were ignorant of our . . . activity and therefore regarded as an alien object what had been constructed by ourselves' (Kuno Fischer in his account of Kant's theory of knowledge).[8]

This is an excellent example of the circularity of philosophical argumentation even in those cases where such argumentation is based upon what seem to be uncontrovertible facts of nature ('inner' nature, that is). This example is a warning that we should not be too impressed by empirical arguments but that we should first investigate the source of their apparent success. Such an investigation may discover a fatal circularity and thereby destroy the force of the argument. It is quite obvious that a circularity of this kind cannot be removed by considering further *empirical* evidence. But it can be removed by an examination of the *methodological* tenability of the procedure described. I shall now give a brief outline of such an examination.

21. There are some philosophers who agree that the *fact* of acquaintance cannot be used as an argument against the materialist (or any other kind of 'internal realist'). Their reasons are not those given above but rather the realization that none of the situations described in the ordinary idiom, in any ordinary idiom, can be known by acquaintance. Realizing this they will look for arguments which remain valid in the face of adverse facts, and they will therefore appeal to norms rather than to facts. They usually suggest the construction of an *ideal language* containing statements of the desired property. In this they are guided by the idea that our knowledge must possess a solid, that is an incorrigible foundation. The construction of such a language has sometimes been represented as a task of immense difficulty and as worthy of a great mind. I submit that this means vastly overestimating it. Of course, if this task is meant to be the discovery of *already existing* statements of the ordinary language which possess the desired property (Russell's 'canoid patch of colour' indicates that he conceived his task in

---

[8] *Immanuel Kant und seine Lehre* (Heidelberg, 1889), 10.

this fashion), then it is perhaps impossible to carry it out. It may also be impossible to give an account of complex perceptions in terms of simple sensible elements (the investigations of the Gestalt school of psychology most definitely indicate that such composition from psychological elements will be an extremely difficult matter). But why should the attempt to find a safe observation language be impeded by such inessential restrictions? What we want is a series of observation statements leading to knowledge by acquaintance. Such statements can be obtained *immediately* by a philosophical laboratory assistant, by taking any observation statement and eliminating its predictive and retrodictive content as well as all consequences concerning public events occurring simultaneously. The resulting string of signs will still be observational, it will be uttered on the same objective occasion as was its predecessor, but it will be incorrigible, and the object described by it will be 'known' by acquaintance. This is how acquaintance can be achieved. Now let us investigate some consequences of this procedure.

22. Such an investigation is hardly ever carried out with due circumspection. What happens usually is this. One starts with a sentence which had a perfectly good meaning, such as 'I am in pain'. One interprets it as a statement concerning what can be known by acquaintance. One overlooks that such an interpretation drastically changes the original meaning of the sentence and one retains in this fashion the illusion that one is still dealing with a meaningful statement. Blinded by this illusion one cannot at all understand the objection of the opponent who takes the move towards the 'given' seriously and who is incapable of getting any sense out of the result. Just investigate the matter in some detail. Being in pain I say 'I am in pain' and, of course, I have some independent idea as to what pains are. They do not reside in tables and chairs; they can be eliminated by taking drugs; they concern only a single human being (hence, when in pain I shall not get alarmed about my dog); they are not contagious (hence, when in pain I shall not warn people to keep away from me). This idea is shared by everyone else and it makes people capable of understanding what I intend to convey. But now I am not supposed to let any one of these ideas contribute to the meaning of the *new* statement, expressed by the same sentence, about the immediately given; I am supposed to free this meaning of all that has just been said; not even the idea that a dreamt pain and a pain really felt are different must now be retained. If all these elements are removed, then what do I mean by the new statement resulting from this semantical canvas cleaning? I may utter it on the occasion of pain (in the normal sense); I may also utter it in a dream with no pain present, and I may be equally convinced that this is the right thing to do. I may use it metaphorically, connecting it with a thought (in the usual sense) concern-

ing the number two; or I may have been taught (in the usual sense of the word) to utter it when I have pleasant feelings and therefore utter it on these occasions. Clearly all these usages are now legitimate, and all of them describe the 'immediately given pain'. Is it not evident that using this new interpretation of the sentence I am not even in principle able to derive enlightenment from the fact that Herbert has just uttered it? Of course, I can still treat it as a *symptom* of the occurrence of an event which in ordinary speech would be expressed in the very same fashion, viz., by saying 'I am in pain'. But in this case I provide my own interpretation which is very different from the interpretation we are discussing at present. And we have seen that according to this interpretation the sentence cannot be taken to be the description of anything definite. It therefore means nothing; it cannot be understood by anyone (except in the sense in which a person looking at someone else's distorted face 'understands' what is going on – but then he does his own interpreting); and it is completely inadequate as a 'foundation of knowledge' or as a measure of factual meaning. Now if the Given were a reality, then this would mean the end of rational, objective knowledge. Not even revelation could then teach us what admittedly cannot be known in principle. Language and conversation, if they existed, would become comparable to a cat-serenade: all expression, nothing said, nothing understood. Fortunately enough, the 'Given' is nothing but the reflection of our own unreason and it can be eliminated by building up language in a more sensible fashion. This finishes our discussion of the argument from acquaintance.

23. To sum up: I have discussed three arguments against materialism. The first argument points out that materialism is not the ontology of ordinary English. I have given the reasons why this argument would be irrelevant even if ordinary English should turn out to be a highly successful *testable* idiom. The second argument refers to results of observation. I have pointed out that results of observation are in need of *interpretation* and that no reason has been given why a materialistic interpretation should be excluded. The third argument relied on the fact of 'acquaintance'. I have shown, first, that this fact is not unchangeable, and second that if it were a *fact*, knowledge would be impossible. I am not aware of any other philosophical arguments against materialism (clearly all considerations of synonymy or co-extensionality belong to what I have above called the first argument). There is, therefore, not a single reason why the attempt to give a purely physiological account of human beings should be abandoned, or why physiologists should leave the 'soul' out of their considerations.

24. A common feature of all the arguments discussed is this: they try to criticize a theory *before* this theory has been developed in sufficient detail to

be able to show its power. And they make established modes of thinking and of expression the basis of this criticism. I have pointed out that the only way of discovering the faults of established modes of thinking is by resolutely trying out a different approach. It would seem to me that the task of philosophy, or of any enterprise interested in the advance rather than the embalming of knowledge, is to encourage the development of such new modes of approach, to participate in their improvement rather than to waste time in showing, what is obvious anyway, that they are different from the status quo.

I I

Realism and instrumentalism

Comments on the logic of factual support

1. EXPLANATION OF CONCEPTS

Realism and instrumentalism provide two alternative interpretations of science and of factual knowledge in general. According to realism such knowledge is descriptive of (general or particular) features of the universe. According to instrumentalism even a theory that is wholly correct does not describe anything but serves as an instrument for the prediction of the facts that constitute its empirical content. Thus, considering Newton's theory of gravitation, a realist would remark that it teaches us of the existence, in addition to physical objects and their spatiotemporal behaviour, of entities of an altogether different kind which cannot be directly seen, heard, or felt, but whose influence is still noticeable enough, viz. forces. An instrumentalist, on the other hand, will take the position that there are no such entities and that the function of words like 'gravitation', 'force' and 'gravitational field' is exhausted by their giving an abbreviated description of the spatiotemporal behaviour of physical objects. He may even deny the existence of these objects and regard object words, too, as instruments, usable for the ordering and predicting of sense data. In this paper I shall argue that realism is preferable to instrumentalism.

2. THE DISTINCTION IS NOT PURELY VERBAL

Such an argument is of interest only if the issue between realism and instrumentalism is more than just a quarrel about words. Some philosophers deny that it is: Nagel, for example, holds that 'the opposition between these issues is a conflict over preferred modes of speech' which cannot be resolved in an objective manner.[1] I do not doubt for a second that there are versions of the problem which do possess this degenerate character. At the same time it seems to me that the instrumentalist position of Proclus, of some astronomers of the early seventeenth century, and of Niels Bohr is prompted by much more substantial motives than the predilection for certain modes of speech. These thinkers offer *physical* arguments for their point of view. They attempt to show that a realistic interpretation of certain

[1] *The Structure of Science* (New York, 1961), 152.

theories is bound to lead to results which are incompatible with observation and highly confirmed physical laws. Now if they are correct in this – and it will soon emerge that they are – then a realist cannot rest content with the general remark that theories just *are* descriptions and not merely instruments. He must then also revise the accepted *physics* in such a manner that the inconsistency is removed; i.e. he must actively contribute to the *development* of factual knowledge rather than make comments, in a 'preferred mode of speech', about the *results* of this development. In addition he must offer methodological considerations as to why one should change successful theories in order to be able to accommodate new and strange points of view. An excellent example of this situation is provided by the arguments against the realistic interpretation of the Copernican hypothesis and by the attempts that were made in order to overcome these arguments.

3. ARISTOTELIAN DYNAMICS

According to the Aristotelian philosophy, which was the accepted basis for physical reasoning throughout the later Middle Ages, motion was to be understood as the actualization of a potentiality inherent in an object.[2] This theory resolved the difficulty of *monism*, first exhibited by Parmenides, which consisted in the fact that change is impossible in a monistic universe. For now we are dealing with at least two different kinds of being, potential and actual. It also implied some very plausible assumptions concerning the circumstances under which change might occur. Actualization of a potentiality is possible only with the help of a form that corresponds to the properties exhibited by the object after the change is over. Forms do not exist by themselves; they can be separated from matter in thought, but not in reality. Hence, the occurrence of change in an object requires the presence of another object which possesses the appropriate form. Everything that is moved is moved by something else. Every motion needs a mover, and this mover must be present in the close neighbourhood of the changing thing, as action at a distance is impossible. Conversely, the state of an object that is not under the influence of forces is a state of rest. This is the Aristotelian 'law of inertia'.[3]

---

[2] I shall develop this point of view in the form which it received in the later Middle Ages and which is in some respects different from what is found in the Aristotelian opus itself. For such a later account concerning the *specific* theory of motion (see n.3, below), see document 7.1 in M. Clagett, *The Science of Mechanics in the Middle Ages* (Madison, Wis., 1957); cf. also Clagett's own summary, pp. 421ff.

[3] This *general theory* of motion and its law of inertia must be distinguished from the *specific theory* that deals with the motions, natural or forced, which actually occur in the universe. In the specific theory motions are called 'natural' when there is no visible *outer* agent that can be used for explaining their occurrence. Adherence to the law of inertia stated in the text, above, makes it necessary in this case to introduce an 'intrinsic form' such as the impetus, or the gravity of the moving object, or celestial intelligences which maintain the circular motion of

It is worth pointing out that this result is confirmed by our everyday experience: physical objects do not move unless they are hit, or pushed by other objects. Their natural state is indeed a state of rest. Note also the quasi-empirical character of some of the assumptions made in the course of the argument. This applies to the assumption of contact action and to the denial of the existence of separate forms. Empirical success (actual motions), theoretical success (solution of the 'Parmenidean problem'), comprehensiveness (applicability to any kind of change), consideration of details (in this respect the Aristotelian theory was superior to the considerations of the atomists) – these are weighty arguments in favour of the Aristotelian point of view. There were undesirable aspects, too, such as the linguistic character of many Aristotelian arguments which make them irrelevant to the solution of problems of fact. However, this need not deter us from properly appreciating the *result* of these arguments. And the result was a very interesting and successful empirical theory.[4]

the celestial spheres. From the point of view of the general theory, therefore, the 'natural motion' of the specific theory is still a motion that occurs under the influence of forces; these forces, however, are frequently left unspecified (although not unnamed).

[4] Another objection might arise from the fact that the theory failed to give a satisfactory account of the motion of projectiles and of falling objects. Two points ought to be remembered in connection with this objection. First, that it *was* possible to account for both kinds of motion within the framework of the general Aristotelian theory of motion. The impetus theory and the theory of antiperistasis gave such an account for the motion of projectiles; the theory of the inherent gravity of the heavy elements combined with the impetus theory explained the motion of falling objects. Second, the initial difficulties of the Aristotelian theory must not be taken as an indication of its 'unscientific' or 'metaphysical' character. *There is no single physical theory that is not beset by similar difficulties* (unless, of course, its defenders refrain from comparing it with the facts). Take Newton's theory of gravitation. It was about a century before the great inequality of Jupiter and Saturn and the secular acceleration of the mean motion of the moon were shown to be in accordance with Newton's law. And there exist still phenomena which resist explanation by the theory although they do not belong to the domain where relativistic effects become relevant. This is true of all theories: they are successful in a number of cases and will be regarded as revolutionary if these cases have been troublesome for a considerable time. But there always exist *other* cases which are *prima facie* refuting instances of the theory but which are put aside, for the time being, in the hope that a favourable solution (i.e. favourable to the theory under consideration) will be forthcoming. Now, if we postulate that a theory which is problematic because of the existence of *prima facie* refuting instances must not be used in cosmological arguments regarding the existence of non-existence of certain situations, then we shall thereby eliminate not only the Aristotelian point of view *but every succeeding physical theory as well*. For example, we shall be unable to use the theory of relativity in arguments concerning the nature of space and time, we shall be unable to use the quantum theory in arguments concerning determinism, and so on. Conversely, if we agree to base our arguments on the best theory available at a given time, then we cannot escape admitting the validity of the considerations to be outlined in section 4.

This problematic character of *every* scientific theory is very often kept from the eyes of the public, and even of students of the subject. Both popular presentations and textbooks dwell at length on the successes of a theory and hardly ever mention the much more interesting difficulties it faces. Some thinkers assert that this is a necessary evil, as only people who are firmly committed to a theory will be able to work strenuously at overcoming whatever difficulties it may possess. This is plain nonsense. It amounts to saying that only those who have first been given an incorrect account of the theory will be able to show that the theory is

4. CONSEQUENCES FOR THE MOTION OF THE EARTH

This theory has immediate implications regarding the motion of the earth. If it is correct, and if we also take into account some very simple facts, then we must conclude that the earth is at rest, i.e. that it neither rotates nor displaces itself in space. The reason is that only those things which are in direct contact with it, such as houses and human beings resting on its surface, would be carried along by the motion; whereas anything disconnected, such as birds, clouds, human beings jumping, would immediately assume their natural state of motion, namely, rest, and would therefore be left behind.[5] Considering that the birds are still with us, and that cannon balls unfortunately are not lost,[6] but hit their target most accurately, we must conclude that the earth cannot possess any motion whatever.

It is with reference to this argument that Ptolemy[7] criticizes 'certain thinkers' who

> have concocted a scheme which they consider more acceptable and they think
> that no evidence can be brought against them if they suggest for the sake of

true after all. And what happens if the theory should break down? Who will then be able to overcome his conditioning to this one theory and suggest something different?

[5] It would be unhistorical at this place to refer to the relativity of location, velocity and, perhaps, of all motion. Place, or position, in Aristotle has physical properties: 'the typical locomotions of the elementary bodies . . . show not only that place is something, but also that it exerts a certain influence. Each is carried to its own place, if it is not hindered, the one up, the other down' (*Physics* 208b, quoted from the Ross edition (Oxford, 1930)). These different properties of different locations enable us to distinguish them absolutely, and not only in relation to objects occupying them. The idea that the observed motions are prompted by *objects* in space (such as the earth) rather than by *positions* in space (such as the centre of the closed universe) is an alternative theory whose advantages were realized only after the Copernican point of view had been generally accepted. To a certain extent general relativity implies a return to the Aristotelian notions.

[6] The cannon argument was frequently used. For a discussion from the point of view of a new, and not yet existing dynamics, see Galileo, *Dialogues Concerning the Two Chief World Systems*, tr. by Stillman Drake (Berkeley and Los Angeles, 1953), 126f. For a very clear statement of the Aristotelian position, see Buridan's 'Questions on the Four Books on the Heavens and the World of Aristotle', Book II, question 22, section 9, quoted from Clagett, *The Science of Mechanics in the Middle Ages*, document 101: 'But the last appearance [that must be adduced against a rotation of the earth] is more demonstrative in the question at hand. This is that an arrow projected from a bow directly upward falls again in the same spot of the earth from which it was projected. This would not be so if the earth were moved with such velocity. Rather, before the arrow falls the part of the earth from which the arrow was projected would be a league's distance away. But still the supporters would respond that it happens so because the air, moved with the earth, carries the arrow, although the arrow appears to us to be moved simply in a straight line motion because it is being carried along with us. Therefore we do not perceive that motion by which it is carried with the air. But this evasion is not sufficient because the violent impetus of the arrow in ascending would resist the lateral motion of the air so that it would not be moved as much as the air. This is similar to the occasion when the air is moved by a high wind. For then an arrow projected upward is not moved as much laterally as the wind is moved, although it would be moved somewhat.' Can there be any doubt of the empirical character of this argument?

[7] Quoted from M. R. Cohen and I. E. Drabkin, *Source Book in Greek Science* (New York, 1948), 126ff.

argument that the heaven is motionless, but that the earth rotates about one and the same axis from West to East, completing one rotation approximately every day, or alternatively that both the heaven and the earth have a rotation of a certain amount, whatever it is, about the same axis, as we said, but such as to maintain their *relative* situations. These persons forget however that, while, as far as appearances in the stellar world are concerned, there might perhaps be no objection to this theory in the simple form, yet to judge by the conditions affecting ourselves and those in the air above us such a hypothesis must seem to be quite ridiculous.[8]

Ptolemy here distinguishes, as is also done in classical physics, between the purely *kinematic* aspects of motion on the one side and the phenomena of *inertia* which are brought to light by motion on the other. *Kinematically* the motion of the earth is indistinguishable from a situation where the earth is at rest and the stellar sphere rotates in the opposite direction (we are now considering rotation only). However, the motion of the earth, if it occurs, will also lead to *dynamical* phenomena of a kind that can be well described in advance. These phenomena do not occur. Hence, the earth does not move. One should note that this argument against the dynamical, or, as one could also call it, the absolute motion of the earth has exactly the same structure as the argument *for* the absolute rotation of the earth as derived from Foucault's pendulum and the variation of pendulum clocks from the

---

[8] Another proof derives from the doctrine of natural places (which is part of the specific theory of motion explained in n.3, above). According to this doctrine, which is again very closely related to experience, the elements of the universe are distinguished by the places to which they tend to move: the earth moves toward the centre, fire moves toward the circumference, water and air move to intermediate places. Sometimes these purely dynamical properties seem to be regarded as the sole *defining properties* of the elements. Earth is distinguished from fire not by its appearance, nor by the fact that the latter burns and the former cools, *but solely by the fact that it moves down whereas fire moves toward the circumference.* (For details see F. Solmsen, *Aristotle's System of the Physical World* (New York, 1960), chs. 11ff.) It has already been pointed out (n.5, above) how this doctrine gives physical content to the notion of position and thereby to (a finite) absolute space. The proof against the motion of the earth derived from it (Ptolemy, quoted from Cohen & Drabkin, *Source Book in Greek Science*, 126), runs as follows: 'So far as the composite objects in the universe, and their motion on their own account and in their own nature are concerned, those objects which are light, being composed of fine particles, fly towards the outside, that is, towards the circumference, though their impulse seems to be towards what is for individuals 'up', because with all of us what is over our heads, and is also called 'up', points towards the bounding surface; but all things which are heavy, being composed of denser particles, are carried towards the middle, that is to the centre, though they seem to fall "down", because, again, with all of us the place at our feet, called "down", itself points towards the center of the earth, and they naturally settle in a position about the center, under the action of mutual resistance and pressure which is equal and similar from all directions. Thus it is easy to conceive that the whole mass of earth is of huge size in comparison with the things that are carried down to it, and that the earth remains unaffected by the impact of the quite small weights (falling on it), seeing that these fall from all sides alike . . . But, of course, if as a whole it had a common motion, one and the same with that of the weights, it would, as it was carried down, have got ahead of every other falling body, in virtue of its enormous excess of size, and the animals and all separate weights would have been left behind floating in the air, while the earth, for its part, at its great speed, would have fallen completely out of the universe itself. But indeed this sort of suggestion has only to be thought of in order to be seen to be utterly ridiculous.'

equator to the pole.[9] The only difference lies in the law of inertia used. According to Aristotle a thing left to itself will remain at rest.[10] According to Newton it will move on a straight line with constant speed. At the time of Ptolemy the limitations of the Aristotelian physics had not yet been exhibited in an unambiguous manner. We have to conclude, then, that apart from the hypothetical character of *any* argument from physical principles the argument for the unmoved earth referred to by Ptolemy was impeccable.

5. THE INSTRUMENTALIST INTERPRETATION OF THE COPERNICAN THEORY

It is this situation which we must keep before our eyes when embarking upon the evaluation of the controversy over the Copernican hypothesis. Seen in the light of the argument given above, the attempt to regard this hypothesis as a correct account of the actual situation in the universe amounts to upholding an unsupported conjecture in the face of fact and well-supported theory.[11] True, the heliocentric point of view gave a simpler explanation of the second inequality of planetary motion (the loops) than did the geocentric scheme. However, it did not on that account alone lead to better predictions. Epicycles were still needed for the first inequality. The specific way in which Copernicus introduced these epicycles might improve the empirical adequacy of the theory. But *these* details were not bound to the heliocentric scheme. They were a mathematical technique which, like the technique of Fourier decomposition,[12] could be applied under the most

[9] As is well known, Newton distinguished relative or apparent motion from absolute or true motion, and he also pointed out that the latter can be recognized through its dynamical effect (bucket experiment). The argument between the defenders of a relational account of space (Leibniz was one of them) and the absolutists is the exact parallel to Ptolemy's argument in the text. Ptolemy points out that although there may be kinematic equivalence between two situations involving motion, there is yet no dynamical equivalance. *This belief is shared by the Aristotelians and by the supporters of absolute space.* The only difference is that, because of the different laws of motion, the Newtonians and the Aristotelians regard different phenomena as indicating absolute motion.

[10] In n.3, above, I have noted some difficulties encountered by this law and have mentioned the theory of impetus as one of the possible ways out. Would it not be natural, therefore, to replace the Aristotelian law of inertia by the corresponding law of the impetus theory and thereby remove one of the most decisive obstacles to the motion of the earth? This would indeed be a possible procedure. But this emphasizes rather than weakens another point I shall make: that in the case of the Copernican hypothesis the realistic position was not a matter of pure philosophy and still less a matter concerning 'preferred modes of speech'. A realist had to change physics as well.

[11] For 'well-supported' see n.4, above. Copernicus, of course, was well aware of the dynamical difficulties connected with the motion of the earth and he therefore tried to introduce a dynamics of his own. The same is true of Galileo, whose main work may be described as the attempt to show that the motion of the earth was not only dynamically possible, but even required. See his arguments in the Second Day of the *Dialogues*.

[12] If I remember correctly it was Norbert Wiener who has pointed to the similarity, from a

varied circumstances. For example, they could also be added to the geo-centric hypothesis, where they had originated in the first place. Hence, 'if the tables newly to be calculated should prove to be superior to the Alphonsine tables, which were based on the Ptolemaic system, this would not ... be due to the heliocentric hypothesis as such, but only to the superior quality of the details of the new system'.[13] There was no indepen-dent evidence in favour of the heliocentric theory; this theory was, at least initially, a conjecture that had no foundation in empirical fact.[14] The only favourable remark that could be made was that it somewhat simplified

mathematical point of view, of the technique of epicycles and the technique of the Fourier decomposition.

[13] E. J. Dijksterhuis, *The Mechanization of the World Picture* (Oxford, 1961), 249. Cf. also T. S. Kuhn, *The Copernican Revolution* (New York, 1957), 169: 'When Copernicus had finished adding circles, his cumbersome sun-centered system gave results as accurate as Ptolemy's but it did not give more accurate results. Copernicus did not solve the problem of the planets.' Kuhn's book contains an excellent semi-technical account of the comparative efficiency of the Ptolemaic and Copernican systems.

[14] This, more than anything else, should exhibit the error in the assertion that science started when people stopped being impressed by theories and turned to observations instead. Galileo especially is often represented as a thinker who started from observations and strictly followed the Baconian method. According to J. Herschel, *The Cabinet of Natural Philosophy* (Philadelphia, 1831), 85, Galileo 'refuted the Aristotelian dogmas respecting motion, by direct appeal to the evidence of sense, and by experiments of the most convincing kind'. As regards Copernicus Herschel has this to say: 'By the discoveries of Copernicus, Kepler, and Galileo the errors of the Aristotelian philosophy were effectually overturned on a plain appeal to the facts of nature.' Now we have seen that there were no 'facts of nature', there was no 'evidence of sense' to which Copernicus *could* have appealed. Even worse, the 'evidence of the senses' was against him and his theory. It was the Aristotelians who could quote 'nature' in their favour. Also 'one outstanding fact about the scientific revolution is that its initial and in a sense most important stages were carried through before the invention of the new measuring instruments, the telescope, and the microscope, thermometer and accurate clock, which were later to become indispensable for getting the accurate and satisfactory answers to the questions that were to come to the forefront of the sciences' (A. C. Crombie, *Mediaeval and Early Modern Science* (New York, 1959), II, 122). The 'Galileo myth' too, according to which Galileo busily rushed around making experiments and 'climbed the leaning tower of Pisa with one one-hundred-pound cannon ball under one arm, and a one-hundred-pound cannon ball under the other" (ironical remark in H. Butterfield, *The Origins of Modern Science* (London, 1957), 81) has been refuted by historical research. 'In general', writes Dijk-sterhuis, 'one has always to take stories about experiments by Galileo as well as by his opponents, with some reserve. As a rule they were performed only mentally, or they are merely described as possibilities' (*Mechanization of the World Picture*, 338). More especially there is evidence which proves 'the complete baselessness of the belief tenaciously main-tained by the supporters of the Galileo myth, namely that he discovered the law of squares by performing with falling bodies a number of measurements of distance and time, and noting in these values the constant ratio between the distance and the square of time' (*ibid.*, 340). All this of course cannot prevent an inductivist like Dingle from repeating (in the 1961 edition of the *Encyclopaedia Britannica*. XIX, 95) that 'Galileo discovered the law of falling bodies by measuring how the space covered varied with the time of fall'. Which shows how difficult it is for an empiricist to accurately represent the facts of history.

It is also surprising that this feature of the origin of modern science (its original *incompatibility* with facts and well-supported theory) has not yet been taken into account by scientific methodology. None of the methodologies in existence today would have permitted Coperni-cus to interpret his theory in a realistic fashion. But of this later.

calculations by a suitable coordinate transformation. This is a phenomenon well known from mathematical physics: there are many problems which admit of immediate solution once a proper choice of coordinates has been made and whose solution is very cumbersome in different coordinates. Such a choice and the resulting mathematical success does not imply that the coordinate system chosen has any *dynamical* preference over other coordinate systems; for example, it does not imply that it is an inertial system. After all, the solubility of a problem, the circumstances under which it can be easily calculated, depend as much upon the mathematical formalism used as upon nature. There may exist asymmetrics in the formalism which do not exist in nature. The fact that the problems of positional astronomy can be dealt with in a more simple manner in a coordinate system in which the sun is at rest therefore does not imply that the sun is actually at rest and that the earth moves.

Combining this with the argument in section 4 above, we arrive at the result that if the Copernican hypothesis has any merit at all (judged from the point of view of the contemporaries of Copernicus) it lies in the fact that it allows for a more effective *calculation* of the position of the planets. It does not lie in its giving a new *and true* account of what goes on in reality. This is one of the ways in which the instrumentalist interpretation of Copernicus was introduced. Considering the dynamical arguments against the motion of the earth, this interpretation would seem to have been unassailable.[15]

6. PHILOSOPHICAL ARGUMENTS FOR THIS INTERPRETATION ARE NOT THE ONLY ONES

At this point it is very important to emphasize that the argument does not at all depend on a general philosophical position considering the nature of our knowledge. No such sweeping assumption is implied or presupposed. The argument is concerned with a *specific theory*, viz. the heliocentric hypothesis. It is based upon facts and physical laws. It shows that, in view of these facts and these laws, the hypothesis cannot be *true*; that it can at most be an *instrument* of prediction. It is arguments of this specific kind with which we shall be concerned in the present paper (which means, of course, that we shall have to find a method of justifying the invention of unsupported conjectures in the face of fact and well supported theory).

It is well known that arguments of this kind are not the only ones which occur in connection with this issue between realism and instrumentalism.

---

[15] To put it in different words: the dynamical arguments amount to a straightforward *refutation* of the Copernican hypothesis. If we take these arguments at their face value, then we must regard this hypothesis as false. This, of course, does not prevent the hypothesis from giving a correct account of some facts of astronomy. It is therefore still a good instrument of prediction.

The above refutation of the heliocentric hypothesis was not even the most popular one. A great deal of the opposition to Copernicus derived from the difficulty in making his ideas agree with the scriptures as interpreted by the Church fathers (which interpretation had become binding after the Council of Trent). There were also more philosophical considerations which were closely connected with the belief that only those theories whose validity had been established by a *proof* implying necessity could be said to be descriptive of reality. No such proof was available in the case of the Copernican hypothesis, nor did it seem likely that it would ever be forthcoming. Hence the hypothesis could at most be regarded as an instrument of prediction. This is the train of reasoning on which Bellarmine seemed to base *his* evaluation of the case.

> If there were one real proof [he writes to Father Foscarini] that the sun is in the center of the universe, that the earth is in the third heaven, and that the sun does not go round the earth but the earth round the sun, then we should have to proceed with great circumspection in explaining passages of scripture which appear to teach the contrary . . . But as for myself I shall not believe that there are such proofs until they are shown to me. Nor is it a proof that, if the sun be supposed at the center of the universe and the earth in the third heaven, everything works out the same as if it were the other way round.[16]

Later historians have repeated the argument.

> Logic was on the side of Osiander and Bellarmine [writes P. Duhem] and not on that of Kepler and Galileo; the former had grasped the exact significance of the experimental method, while the latter had been mistaken . . . Suppose that the hypotheses of Copernicus were able to explain all known appearances. What can be concluded is that they may be true, not that they are necessarily true, for in order to make legitimate this last conclusion it would have to be proved that no other system of hypotheses could possibly be imagined which could explain the appearances just as well.[17]

There is a very definite epistemology connected with this attitude. Different kinds of knowledge are distinguished and different claims to reality allotted to each kind. Physics deals with causes, with substance, and the real constitution of things, and it is capable of *proving* the truth of its assertions; astronomy is concerned with prediction only, and may for this purpose introduce hypotheses which are actually false.[18] Predictive success in astronomy is therefore no indication of truth and of factual relevance. Proof alone is. It is impossible to explain here why this position is both untenable and undesirable. However, it will perhaps be admitted that such an explanation will be of a general kind, that it will be a matter of pure philo-

[16] Letter of April 12, 1615, quoted in G. di Santillana, *The Crime of Galileo* (Chicago, 1955), 99ff.
[17] Quoted in Santillana, *The Crime of Galileo*, 107.
[18] For a more detailed account see Simplicius, *Commentary on Aristotle's Physics*, quoted from T. L. Heath, *Aristarchus of Samos* (Oxford, 1913), 275–6. Cf. also P. Duhem, *Aim and Structure of Physical Theory* (Princeton, 1934).

sophy, and that it will concern not only *one* theory but *any possible* theory: the reasons which prompted Bellarmine and Duhem to regard the instrumentalism of Osiander as 'logically superior' to the realism of Galileo can be refuted by philosophical considerations.[19] *However, it would be a great mistake to assume that thereby all the objections to the heliocentric point of view have been done away with.* The physical arguments of sections 4 and 5 still stand unrefuted and require an answer.

It is very important to realize this complex character of the situation, for otherwise one will be satisfied too early and too easily. Thus a thinker who is acquainted with the epistemological arguments only will regard a refutation of these arguments, and the construction of an alternative epistemology which allows hypothetical statements to have realistic implications, as the completion of his task, and he may even be seduced into thinking that his epistemology once and forever settles the issue between realism and instrumentalism in favour of the former. Quite obviously such an attitude will not impress the 'physicists' whose arguments have not even been touched, and it will thereby either create, or further contribute to, a very undesirable split between physics and philosophy. And this is precisely what happens today in microphysics.[20] There exist some very impressive physical arguments against a realistic interpretation of quantum theory. There are also more philosophical arguments trying to establish the same result, viz. that the quantum theory is an instrument of prediction from which no realistic consequences can be drawn. These philosophical arguments proceed from the assumption that only observational terms are candidates for a realistic

[19] For an analysis and criticism of this position see K. R. Popper, *Conjectures and Refutations* (New York, 1962), 97ff. But Popper overlooks the physical arguments which were used in addition to the general philosophical theory and thus is content too soon. Cf. ch. 1, n.8.
[20] Other examples are the kinetic theory of matter and the theory of relativity. The kinetic theory of matter of the late nineteenth century was attacked *both* by philosophical arguments which tried to show the undesirability, from an empirical point of view, of the abstract and unobservable notions introduced by that theory, *and* by physical arguments which pointed out that any kinetic theory will be inconsistent with the laws of the phenomenological theory, especially the second law (reversibility objection of Loschmidt; recurrence objection of Poincaré–Zermelo). A purely philosophical defence of atomism is therefore insufficient. In addition it must be shown how the inconsistency can be circumvented and to what extent it is really an objection. When the theory of relativity first became known to a wider circle of people, including philosophers, it appeared that its main conclusions were a result of the positivistic doctrine that things that cannot be measured do not exist. It was therefore immediately attacked by realists who believed that their defence of realism was at the same time a defence, and a complete rehabilitation, of the notions of absolute space and time. Again, it was overlooked that relativity rests on consideration of a much more substantial nature than a positivistic theory of knowledge can offer. But it must be admitted that in this case it was *the physicists themselves* who created the confusion. Thus Bridgman's widely read *Logic of Modern Physics* (Princeton, 1936) represents the theory of relativity as the transition to a new era where considerations of objective existence are replaced by considerations of measurability, and Niels Bohr has interpreted the theory in the same manner. The most outstanding example of the confusion referred to in the text, however, is presented by the discussions of the quantum theory.

interpretation, an assumption which can be refuted once and for all by philosophical reasoning.[21] Thinkers who are conversant with philosophy only will assume that this settles the matter – which is far from being true.

> This situation [I wrote when discussing the interpretation of the quantum theory] accounts for the strangely unreal character of many discussions on the foundations of the present quantum theory. The members of the Copenhagen school are confident that their point of view with whose fruitfulness they are well acquainted is satisfactory and superior to a good many alternatives. But when writing about it, they do not draw sufficient attention to its physical merits but wander off into philosophy and especially into positivism. Here they become an easy prey to all sorts of philosophical realists who quickly . . . exhibit the mistakes in their arguments without thereby convincing them of the invalidity of their point of view – and quite justly so, for this point of view can rest on its own feet and does not need support from philosophy. So the discussion between physicists and philosophers goes back and forth without ever getting anywhere.[22]

It is imperative to avoid a vicious circle of this kind and to *attack the instrumentalist position where it seems to be strongest; and that is where it is based upon specific factual argument rather than upon general philosophy.* However, before doing so I shall introduce the example of the quantum theory in addition to the example already discussed.

7. THE QUANTUM THEORY: BOHR'S HYPOTHESIS

Not long after Planck had introduced the quantum of action[23] it was realized that this innovation was bound to lead to a complete recasting of the principles of motion of material systems. It was Poincaré who first pointed out that the idea of a continuous motion along a well-defined path could no longer be upheld and that what was needed was not only a new *dynamics*, i.e. a new set of assumptions about the acting forces, but also a new *kinematics*, i.e. a new set of assumptions about the kind of motion initiated by these forces.[24] Both Bohr's older theory and the dual nature of light and matter further accentuated this need. One of the problems arising in the older quantum theory was the treatment of the interaction between two mechanical systems.[25] Assume (fig. 1) that two systems, A and B, interact in

---

[21] An attempt at such a refutation is contained in Popper, 'Three Views'. For a different account, see my paper, 'Das Problem der Existenz theoretischer Entitaeten', *Probleme der Wissenschaftstheorie* (Vienna, 1960), 35–72. Cf. also ch. 1, n.8.

[22] For this quotation and a more detailed account of what follows, see my 'Problems of Microphysics', *Pittsburgh Publications in the Philosophy of Science* (Pittsburgh, 1963), I.

[23] I am here referring to what is known as Planck's First Theory in which both absorption and emission were regarded as discontinuous processes (*Verh. phys. Ges.*, 2 (1900), 237ff) and which also implied discontinuities in space (cf. E. T. Whitaker, *History of the Theories of Aether and Electricity* (Edinburgh, 1953), II, 103).

[24] 'Sur la théorie des quanta', *J. Phys.* (1912), 1.

[25] Niels Bohr, *Atomic Theory and the Description of Nature* (Cambridge, 1932), 65.

such a manner that a certain amount of energy, E is transferred from A to B. During the interaction the system $A + B$ possesses a well-defined energy. Experience teaches that the transfer of E does not occur immediately, but takes a finite amount of time. This seems to suggest that both A and B change their state gradually, i.e. A gradually falls from 2 to 1, while B gradually rises from 1 to 2. However such a mode of description would be incompatible with the *quantum postulate* according to which a mechanical system can only be in either state 1 or state 2 (we shall assume there are no admissible states between 1 and 2), and is incapable of being in an intermediate state. How shall we reconcile the fact that the transfer takes a finite amount of time with the non-existence of intermediate states between 1 and 2?

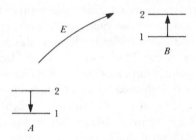

Fig. 1

This difficulty was resolved by Bohr,[26] on the basis of the assumption that during the interaction of A and B the dynamical states of both A and B cease to be well defined so that it becomes *meaningless* (rather than *false*) to ascribe a definite energy to either of them.[27]

This simple and ingenious hypothesis has so often been misrepresented that a few words of explanation are needed. First of all it must be pointed out that in the above formulation the term 'meaning' has not entered, as has been asserted by various critics, because of some connection with the now customary attitude of preferring semantical analysis to an investigation of physical conditions.[28] After all, there are well known classical examples of

[26] I shall not contend that this is the only way of getting around the difficulty, but it is a very reasonable physical hypothesis which has not yet been refuted by any of the arguments aimed against it.

[27] By the expression 'dynamical state' we refer to 'quantities which are characteristic of the motion' of the system concerned (such as the positions and the momenta of its components), rather than those quantities which, like mass and charge, serve as a characteristic of the kind of system it is. For this explanation see L. Landau and E. M. Lifshitz, *Quantum Mechanics* (London, 1958), 2, as well as N. Bohr, *Atomic Physics and Human Knowledge* (New York, 1956) 90; cf. also H. A. Kramers, *Quantum Mechanics* (New York, 1957), 62.

[28] It is to be admitted, however, that most derivations of the uncertainties, and especially those based upon Heisenberg's famous thought experiments, *do* make use of philosophical theories of meaning. Usually these arguments (and other arguments which proceed from the

terms which are meaningfully applicable only if certain physical conditions are first satisfied and which become inapplicable, and therefore meaningless, as soon as these conditions cease to hold. A good example is the term 'scratchability' (Mohs scale), which is applicable to rigid bodies only and which loses its significance as soon as the bodies start melting. Secondly, it should be noted that the proposed solution does not contain any reference to *knowledge*, or *observability*. It is not asserted that during the time of transfer *A* and *B* may be in some state which is unknown to us, or which cannot be observed. For the quantum postulate does not merely exclude the knowledge of, or the observability of, the intermediate states; it excludes these intermediate states themselves. Nor must the argument be read as asserting, as is implied in many presentations by positivist-minded physicists, that the intermediate states do not exist *because* they cannot be observed. For it refers to a postulate (the quantum postulate) which deals with existence, and not with observability. It is at this point that the most misleading arguments occur. Physicists who have adopted the positivistic principle that things which cannot be observed do not exist try to justify the indefiniteness of state descriptions by a combined reference to the fact that they cannot be observed and to this principle. Philosophers immediately expose the fallacy of the argument (if they are anti-positivists, that is) and think that they have thereby shown the existence, or at least the physical possibility, of sharp states. This is, of course, not correct, for from the fact that a certain argument has been found to be fallacious it does not follow that a better argument does not exist. But this better argument is hardly ever used by the physicists, which creates the impression that positivism is indeed the only source of the peculiar features of the present quantum theory.[29]

The emphasis upon the absence of *predictability* is not satisfactory either. For this way of speaking would again suggest that we could perhaps predict

commutation relations of the elementary theory) only establish that inside a certain interval *measurements cannot be carried out*, or that the products of the mean deviations of certain magnitudes *cannot be ascertained* below Planck's constant h. The transition from this stage of the argument to the assertion that it would be *meaningless* to ascribe definite values to the magnitudes in the interval is then achieved on the basis of the principle that what cannot be measured cannot be meaningfully asserted to exist. This argument is, of course, unacceptable because the principle on which it is based is unacceptable. Moreover, it is liable to lead to a dogmatic belief in the result. For whereas a physical hypothesis such as the one discussed in the text will be accepted with caution, it is customary to assume that philosophical considerations, and especially considerations flowing from a meaning criterion, possess a much greater argumentative force.

[29] A writer who holds this belief is M. Bunge, who asserts in his *Causality* (Cambridge, Mass., 1959), 328, that 'the empirical indeterminacy characterizing the usual interpretation of the quantum theory is a consequence of its idealistic presuppositions'. Similar sentiments have been expressed occasionally by Bohm, Kaila, Landé and Popper. For a criticism, see again my 'Problems of Microphysics', as well as the final section of my review of Bunge's book in *Phil. Rev.*, 60 (1961), 396–405.

better if we only knew more about the things that exist in the universe, whereas Bohr's suggestion denies that there *are* things whose detection would make our knowledge more definite.

The third point concerns a suggestion for getting around the kinematics of ill-defined states which has often been made in connection with wave mechanics and which will be discussed in detail later in the present paper. According to this suggestion the difficulties which arise when we try to give a rational account of processes of interaction are due to the fact that classical point mechanics is not the correct theory for dealing with atomic systems, and state descriptions of classical point mechanics are not adequate means for describing the state of systems upon the atomic level. According to this suggestion we ought not to retain the classical notions, such as position and momentum, and make them less specific. What we ought to do is to introduce completely new notions which are such that when *they* are used states and motions will again be well defined. Now if any such new system is to be adequate for the description of the quantum phenomena, then it must contain means for expressing the quantum postulate, which is one of the most fundamental microlaws; and it must therefore also contain adequate means for expressing the concept of energy. However, once this concept has been introduced, all our above considerations immediately apply again with full force: while being part of $A + B$, neither A nor B can be said to possess a well-defined energy; whence it follows at once that the new and ingenious set of concepts will also not lead to a well-defined and unambiguous kinematics. Now if the new formalism should happen to work with functions, operators, and other mathematical tools which are unambiguous and precise from a mathematical point of view, then we shall have to conclude that this definiteness and absence of ambiguity has no correlate in the real world. In other words, *we shall have to interpret these mathematical tools in a purely instrumentalist manner.* 'It would [therefore] be a misconception to believe', writes Niels Bohr,[30] 'that the difficulties of the atomic theory [i.e. the indefiniteness of state descriptions demanded by the features of processes of interaction as well as by duality] may be evaded by eventually replacing the concepts of classical physics by new conceptual forms.' This last remark will be of great importance in connection with the interpretation of Schrödinger's wave mechanics in section 9.

The empirical adequacy of the proposed solution is shown by such phenomena as the natural line breadth, which in some cases (such as in the absorption leading to states preceding the Auger effect) may be quite considerable.

Its consequence is, of course, the *renunciation of the kinematics of classical physics* and an instrumentalist interpretation for any future quantum theory

[30] *Atomic Theory and the Description of Nature*, 16.

that works with state descriptions which are well defined from the mathematical point of view. For if during the interaction of A and B neither A nor B can be said to be in a well-defined state, then the change of these states, i.e. the *motion* of both A and B, will not be well defined either. More particularly, it will no longer be possible to ascribe a definite trajectory to any one of the elements of either A or B. If, on the other hand, the state function of some quantum theory should happen to develop in a well-defined fashion, then this development cannot have any real significance, it cannot correspond to any process in nature; that is, it can at most be regarded as an instrument for the prediction, different at different times, of observational results. This is a very forceful argument indeed in favour of instrumentalism, and it can be further supported by a detailed investigation of the properties of wave mechanics. Popper's assertion to the effect that 'the view of physical science founded by Cardinal Bellarmino and Bishop Berkeley has won the battle without a further shot being fired'[31] therefore shows only that he knows neither the anti-Copernican arguments, nor the better arguments about the interpretation of the quantum theory.[32] Not only was the instrumentalist position at the time of Copernicus supportable by arguments which were much stronger, at least for a contemporary thinker, than were the arguments flowing from Bellarmine's Platonistic epistemology. But modern physics has found *new* physical reasons why its own most important theory, the quantum theory, cannot be anything but an instrument of prediction. These reasons are of precisely the same character as were Ptolemy's: a realistic interpretation of the quantum theory is bound to lead to incorrect predictions. Admittedly, in the arguments usually presented these physical difficulties are almost buried beneath an unacceptable positivism. However, this does not mean that they do not exist and that no 'further shot had been fired' since the time of Bellarmine and Berkeley.

One may now attempt to retain the idea of a well-defined motion and merely make indefinite the relation between the energy and the parameters characterizing this motion. The considerations in the next section show that this attempt encounters considerable difficulties.

---

[31] 'Three Views Concerning Human Knowledge', 2ff.

[32] I agree with Popper in his attitude towards instrumentalism. I also agree that one should not rest content with a theory which at most admits of an instrumentalist interpretation but becomes false when interpreted realistically. Thirdly, I agree that in the case of the quantum theory such an attitude is usually supported by the view that in any case theories are nothing but instruments of prediction. However, I also believe that within the quantum theory the instrumentalist position has been forced upon the physicist by the realization that the current theory interpreted realistically must lead to wrong results; it is not merely a repetition of the philosophical idea that all theoretical thinking is of instrumental value only.

8. IN THE QUANTUM THEORY, TOO, PHILOSOPHICAL ARGUMENTS FOR
   INSTRUMENTALISM ARE NOT THE ONLY ONES

The reason for the difficulty is that the *duality of light and matter* provides an even more decisive argument for the need to replace the classical kinematics by a new set of assumptions. It ought to be pointed out, in this connection, that dealing with both light and matter on the basis of a single general principle, such as the principle of duality, may be somewhat misleading. For example, whereas the idea of the position of a light quantum has no definite meaning,[33] meaning can be given to the idea of the position of an electron. Also, no account is given in this picture of the coherence length of light. Omitting these details, however, we can now argue as follows.

It has been asserted that the interference properties of light and matter and the duality resulting from them are but an instance of statistical behaviour in general,[34] which latter is then thought to be best explainable by reference to such classical devices as pin boards and roulette games. According to this assertion, elementary particles move along well-defined trajectories and possess a well-defined momentum at any instant of their motion. It is sometimes admitted that their energy may occasionally undergo sudden and perhaps individually unexplainable changes. But it is still maintained that this will not lead to any indefiniteness of the *state* that experiences these sudden changes.

I shall now try to show that this assumption cannot give a coherent account of the wave properties of matter and of the conservation laws. It is sufficient, for this purpose, to consider the following two facts of interference: (1) Interference patterns are independent of the number of particles which at a given moment are dwelling in the apparatus; for example, we obtain the same pattern on a photographic plate whether we use strong light and short time of exposure or very weak light and a long time of exposure.[35] (2) The two-slit interference pattern is not simply the arithmetical sum of the patterns created by each single slit. It is quite possible for the two-slit pattern to possess a minimum in a place, say P, in which the one-slit pattern shows a finite intensity (see fig. 2). The first fact allows us to neglect the mutual interaction (if any) between the particles.

[33] See, e.g. E. Heitler, *Quantum Theory of Radiation* (Oxford, 1957), 65; D. Bohm, *Quantum Theory* (Princeton, 1951), 97ff.

[34] An example is A. Landé, 'From Duality to Unity in Quantum Mechanics' in *Current Issues in the Philosophy of Science*, ed. G. Maxwell (New York, 1961), 350ff.

[35] With respect to light this was shown by Janossy. See the booklet edited by the *Hungarian Academy of Sciences* (1957), where previous experiments are reported, as well as L. Janossy, 'On the Classical Fluctuations of a Beam of Light', *Nuovo Cimento*, 6 (1957), 111ff, and A. Adam, L. Janossy and P. Varga, 'Coincidences between Photons contained in Coherent Light', *Acta Phys. Hung.*, 4 (1955) 301ff.

Considering the second, we may now reason as follows: if it is correct that each particle always possesses a well-defined trajectory, then the finite intensity in P is due to the fact that some particle E wandered along b and ended up in P. As long as slit 2 remains closed there will always be some particles which, having passed slit 1, will travel along b. Now consider such a particle E and assume that it is about to enter slit 1. If we open slit 2 at this very moment we have thereby created conditions which are such that E must not arrive at P. Hence, the process of opening slit 2 must lead to a change in the path of E. How can this change be accounted for?

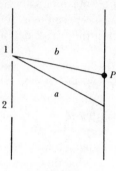

Fig. 2

It cannot be accounted for by assuming action at a distance. There is no room in the conservation laws (which remain valid also in the quantum theory) for energies deriving from such action. Furthermore,[36] the alleged action works not everywhere in space but only along those surfaces which in the wave picture are surfaces of equal phase, and reference to it is therefore nothing but a misleading way of bringing in the wave picture.

According to Popper[37] and Landé[38] the change of path of the *individual* particle is not in need of explanation. What *can* be explained, by reference to the change of physical conditions (opening of slit 2) is the emergence of a new stochastic process that leads to a new interference pattern. This position is indeterministic as it admits the existence of spontaneous individual changes and its indeterminism is about as radical as that of the Copenhagen point of view. It also shares with that point of view an

---

[36] The idea of action at a distance has been discussed, and regarded as a possible explanation by Hans Reichenbach. See his *Philosophic Foundations of Quantum Mechanics* (Berkeley and Los Angeles, 1945), section 7. Action at a distance is not the solution Reichenbach himself adopts. For an evaluation of Reichenbach's analysis and of his own solution (three-valued logic), see ch. 15.

[37] *Observation and Interpretation*, ed. S. Körner (London, 1957), 65ff.

[38] Landé's suggestions are in many respects similar to those of Popper. Originally, in *Quantum Theory, A Study of Continuity and Symmetry* (New Haven, 1955), esp. 24ff) Landé adopted the assumption of the indefiniteness of state descriptions. In his later publications he dropped this assumption.

emphasis on the importance of the experimental situation: predictions are valid only under certain experimental conditions; they are not unconditionally valid. It differs from the Copenhagen Interpretation in that it works with well-defined trajectories. This being the case it must, and does,[39] admit that the conservation laws are valid only for large ensembles of particles in a certain situation and that they may be violated in the individual case. It is here that the difficulties arise. For energy and momentum are conserved in each individual case of interaction also for elementary particles.[40] This position must therefore be rejected *unless* it is developed in such detail that an account can be given of all those experiments which have convinced the physicists of the individual validity of the conservation laws. Until such a detailed account can be given (and nobody can say in advance that it is impossible) we must again regard the assumption of the indefiniteness of state descriptions as the only satisfactory account. Let it be noted, by the way, that this restriction applies to all the arguments we are going to develop in favour of the assumption of the indeterminateness of state descriptions and the instrumental character of any quantum theory which describes states with the help of functions which are well defined from the mathematical point of view. All these arguments presuppose the validity of certain experimental findings such as the quantum postulate, the laws of interference, and the individual validity of the conservation laws – and the arguments shows that *given* these experimental findings we are forced to adopt instrumentalism.

This demonstrates what has been emphasized before, that the instrumentalism in the quantum theory is not a purely philosophical affair that can be disputed away by general arguments in favour of realism. The argument 'Quantum theoretical instrumentalism is a result of positivism; positivism is false; hence, we must interpret the quantum theory in a realistic fashion',[41] is completely irrelevant and also very misleading. It is very misleading because it suggests that a realist can at once interpret the ψ-function realistically, and that the reason why it was not so interpreted was only philosophical prejudice. And it is irrelevant because it does not proceed a single step on the way to resolving the *physical* difficulties which are connected with the realistic position in microphysics. A realistic alternative to the idea of complementarity is likely to be successful only if it implies that certain experimental results are not strictly valid. It therefore demands the construction of a *new theory*, as well as demonstration that this new theory is experimentally at least as valuable as the theory that is being used at present. This is a formidable task indeed, and a task that is not even

[39] K. R. Popper, private communication.
[40] This was established by the experimental follow-up of the Compton-effect, and especially by the experiments of Bothe and Geiger, and Compton and Simon.
[41] See n.29, above.

recognized by the purely philosophical champions of realism in microphysics.

But the situation is even more complex. We have already pointed out that physicists, too, use the very same philosophical arguments which philosophers think to be the only reasons for their having accepted instrumentalism. It is as if they did not really trust their very forceful physics and as if they needed support from something more 'fundamental'. Moreover it is the belief of many followers of the 'orthodox' point of view that a theory which provides a realistic alternative to the present quantum theory is impossible, either for logical reasons or for empirical reasons. They therefore not only suggest an interpretation of the known experimental results in terms of indefinite state descriptions and instrumentalism. They also suggest that this interpretation *be retained forever*, and that it be the foundation of any future theory of the microlevel. This is another mistake.[42] Small wonder that philosophers find it difficult to unearth the valuable core of their argument from beneath all the philosophical rubbish that conceals it.

9. THE INTERPRETATION OF WAVE MECHANICS

What we have said so far has immediate application to the interpretation of wave mechanics. The older quantum theory, although experimentally very successful and also extremely useful in its power to unite a host of otherwise disconnected facts, was nevertheless always regarded as unsatisfactory by many physicists. Its main fault was seen to lie in the manner in which it combined classical and non-classical assumptions, making a coherent interpretation impossible. For many physicists it was therefore nothing but a stepping stone on the way to a really satisfactory theory, i.e., to a theory which could give us not only correct predictions but also some insight into the nature and dynamics of microscopic entities. It is quite true that Bohr, Kramers, Heisenberg and others worked along very different lines. Their main objective was not the construction of a new physical theory about a world that existed independently of measurement and observation. Rather, they sought to construct a logical machinery for the utilization of those parts of classical physics which could still be said to lead to correct predictions. The inspiration for this lay no doubt in the surprising fact that many classical laws remained *strictly valid* even on the quantum level (for example, the laws of interference discussed in section 8). This suggested that what was needed was not the elimination and complete replacement of classical physics, but rather a modification of it. However that may be, the philosophical spirit behind the *Korrespondenzdenken* was not shared by everybody.

---

[42] For a more detailed account of this mistake, see section 7 of my 'Problems of Microphysics'.

De Broglie and Schrödinger tried to develop an entirely new theory for the description of the nature and the behaviour of atoms, molecules, and their constituents. When this theory was finished it was hailed by many as the long-expected coherent account of the microlevel. The hypothesis of the indefiniteness of state descriptions, so it was thought, had only reflected the indefiniteness and incompleteness of the early theory, and it was no longer necessary. More especially, it was assumed either that the states were now new, but well defined entities (the ψ-waves), or it was assumed that whatever incompleteness occurred was due to the statistical character of the theory, i.e. to the fact that wave mechanics was 'primarily a variety of statistical mechanics, similar to the classical statistical mechanics of Gibbs'.[43] These two interpretations still survive. I hope that our arguments in the preceding section have made it clear that such interpretations of wave mechanics are bound to lead to inconsistencies. The only presupposition of the hypothesis of indefinite state descriptions is the quantum postulate and the dual nature of light and matter (taken together with the individual conservation of energy and momentum). Both these facts are contained in the wave mechanics, which will therefore be equally in need of the said hypothesis and which will at most admit of an instrumentalist interpretation. A closer analysis of the two main alternatives shows that this is indeed correct.[44]

In summary: any attempt to give a realistic account of the behaviour of the elementary particles is bound to be inconsistent with highly confirmed theories. Any such attempt therefore amounts to introducing unsupported conjectures in the face of fact and well-supported physical laws. This is the main objection which is used today against the theories of Bohm, Vigier, de Broglie and others. It is similar to the objections which were raised, at the time of Galileo, against the idea that Copernicus should be understood realistically.

10. COMMON FEATURES OF THE COPERNICAN CASE AND THE QUANTUM CASE

To repeat: the objections to a realistic interpretation of the Copernican hypothesis and the objections to a realistic interpretation of quantum theory have this in common: they point out that such an interpretation, quite apart from not having any factual support of its own, is actually inconsistent with observation and well-confirmed physical laws. Now, in the case of the Copernican theory a new dynamics was found which was not only better than the Aristotelian dynamics – more detailed, and allowed for

---

[43] E. C. Kemble, *The Fundamental Principles of Quantum Mechanics* (New York, 1937), 55.
[44] For this, see section 3 of my 'Problems of Microphysics'.

the motion of the earth – but even led to dynamical arguments *in favour* of at least part of earth's motion (rotation). The whole development was accompanied by the discovery of further difficulties for the Aristotelian scheme (sun spots, new stars, path of comets, etc.). Thus the persistence of the Copernicans was finally rewarded and the belief in the basic correctness of their point of view justified. The realistic position triumphed as the result of laborious research and seemed thereby to be proved as essentially correct. Is this not a splendid argument for realism? Does it not show that the realistic position encourages research and stimulates progress, whereas instrumentalism is more conservative and therefore liable to lead to dogmatic petrifaction? This consequence has been drawn by many thinkers. This is the philosophy that allowed Boltzmann and the other proponents of the kinetic theory to remain persistent in the face of the sometimes quite formidable objections of their opponents; and this is also the philosophy that inspires the contemporary critics of the Copenhagen Interpretation. It is a very positive philosophy and a very optimistic philosophy. Personally I am very much inclined to accept it. At the same time there are grave difficulties, and even absurdities arise if one tries to think it through consistently. It is now time to face these difficulties, and perhaps to remove them.

11. THE FORCE OF EMPIRICAL OBJECTIONS

It is clear that the final success of the Copernican theory could not have been foreseen in the beginning. No set of methodological principles can ever guarantee the essential correctness of a theory that has just been introduced; this follows from Hume's investigations. And if the theory *contradicts* the accepted laws and facts then its future success would seem to be even less certain. The belief that such success is none the less bound to occur, which inspired the Copernicans, can therefore be based neither on methodological considerations nor on factual argument. It is a metaphysical belief. So if it was legitimate for the Copernicans to act on such a belief, then more recent metaphysical assumptions cannot be excluded either. For example, there is no reason why one should not now reintroduce Aristotelianism and hope for the best. The retort that this theory has already been given its chance does not count; the same was true of the hypothesis of the motion of the earth which was well known in antiquity and which was given up in view of the arguments of Aristotle and his followers. However, does not such an admission open the door to all sorts of wild speculations such as the hollow earth theory, Wilhelm Reich's Orgonomy, Dianetics,[45] astrology, and other crazy ideas? Is this not making illegitimate use of the success of

[45] For an amusing description of these and other odd theories, see Martin Gardner, *Fads and Fallacies* (New York, 1952).

Copernicus (and, so one might add, of Boltzmann and Schrödinger?)[46] Is there not a decisive difference between the Copernican conjecture and astrology? I agree: Copernicus has been very successful, astrology has not been quite so successful. But what I am talking about now is the attitude to be adopted *before* a theory has proved its fruitfulness. The objection assumes that the final success of Copernicus *could somehow have been foreseen* and that we know *in advance* that Orgonomy is completely fanciful, and hopelessly out of touch with reality. But how could we possibly possess such knowledge? Because the existence of Orgon is *inconsistent with contemporary physics?* Copernicus was inconsistent with the physics of his time in a very simple and straightforward manner. The case of Orgon, its relation to the rest of physics, is much more complicated and doubtful. Or shall we reject the Orgon idea or, to take an even better example, the hollow earth theory (the earth is hollow and we live in its inside) because it is *absurd?* Copernicus was regarded as absurd; read Luther, Sir Francis Bacon and the professional astronomers of the time. From all this it seems to emerge that from the point of view of their status *before* their success (or failure) there is not much to choose between the hypothesis of the moving earth in the infinite universe on the one hand, and the hollow earth theory, astrology, and Ehrenhaftian physics on the other. This is the difficulty inherent in the optimistic and naive realism which we have described above. It does not allow us to separate the real from the fanciful, the fruitful hypothesis from the products of the crank. How can this difficulty be resolved?

12. CONTRADICTION OF OLD FACTS AND NEW IDEAS IS NO ARGUMENT AGAINST THE LATTER

I think it can be resolved only if we are prepared to give up certain very deep-rooted prejudices concerning the nature of empirical support. It is not at all difficult to give up these prejudices once the matter is put in the right light. In order to be able to do this we must first abandon the attitude that the Aristotelian physics was just a heap of rubbish which no careful thinker would ever have supported, and that the inconsistency between it and the heliocentric hypothesis could therefore not at all be counted as an argument against the latter. It must again be emphasized that from the point of view of empirical method the Aristotelian physics was as good as any theory that could have been devised at that early time. It was partly supported by the evidence then available, it was confronted with certain difficulties – and it was in all these respects very similar to the more detailed theories we possess at present.[47] It was therefore completely legitimate to use it in the

---

[46] For the very interesting history of Schrödinger's early attempts to solve the problem of atomic spectra see Dirac's letter in *Sci. Mon.*, 79 (1954), no. 4.
[47] See n.4, above.

empirical arguments purporting to show the untenability of the Copernican hypothesis. At the same time the eventual success of the latter hypothesis shows what empirical arguments of such a kind are worth. It shows that such arguments are not at all final and irrevocable and that it is possible to revise their premises in the light of further research. And this is not surprising. After all, the laws which are used in such arguments (the laws of the Aristotelian dynamics in the case of the Copernican hypothesis; the second law of thermodynamics in the case of the kinetic theory; the conservation laws, the quantum postulate, and the interference laws in the case of the quantum theory) always go far beyond what could have been shown by experience. They use precise concepts where experience can at most give imprecise information; and they are general statements where experience can at most give rise to a finite number of singular statements of observation.[48] A new point of view which contradicts these laws therefore need not on that account alone be factually inadequate, provided the inconsistency occurs inside the domain of imprecision and outside the class of known facts. Even a singular statement of observation need not be regarded as final and irrevocable. Any such statement will use terms which are part of a fairly comprehensive conceptual system whose postulates have either been explicitly formulated or function implicitly as the 'rules of usage' according to which the terms are habitually used. Now a term which is used for describing the result of a (direct) observation will obtain its meaning partly from the impression created by the observed situation, partly from the postulates of the conceptual system to which it belongs. Hence, although it may adequately express what is observed it may still be inadequate if some of those postulates have been found to be incorrect. This being the case, not even a direct observational report is exempt from criticism and reformulation.

Considering all these possibilities of change, the discrepancy between a new point of view on the one hand and accepted theories and observations on the other does not at all decide the fate of the new point of view – and this even if the accepted theories should be as well confirmed and as precisely formulated as is the second law of thermodynamics or the law of conservation of energy. Any such discrepancy creates a problem that must be further investigated in order to ascertain whether it is indeed a contradiction between fact and theory that is here being revealed and not, rather, a contradiction between one theory and the as yet untested part of another theory, or between one theory and the as yet untested part of a principle contributing to the meaning of a key term in an observational statement. Nobody can say in advance where such further investigation will lead and nobody should therefore allow a controversy between the point of view he

[48] For a more detailed account, see ch. 4.6 and ch. 12, as well as section 4 of my paper 'How to Be a Good Empiricist', *Delaware Studies in the Philosophy of Science* (New York, 1963), 1.

likes on one side and facts plus well established theories on the other to deter him from carrying out such an investigation.

13. OBJECTIONS AGAINST A CERTAIN WAY OF TREATING THE CONTRADICTION

It is here, by the way, that the distinction between 'respectable' people and cranks must be drawn. The distinction does not lie in the fact that the former suggest what is plausible and promises success, whereas the latter suggest what is implausible, absurd, and bound to fail. It *cannot* lie in this because we never know in advance which theory will be successful and which theory will fail. It takes a long time to decide this question, and every single step leading to such a decision is again open to revision. Nor can the absurdity of a point of view count as a *general* argument against it. It is a reasonable consideration for the choice of one's own theories to demand that they seem plausible to oneself. This is one's private affair, so to speak. But to declare that only plausible theories should be considered is going too far. No, the distinction between the crank and the respectable thinker lies in the research that is done once a certain point of view is adopted. The crank usually is content with defending the point of view in its original, undeveloped, metaphysical form, and he is not at all prepared to test its usefulness in all those cases which seem to favour the opponent, or even to admit that there exists a problem. It is this further investigation, the details of it, the knowledge of the difficulties, of the general state of knowledge, the recognition of objections, which distinguishes the 'respectable thinker' from the crank. The original content of his theory does not. If he thinks that Aristotle should be given a further chance, let him do it and wait for the results. If he rests content with his assertion and does not start elaborating a new dynamics, if he is unfamiliar with the initial difficulties of his position, then the matter is of no further interest. However, if he does not rest content with Aristotelianism in the form in which it exists today but tries to adapt it to the present situation in astronomy, physics, and microphysics, making new suggestions, looking at old problems from a new point of view, then be grateful that there is at last somebody who has unusual ideas and do not try to stop him in advance with irrelevant and misguided arguments.

14. TREATED CORRECTLY, THE CONTRADICTION CAN BE MAINTAINED FOR A CONSIDERABLE TIME

I think it is clear now *that there is no harm* in proceeding as Copernicus did, and as Bohm does, in introducing unfounded conjectures which are inconsistent with facts and accepted theories and which, moreover, give the impression of absurdity – provided the suggestion of such conjectures is followed up by detailed research of the kind outlined in the preceding section. There is no harm in proceeding in this manner. But we have not yet given a single reason why one *should* proceed in this manner. After all, there are many activities which are not harmful but on which nobody would want to waste his time. We can, of course, try to defend such a procedure by reference to its *possible* success. But the accepted theory might also continue to succeed in the future. It might survive all difficulties, and if it does then there is no need to introduce new ideas and get involved in the laborious investigations connected with such a procedure. Hence, the best method would seem to be to wait until the current theory gets into difficulties and only *then* to start looking for new theories.

This opinion is widely accepted on account of its apparent reasonableness. It has a lot to recommend it *provided the difficulties which may beset a theory will always be discovered without any help from other theories*. It is only then that the advice to wait until the accepted point of view has collapsed will be feasible. It can be shown, however, that *there exist potential difficulties for any theory that can be detected only with the help of further theories*. If this is correct, then the development of such further theories is demanded by the principle of testability, according to which it is the task of the scientist relentlessly to test whatever theory he possesses, and it is also demanded that these further theories *be developed in their strongest possible form, i.e. as descriptions of reality rather than as mere instruments of successful prediction*. Even where it contradicts the accepted views and facts, realism can be justified by methodological considerations. In the next section we shall show how development of additional theories may increase the testability of the accepted point of view.

15. AN ARGUMENT FOR MAINTAINING THE CONTRADICTION

The argument is very simple. Consider a theory T which makes predictions P in a domain D, and assume also that the actual state of affairs P' is different from P but to such a small extent that the difference is far below the experimental possibilities. In this case T is incorrect without our being able to discover this incorrectness. One may now hope that development of new experimental methods will eventually reveal that P' obtains, and not P. Now when the difference between P' and P is small enough this hope would seem to be as unrealistic (or as realistic) as the hope that invention of a new theory might lead to giving a better account of what is going on in nature. After all, the development of instruments of measurement is guided by the ideas and interests of scientists, and it is very unlikely that it will automatically lead to the discovery of all the shortcomings of the theories we possess. Moreover, there are cases where construction of instruments for the direct detection of differences between P and P' *is excluded by laws of nature* and therefore impossible. The behaviour of electrons inside an atom can never be investigated by a direct test, i.e. by a test of the kind one would carry out if Newton's celestial mechanics and Coulomb's law were the only theories in existence and the ideas connected with these theories the only ideas we possessed. Nor is it possible by a direct test to discover that Brownian motion is due to a transfer of heat from the embedding fluid to the particle, which means that the difficulties presented for the second law of the phenomenological theory of thermodynamics by the existence of the Brownian particle would in this way never be revealed.[49] Thirdly, it is very unlikely that the discovery of a discrepancy would at once lead to its correct interpretation. Many such discrepancies, if they were small and irregular enough, would be regarded as oddities much in the same fashion as Ehrenhaft's astounding effects are today regarded as oddities rather than as refuting instances for part of contemporary physics. All these circumstances work in the same direction – they tend to hide from us the weaknesses of a theory in which we believe.

Assume now that we introduce alternative theories T', T'', etc., which are inconsistent with T inside D and which predict P' rather than P. Now if we succeed in elaborating one of these theories in such detail that it can be compared with T as regards simplicity and effectiveness, if this theory is confirmed where T was confirmed, if it solves some cases which belonged to the class of unsolved problems of T (see n.4 above), if it makes predictions which are not made by T, and if these predictions are confirmed as well then we shall take T' as our measure of truth and regard T as refuted – and this despite the fact that no *direct* refuting instance has as yet been found for T. This is why the invention of new theories which are inconsistent with the accepted point of view is demanded by the principle of testability, and this is also the promised methodological justification for realism.

16. REALISM IS ALWAYS PREFERABLE TO INSTRUMENTALISM

To sum up: the issue between realism and instrumentalism has many facets. There are arguments of a philosophical kind which demand that theories be regarded as instruments of prediction only and not be used for inferences concerning the structure of our universe. In the present paper such arguments were only mentioned; they were not discussed. There are other arguments for instrumentalism which concern specific theories such as the quantum theory or the heliocentric hypothesis and which are based upon specific facts and well confirmed theories. It was shown that to demand realism in *these* cases amounts to demanding support for implausible conjectures which possess no independent empirical support and

[49] For details, see ch. 4.7.

which are inconsistent with facts and well confirmed theories. It was also shown that this is a plausible demand which immediately follows from the principle of testability. Hence realism is preferable to instrumentalism even in these most difficult cases.

12

A note on the problem of induction

The so-called 'problem of induction' comprises many different but related questions. This variety is in part due to the difficulty of finding a satisfactory solution: older formulations are given up and are replaced by weaker problems in the hope that what is impossible to prove in the strong case might perhaps yield to proof in a weaker case. Roughly, the development is as follows.

Originally it was believed that the conjunction $P(a_1)$ & $P(a_2)$ ... & $P(a_n)$, or $P(n)$ for short, could in some way guarantee the *truth* of $(x)P(x)$. (The predicate 'P' occurring here may be expressed in ordinary English ('being black, provided one is a raven'); it may be expressed in terms of physics ('moving on a straight line with constant speed with no forces present'); or in terms of some other discipline. This way of defining P allows us to state any theory in the form $(x)P(x)$.) This assumption, which I shall call the *simple generalization hypothesis*, leads to this programme: to discover, and to state explicitly, the specific inferences according to which $(x)P(x)$ can be obtained from $P(n)$. The hypothesis was refuted by Hume, who also showed that the corresponding programme could not be carried out.

Next, the simple hypothesis was replaced by the assumption that $P(n)$ might guarantee a *high probability* (in the objective sense) of $(x)P(x)$. Hume's argument refutes this hypothesis also (the disproof was provided already by Hume himself).

The breakdown of both the simple and the probabilistic generalization hypotheses led to the search for a formulation of the key statement of the problem of induction which would be weak enough to escape refutation by Hume's arguments. The formulation which provides the background of much contemporary thought on the problem and which we shall call the *modified generalization hypothesis* asserts that, given $P(n)$, it is *reasonable* to adopt $(x)P(x)$. I shall now make a few comments on this modified hypothesis.

First, it is important to point out that this hypothesis is no longer concerned with the truth, or even with the probability (in the objective sense) of the generalization whose use it recommends. It asserts that it is *reasonable* to generalize a predicate that has been found to be instantiated in

a finite number of cases. It does *not* assert that the result of the generalization will be *true*, or even *highly probable*. It is particularly important to make
this point. The modified hypothesis has often been misunderstood as
justifying an expectation of *success*. Even the most careful thinkers are
sometimes found to believe that proof of the modified hypothesis gives them
the right to expect success, or success in the long run. This of course means
regressing to the simple generalization hypothesis or to the probabilistic
generalization hypothesis, which have both already been refuted.

Secondly, it should be realized that the modified hypothesis is behind
almost all recent attempts to solve the problem of induction (this
includes Popper's theory). All these attempts in one way or another aim
to prove the hypothesis. This is certainly true of the argument that the
modified hypothesis agrees with commonsense. But it is also true of the
more complex attempt to show that, given $P(n)$, $(x)P(x)$ has a higher degree
of confirmation than any one of its alternatives (where by an alternative to
$(x)P(x)$ we mean a statement that is as general as $(x)P(x)$ but entails
$(\bar{x})P(x)$). The concentration upon $(x)P(x)$ and the neglect of alternatives is
found even in Hume. One almost never starts with $P(n)$ and asks *what*
generalization should be adopted. One *takes it for granted* that adopting
$(x)P(x)$ is the right thing to do, and one looks for some plausible argument
supporting this belief.

Thirdly, it is clear that the modified hypothesis has a much greater
chance of succeeding than any one of its more demanding predecessors.
What is reasonable or not is a notoriously vague affair, and it is easy first to
define (either by explicit stipulation or by uncritical adoption of commonly
accepted standards) a 'reasonable procedure' in terms of direct generalization from finite evidence, and then to obtain the modified generalization
hypothesis by an analysis of the procedure thus defined. If definition and
analysis are separated by many steps, or by many years, then the circularity
will not be at all disturbing; it will rather create the impression that a
particularly solid foundation has been found for the hypothesis. It needs of
course only little thought to realize that, circularity or no, the standards
implied in common behaviour are themselves open to criticism and that it is
the task of the philosopher to provide such criticism, and not to be satisfied
with popularity.

Assume, now, fourthly, that the modified hypothesis is *false;* i.e. assume
that given $P(n)$ it is *not* reasonable to choose $(x)P(x)$ over all its possible
alternatives and that it *is* reasonable to consider at least some of those
alternatives. Such a result would refute the simple generalization hypothesis and the probabilistic generalization hypothesis and would also be
stronger than Hume's disproof of them. The first point follows from the fact
that it is desirable, or reasonable, to choose what is known to be true or
highly probable over what is known to be false or of low probability. For if it

is *not* desirable, or reasonable, to choose $(x)P(x)$ over any one of its alternatives, then the truth or the high probability of $(x)P(x)$ cannot have been
established, as this would mean that it was desirable to choose what is
known to be false in addition to what is known to be true, and to choose
what is known to have a low probability in addition to what is known to
have a high probability. The second point becomes clear if we consider that
Hume showed the *impossibility* of obtaining the truth or the high probability
of $(x)P(x)$ given $P(n)$, whereas refutation of the modified generalization
hypothesis would in addition show the *undesirability* of such an achievement. It would not only show that the problem of induction *cannot* be solved;
it would also show that it *should not* be solved.

Having explained the role of the modified generalization hypothesis, I
now proceed to show that this hypothesis can indeed be refuted and that the
procedure suggested by it can be shown to be undesirable.

The argument will be in three steps. I shall first discuss abstractly an
individual case of a rather peculiar nature. I shall then give an example that
exhibits all the features described in the abstract discussion. Finally, I shall
show that the defender of *any* theory must act as if his theory possessed these
features. This final step will refute the modified generalization hypothesis.

The abstract discussion is as follows. Consider a theory T (expressed, as
indicated in the beginning, by universalizing the property P) which entails
that F. Assume that actually F' (where 'F takes place' is inconsistent with
'F' takes place'). Assume also that the laws of nature forbid the existence of
equipment for distinguishing F and F'. The theory T is then obviously false;
only we shall never be able to discover this by a consideration of 'the facts'
only.

An example that vividly illustrates this situation is the phenomenon of
Brownian motion. This phenomenon refutes the second law of the phenomenological theory of thermodynamics: the Brownian particle is a
machine that achieves what the second law says should not occur. It
absorbs heat from the surrounding fluid and transforms it directly into (its
own) motion. Still, it is impossible to show in a direct way that a refutation
is taking place. The reasons are as follows. It is impossible to follow in detail
the path of the particle in order to detect the amount of work done against
the fluid (this is connected with the fact that the Brownian motion obeys an
uncertainty relation very similar to the relations known from quantum
theory, the diffusion constant of the medium replacing the quantum of
action). And it is also impossible to measure the amount of heat lost by the
fluid (this is due to the fact that the fluctuations to be measured will
inevitably be overlaid by the fluctuations of the thermometer. This 'noise'
depends on the temperature only and cannot be eliminated). Result:
Brownian motion violates the second law, but the facts are such that they
do not allow us to discover the violation. There are many other cases which

show exactly the same features (an example would be the motion of an electron in the shell of an atom). What is the solution? To explain it, we return to the abstract argument.

Assume that, in addition to T, we introduce another theory T' (expressed by universalizing a property whose universalization entails $(\bar{x})P(x)$) which covers the facts supporting T, makes successful additional predictions A, and entails that F'. The test of the additional predictions may be regarded as an indirect proof that F' and, thereby, as an indirect refutation of T.

In the case of our example it was Einstein who resolutely used the kinetic theory of matter (which plays the role of the alternative theory T') for calculating those properties A of the Brownian motion which can be checked by experiment (the main prediction was the proportionality between time and the mean-square displacement of the particle) and who thereby made an indirect refutation of the phenomenological theory possible.

It is clear that the existence of cases like the one we have just discussed cannot be ascertained in advance. *Any* theory T under consideration (including that zero-case of theory construction 'All ravens are black') may be inconsistent with facts which are accessible only indirectly, with the help of an alternative T'. Now it is surely reasonable to demand that the class of refuting instances of a given theory be made as large as possible and that especially those facts which belong to the empirical content of the theory, which refute the theory but which cannot be distinguished from similar, but confirming facts, be separated from the latter and be thus made visible. However, this means that, given $P(n)$, it is reasonable to use not only $(x)P(x)$, *but as many alternatives as possible*. This is the promised refutation of the modified generalization hypothesis.

As we have said above, this refutation demands of us a completely new attitude towards the so-called 'problem of induction'. The fact that the problem is so difficult to solve need not worry us any longer. As a matter of fact, we should rejoice that we are not restricted, by some proof, to the use (given $P(n)$) of *one* generalization only and are thus able to discover some perhaps decisive shortcomings of this generalization.

13

On the quantum theory of measurement

1. THE PROBLEM

Within classical physics the relation between physical theory and ordinary experience was conceived in the following way: ordinary experience is something that can be described and understood in terms of physics. Such a description involves, apart from physical theory, certain approximations. But the conditons under which those approximations apply (they correspond to the initial conditions of, say, celestial mechanics), taken together with the theory, are supposed to be sufficient for giving a full account of ordinary experience. Consequently, classical theory of measurement (which, like any other theory of measurement, links together terms of ordinary experience and theoretical terms), is a piece of applied physics and all processes which happen during measurement can be analysed on the basis of the equations of motion only.

When we enter quantum mechanics (QM), we are apparently presented with a completely different picture. For according to the current interpretation of elementary quantum mechanics, ordinary experience – and this now means classical physics – and physical theory (and this now means QM), belong to two completely different levels and it is impossible to give an account of the first in terms of the second.[1] Any transition from the quantum level to the level of classical physics must be taken, not as a transition *within* QM from the general to the particular, but as an essentially new element which is incapable of further analysis. Consequently, the quantum theory of measurement, as it has been developed by Bohr, Heisenberg,[2] and in its most elaborate form by von Neumann, involves, apart from the equations of motion, such independent and unanalysable processes as 'quantum-jumps', 'reduction of the wave-packet' and the like.

It is the purpose of this paper to show the inadequacy of this theory of measurement. I shall try to show that it is possible to give an account of the process of measurement which involves nothing but the equations of motion and statements about the special properties of the systems involved, especially statements about the properties of the measuring device; and according to which the theory of measurement is a piece of applied physics,

---

[1] Cf. e.g. [1] ch. 23.   [2] See [12], [13].

just as it was in classical theory. As a satisfactory account of the classical level in terms of Q M is still missing, my suggestions will have to be somewhat sketchy – but they may still be useful, at least as an indication of how a more satisfactory theory of measurement may be built up. The last section of the paper will also develop some more general consequences of the possibility of such a theory.

2. VON NEUMANN'S THEORY OF MEASUREMENT

My point of departure is von Neumann's theory of measurement.[3] The essential point of this theory is that it is about the behaviour of *individual systems*, although only a probabilistic account is given of this behaviour. Probability is introduced in two steps. First, in the usual way, viz. as the limit of relative frequencies within large ensembles.[4] This procedure which has greatly clarified the role of probability within Q M, is suggested (a) by the fact that the Born interpretation, 'the only consistently enforceable interpretation of quantum mechanics to-day',[5] is a statistical interpretation in a straight-forward and classical sense,[6] and (b) by the fact that the statistical properties of large ensembles can be studied by experiments upon small samples, whatever happens during those experiments. The statistical properties of the ensembles are completely characterized by their statistical operators. In the second step which is usually overlooked by those who interpret Q M as a variety of classical statistical mechanics, it is proved (a) that every ensemble of quantum-mechanical systems is either a pure ensemble, or a mixture of pure ensembles, and (b) that the pure ensembles (1) do not contain subensembles with statistical properties different from their own, (2) are not dispersion-free. This proof allows for the application, to individual systems, of probabilities in the sense of relative frequencies.[7] Hence, any operation with an *ensemble* which leads from a statistical operator W to another statistical operator W', can also be interpreted as an operation with an *individual system* (which may or may not be completely known), leading from the state W to the state W' and vice versa, a procedure which would not be possible in the classical case.[8]

---

[3] See [18], chs. 4, 6. Most of von Neumann's results use only the formalism of Q M together with the Born-interpretation. This is especially true of the famous 'Neumann Proof' which contains none of the more philosophical elements characteristic of Bohr's approach. This is also the reason why the proof does not achieve its aim. Cf. ch. 16.8, n.90 and text, as well as n.6 below.

[4] Von Neumann adopts the Mises interpretation of probability. Cf. [18], 289n. There is no objection against this procedure as within physics the Mises approach does not lead to any of the difficulties which have made it untenable for mathematicians and philosophers.

[5] [18], 210.

[6] Cf. also [6].

[7] This is the 'Neumann Proof'.

[8] Attempts have been made (cf. e.g. [20]) to justify such a procedure on the basis of a slightly

The theory of measurement proper rests upon the assumption that the state of a system may undergo two different kinds of changes, viz. changes which are continuous, reversible and in accordance with the equations of motion – those changes happen as long as the system is not observed, however strong its interaction with other systems may be; and changes which are discontinuous, irreversible, not in accordance with the equations of motion and which happen as the result of a measurement. Two arguments are presented for this assumption, an inductive argument, trying to relate the existence of the discontinuous jumps to experience, and a consistency argument which shows how they can be fitted into the theory without leading to contradictions. The inductive argument is invalid and will not be discussed here.[9] The consistency argument which may be said to contain von Neumann's theory of measurement, may be stated thus: assume that the observable R (eigenfunctions $|\varphi_1\rangle$, $|\varphi_2\rangle$, . . . , eigenvalues λ_1, λ_2, . . .; we shall assume that the spectrum is discrete and nondegenerate) is measured in S (initial state $|\Phi\rangle$). If the result of the measurement is taken into account, or, as we shall express ourselves, if a *complete measurement* is performed, then the statistical operator of S will change from P_Φ to $P_{\varphi i}$, λ_i being the eigenvalue found (we assume that all measurements are well designed). If we make a measurement without taking notice of the

modified interpretation of the concept of probability. In this modified interpretation it is possible to apply probabilities also to individual systems. This is done e.g. by ascribing to a single die the *propensity* to exhibit a certain distribution (e.g. 1/6, 1/6, 1/6, 1/6, 1/6, 1/6) when suitably thrown (i.e. when thrown in such a way that the conditions of equiprobability are fulfilled). In this interpretation the statement 'die X has the probability of showing six in 20 per cent of all cases' is no longer equivalent to the statement '20 per cent of all throws with die X are 6' – for conditions may have been realized which led to a different distribution, in spite of the fact that the probability (in the sense of propensity) was as indicated. This interpretation is still not sufficient for the purpose of Q M where we want to use ensembles in order to get information about the *actual state* of a system rather than about its *ability*, or *disposition* to produce such a state with a certain relative frequency, when handled in a suitable way. For within the propensity interpretation we can still assume that a system which possesses a propensity $0 < x < 1$ to be in state A is, in fact, in state A whereas the same probability statement, made with respect to a pure state, *excludes* such an assumption (or its negation, viz. that the system is *not* in state A). Hence, turning frequencies into propensities does not strengthen the connection between statistical assertions and assertions about the individual system, it leaves it unchanged. This change of interpretation of probability has therefore no influence upon the problems of quantum mechanics. Nor is it possible to compare the quantum-mechanical 'reduction of the wave-packet' with the sudden change of of probabilities which occurs whenever the evidence is changed. Cf. ch. 16.5.

It should also be noted that the peculiar property of the ensembles of quantum-mechanical systems which was mentioned in the text does not necessitate a revision of the concept of probability in the sense of relative frequency (or propensity). For the property of being not dispersion-free and yet irreducible is not at variance with the property of being a collective (in von Mises's or in Wald's sense). It is an additional property which is satisfied by special collectives. Cf. ch. 16.8.

[9] The argument is based upon an analysis of the Compton-effect. It is invalid as it takes into account only the classical features of this effect and completely neglects its finer, quantum-mechanical properties. Cf. ch. 17.6.

result, i.e. if we make what we shall call an *incomplete measurement*, then we can only say that λ_i will appear with the probability $|\langle\Phi|\varphi_i\rangle|^2$ which means that an incomplete measurement introduces the transition

$$P_\Phi \to \Sigma_i \mid \langle\Phi|\varphi_i\rangle|^2 P_{\varphi i} = \Sigma_i P_{\varphi i} P_\Phi P_{\varphi i} \tag{1}$$

This process contains only part of what is happening during the measurement. For (1) compares the initial state of S with its state immediately after the end of the interaction, without considering this interaction itself. We shall therefore call the account of measurement which consists in asserting that transition (1) has happened the *direct account*. In order to show the consistency of this direct account with the equations of motion it must now be demonstrated that the interaction between M and S, taken together with a direct account of a measurement of the apparatus-variable R' (performed immediately after the interaction) produces, with respect to S, the same result, as the direct application of (1) to S. It can be shown that for well-designed measurements this problem is equivalent to the problem whether it is possible to choose an H (of $\{S+M\}$) and a T such that

$$\exp\left[(-i/\hbar)\ HT\right]/\Phi\Psi\rangle = |\Phi\Psi\rangle' = \Sigma_i\langle\Phi|\varphi_i\rangle|\varphi_i\rangle|\psi_i\rangle \tag{2}$$

and this problem is easily solved – at least theoretically. This completes von Neumann's theory of measurement.

It should be kept in mind that this theory does not lead to the elimination of process (1). It is not shown that the mixture in (1) emerges from P_Φ on the basis of nothing but the initial states of M and S together with the equations of motion for the combined system $\{S+M\}$. It is only shown that the direct account of a *further* measurement, viz of R', leads, with respect to S, to the same result as if one gave a direct account of the measurement of R in S from the very beginning. Hence the jumps (1) are still an irreducible element of the theory. Are they a necessary element? This question must be answered by a more detailed analysis of the current theory. In order to carry out such an analysis we first present this theory in a suitable form. This will be done in the next section.

3. STAGES OF MEASUREMENT

We start with the following two assumptions: (A) R has only two eigenstates, $|\varphi_1\rangle$ and $|\varphi_2\rangle$, and

$$|\langle\Phi|\varphi_1\rangle|^2 \neq \mid \langle\Phi|\varphi_2\rangle|^2 \tag{3}$$

This assumption is introduced in order to simplify the argument, but (3) will be dropped in section 5; (B) the states of M which correspond to $|\varphi_1\rangle$ and $|\varphi_2\rangle$ (the eigenstates of R') are macroscopically distinguishable. In order to fix our ideas we may assume that M contains a pointer which is capable of only two positions, A and B (for example up and down) and that A and B can be easily registered by some means as taking a photograph, asking somebody else to look, looking oneself, etc.

Using these simplifications and taking into account the interaction between S and M, we may now distinguish three stages of a measurement of R in S.

Stage 1. Interaction between S and M until the combined state of $\{S+M\}$ is of the form (2).

In classical physics this stage is the only interesting one. For according to the classical point of view, S as well as M are now in a well-defined, though unknown, state. The remaining processes (looking at M, etc.), although of importance for the observer (they inform him about the state of M and thereby about the state of S) cannot have any decisive influence upon what is already fixed in M: The pointer is already pointing up, or down, and we need only to look in order to discover what is already there in nature, i.e. in M.

Within QM this classical point of view leads into difficulties as is easily seen from the following considerations: According to (2) the state of $\{S+M\}$ at the end of stage 1 is given by $P_{|\Phi\psi>} = P'$ (for short). On the other hand the assumption that M is already in a well-defined, though unknown, state means, within the formalism of QM, that $\{S+M\}$ is the mixture of $P_M = |\ <\Phi|\varphi_1>|^2 P_{\varphi_1} P_A + |\ <\Phi|\varphi_2>|^2 P_{\varphi_2} P_B$. And as in general

$$P_M \neq P' \qquad (4)$$

this latter assumption must be given up: we must not say that at the end of the interaction M (or S) is in a well-defined, though unknown, state.

This fact has sometimes been used as an argument in favour of the assumption that quantum mechanics is an incomplete theory which must be supplemented by additional parameters in order to become a complete theory.[10] This is the argument: at the end of stage 1 the pointer is already in a well-defined position; P' does not allow us to draw this conclusion. Hence, the description on the basis of P' is incomplete; and as in the situation described QM cannot provide us with a better description, QM is itself incomplete. In this argument it is overlooked that P' is not 'too poor' to provide us with such a description since it explicitly *excludes* such a description. It is therefore not possible to interpret transition (1) or the transition

$$P' \rightarrow P_M \qquad (5)$$

as a substitution of an incomplete description by a more complete description which does not correspond to an change in reality. Transition (5) must

---

[10] The argument is Einstein's. It is frequently used by the more recent opponents of present-day quantum mechanics.

be interpreted as a real process. This is the most decisive argument in favour of the existence of 'quantum-jumps'.

Stage 2. This involves two elements, namely (a) the incomplete measurement of R' and (b) the direct account of this incomplete measurement.

Since R' is a macroscopic variable (a) does not create any technical difficulties. Making an incomplete measurement of R' means measuring R' without taking the result into account; for example making a photograph of M and destroying it; asking an observer to watch M and not listening to his report; looking oneself at M without thinking. Such a measurement is a physical (or biological) process which does not involve any element of consciousness. If we treat this measurement as a further interaction between, say, $\{S+M\}$ on the one side, and light striking a photographic plate on the other, we have, on account of the above argument, to assume that the plate, even if developed, will not contain a definite picture of M. This indefiniteness will spread, the more objects interact with M, and it may even reach the mind of the conscious observer, making it impossible to say that he has received definite information, however certain he himself may feel about it.[11] It is not before we give a direct account of one of the stages described that we can return to attributing definite properties (positions, perceptions) to well-defined objects (measuring instruments, observers). Hence, this step, i.e. transition (5) is necessary for connecting the quantum level with the classical level (the theory with experience – in Heisenberg's discussion of measurement).

Stage 3. Completion of the measurement of R' in M. Result: S (as well as M) is left in a well-defined (and known) pure state.

4. DIFFICULTIES

This theory of measurement is unsatisfactory for the following reasons.

(i) R' is a macroscopic observable. The values of macroscopic observables are fixed independently of any account which is given of their observation. The theory does not yield this result, not even as a first approximation.[12]

(ii) Assume that this first difficulty has been solved, i.e. assume that it is possible to characterize the state of $\{S+M\}$ by P_M already at the end of stage 1. Assume, furthermore, that equation (3) does not hold. Then we are still unable to say the pointer is either pointing up or pointing down as now every linear combination of A and B is compatible with P_M. This shows that

---

[11] It is frequently assumed that M and S both jump into one of the eigenstates of R' and R respectively as soon as the observer *looks* at M (*esse est percipi*). The above considerations show that this is too simple a picture of the situation with which we are faced when trying to give a rational account of the process of measurement.

[12] This difficulty was first discussed by Schrödinger [22], section 5. (Schrödinger's cat.)

stage 3 which looks quite innocuous and which is usually not separated from the other stages, is not always unproblematic.[13]

(iii) Any measurement leads to irreversible changes. For a wide class of changes, including incomplete measurements, the H-theorem can be proved without involving anything but (1) the equations of motion; (2) the specific properties of M; (3) the specific properties of the observer B who registers the state of M. Yet, within von Neumann's theory, irreversibility appears only if we apply the direct account, whereas any process which is in accordance with the equations of motion leaves the entropy unchanged.

A careful analysis of these three difficulties, especially of the last one, leads to the following suggestion: the theory of measurement which was developed in sections 2 and 3 is correct, but incomplete. What is omitted is the fact that M is a macroscopic system and that B cannot discern the finer properties of M. In other words, what is omitted is the fact that stage 2 and stage 3 occur at the macroscopic level. Now the transition from the level of Q M to the level of classical mechanics involves certain approximations.[14] Within a theory of measurement which omits reference to the macroscopic character of both M and B those approximations cannot be justified. Hence, within such an incomplete theory the transition to the classical level will have to be treated as an independent element which cannot be further analysed and which cannot be explained in terms of the equation of motion. We suggest that a complete theory which contains a reference to the macroscopic character of both B and M will allow for such an explanation. And we now proceed to discuss the outlines of such a complete theory.

5. THE CLASSICAL LEVEL

A simple calculation shows that immediately after the interaction is over, $|\psi_1\rangle$ and $|\psi_2\rangle$, the eigenfunctions of R', may be written in the form

$$|\psi_1\rangle' = |c_1\psi_1\rangle \exp(-i\alpha_1) \quad |\psi_2\rangle' = |c_2\psi_2\rangle \exp(-i\alpha_2) \tag{6}$$

where $\alpha_1 = \alpha_1(\lambda_1 H); \alpha_2 = \alpha_2(\lambda_2 H)$ (H being the Hamiltonian of interaction, supposed to be large as compared with H_S and H_M and commuting with R). The demand that $|\psi_1\rangle$ and $|\psi_2\rangle$ should be classically distinguishable implies that the real parts of the exponentials in (6) will have to differ by a great many nodes, i.e. it implies that

$$\alpha_1 - \alpha_2 \gg \pi \tag{7}$$

Now consider the following two assertions:[15] (a) Interaction satisfying (7)

[13] This difficulty has hardly ever been discussed in connection with the process of measurement. Jordan [14] was the first physicist who explicitly drew attention to it; but his treatment is impaired by the fact that he does not distinguish it from difficulty (i).
[14] There is still no satisfactory account available of those approximations.
[15] Cf. [1], ch. 22.

destroys all interference between the eigenfunctions of the observable measured and hence also between the corresponding eigenfunctions of the apparatus variable; (b) immediately after the interaction is over, i.e. already at the end of stage 1, the statistical operator P' can be substituted by P_M.

Assertion (a) follows from equation (7) only if we assume that all values of α_1 and α_2 which are compatible with (7) are equally probable. We shall call this assumption the *principle of equiprobability*. This assumption, which so far has only been used in connection with quantum statistics, is an indispensable part of any complete theory of measurement. As opposed to classical theory of measurement the quantum theory of measurement is essentially a statistical theory, i.e. it is a theory which uses, apart from the equations of motion, also further statistical assumptions. (7), i.e. the assertion that A and B are macroscopically distinguishable, implies, together with the principle of equiprobability, that, for all practical purposes, P' and P_M yield the same results with respect to the properties of S (we omit the simple calculation). This is the first step towards the solution of difficulty (i) in section 4: P' can be substituted by P_M not because the jump (5) happened in nature, but because under the conditions mentioned the difference between P' and P_M is neglibible, i.e. because it is highly improbable that there is a measurement (in S) of a magnitude compatible with R whose result in P' differs appreciably from its result in P_M.

In the second step we make use of the fact that B is not able to discern the finer details of M. More especially, B cannot discern anything like complementarity;[16] or, to be more precise: if R is observable for B then S is observable for B if and only if S commutes with R. This we call the *observer principle*. With respect to the theory of measurement this observer principle has the following consequences.

First: equation (4) implies that there are observables whose expectation is different in P' and in P_M. It can be shown that this does not apply to any observable which commutes with either R or R'.[17] But according to the observer principle an observable which does not commute with either R or R' will not be accessible to a macro-observer. Hence, *for a macro-observer,*

---

[16] From the fact that the classical level does not show the feature of complementarity, it has been concluded that classical objects cannot be described in terms of QM, that QM must be restricted at the classical level, and that this restriction must be regulated by a new axiom (for such a suggestion cf. Jordan [14]). But in such an argument it is overlooked that the absence of complementarity at the classical level is due to the superposition of *two* causes viz. (1) the special properties of macroscopic objects which justify the introduction, into the formalism, of the principle of equiprobability and (2) the special properties of the macroscopic observer who is unable to determine all the properties of these objects: we may, without fear of contradiction, admit that macroscopic systems can be described in terms of wavefunctions if we assume at the same time that a macroscopic observer has never enough information at his disposal to set up such a wave-function.

[17] For this cf. [10] and [11].

systems in the state P' will be indistinguishable from systems in the state P_M. This goes beyond our above result, where it was shown that (on the basis of (7) and the principle of equiprobability) P' coincides with P_M – with respect to the properties of S which are compatible with R. Our present result is that $P' = P_M$ for a macroscopic observer – with respect to any property of P' or P_M. But there is a still more important application of the observer principle which does not follow from (7) and the principle of equiprobability. It is

(*Secondly*) connected with the solution of difficulty (ii): if equation (3) does not hold then any linear combination of A and B will be compatible with P_M. In a state (of M) which is a linear combination of A and B neither A nor B will be diagonal. Such a state is the eigenstate of an observable which is not accessible to a macro-observer. Hence, the observer principle guarantees the consistency, with the formalism, of our third assumption: classical properties are always diagonal (the *diagonality principle*). This solves the second difficulty.

Example: An especially striking example where (3) does not hold has been discussed by Einstein:[18] a macroscopic particle is elastically reflected between the sidewalls of a cubical container of length l. The wave-function corresponding to the stationary process is $C\sin[(n\pi/l)x]\exp[(-i/\hbar)E_n t]$ (E_n being the nth eigenvalue of the energy). The macroscopic character of the process implies that $n\pi/l = \sqrt{2}\,mE_n/\hbar \gg 1$. Therefore $\sin^2[(n\pi/l)x]$ will oscillate rapidly within any classically definable interval Δx and $W(\Delta x)$ (the probability of finding the particle within Δx) will be constant. Now, although this is a macroscopic process we are, on the basis of QM, apparently still unable to say that the particle has a definite, though perhaps unknown, position within the container, for example we cannot say that it is in the right part, or that it is in the left part of the container. This difficulty can now be solved by pointing out that an account of the classical level is an account of classical objects *as observed by classical observers*. Such an account is compatible with the diagonality-principle and hence, also with the assumption that the particle is at a definite, though unknown place within the container. And such an account also disposes of the objections which Einstein derived from his example.

Thirdly: the introduction of a macroscopic observer is of decisive importance for the investigations in connection with the quantum-mechanical ergodic theorem and H-theorem (it corresponds to the introduction of coarse-grained densities).[19] But it was not before quite recently that those

---

[18] [7], 32. Cf. also the discussion by Bohm in the same volume. This discussion is unsatisfactory as it consists in nothing but a detailed exposition of the transition from what we called stage 1 to what we called stage 2. But this amounts to *discussing* the difficulty, not to *solving* it.

[19] Cf. [8] and the literature given there. For a less recent but more detailed discussion cf [27]. [26] contains all the literature in the field.

investigations were connected with the quantum theory of measurement.[20] The result of those more recent investigations is that the observer principle is valid under conditions which are nearly always fulfilled in reality;[21] and that it may be possible to give a proof of the diagonality principle. This result amounts to a consistency proof of the extended theory of measurement as it has been developed in this paper, as well as to a reduction of the diagonality principle to some deeper assumptions. It also amounts to a solution of our third difficulty.[22]

6. CONCLUSION

A theory of measurement can be developed which depends, just as its classical counterpart does, on nothing but the equations of motion and the special conditions (macroscopically distinguishable states; macro-observers) under which those equations are applied. This has the following consequences:

(1) All the processes which happen during measurement can be understood on the basis of the equations of motion only; hence, it is incorrect to say (as Dirac does for example, thereby expressing what is believed to be true by numerous physicists) that a mere interaction between a system S and its surrounding 'is to be distinguished from a disturbance, caused by a process of observation, as the former is compatible with the . . . equations of motion while the latter is not.'[23]

(2) As von Neumann has pointed out,[24] the transitions (5) are the way in which the old idea of 'quantum-jumps' is expressed within the formalism. Hence, our result can also be stated by saying (a) that there are no quantum-jumps and (b) that the idea that there are quantum-jumps has its origin in an incomplete theory of measurement. In this respect we agree with Schrödinger's recent attack against the orthodox interpretation of elementary quantum mechanics.[25]

(3) The theory of section (5) has been criticized; it has been objected that it is not 'exact' as it omits, by a process of approximation, interference terms which are postulated by Q M.[26] Now those interference terms are omitted also within the 'exact' theory. But the 'exact' theory omits them, not on the basis of a rational account, which does not deny their existence and which tries to *explain* why they can be neglected; it just omits them, gives a name to this procedure ('reduction of the wave-packet'; 'cut'; 'decision'; 'quantum-jump') and assumes that this amounts to having discovered a new kind of physical process (process (1) or process (5)).

[20] This was mainly done by Ludwig [17] and his pupil Kümmel [16].
[21] Those conditions are stated very clearly in [16].
[22] In fact this third difficulty was first attacked in von Neumann's attempt at a proof of a macroscopic H-theorem in 1929.   [23] [5], 110.   [24] [18], 218, footnote.
[25] [23]. Cf. also my note in [9].   [26] For this objection cf. e.g. [25].

(4) If we want to understand why this is done we must remember that the current interpretation of quantum mechanics contains the following philosophical thesis: Q M is a tool for producing predictions rather that a theory for describing the world, whereas classical terms have direct factual reference.[27] This thesis implies, of course, that the classical level and the quantum level are entirely distinct and that the transition from the one to the other cannot be further analysed.

(5) Our analysis, if it is correct, shows the classical level cannot be regarded as something which is totally distinct from the quantum level; it is rather a (particular) part of that level. Hence, the philosophical thesis, referred to in the last paragraph, must be revised and replaced by a realistic interpretation of the formalism of Q M.

(6) Apart from leading to the rejection of part of the current *interpretation* of Q M the result of our analysis can also be used for showing the inadequacy of various attacks against the *theory itself*. More especially, it can be used as an argument against all those attacks, which proceed from the difficulties we mentioned in section (4), and which interpret them as an indication that the present theory is incomplete,[28] or subjectivistic. For we have suggested that, and how, those difficulties can be solved *within* the present theory.

(7) Within certain schools of philosophy it was, and still is, fashionable to distinguish the level of everyday experience (or the 'observation language',[29] or the 'everyday language') from the theoretical level, and to assume that the transition from the first level to the second level is totally different from transitions between parts of either the first, or the second level. This view is a generalization of the 'orthodox' view about the relation between classical mechanics and Q M and it may therefore be called 'scientific'. But this only shows that nowadays scientists are committing a mistake which previous philosophers (notably positivistic, or 'scientific' ones) had the privilege to commit alone. On the other hand par. (5) of this section suggests that, quite in general, the everyday level is part of the theoretical rather than something completely self-contained and independent; and this suggestion can be worked out in detail and leads to a more satisfactory account of the relation between theory and experience than is the account given by Carnap, Hempel and their followers on the one side, and some contemporary British philosophers on the other.[30]

---

[27] 'The entire formalism is to be considered as a tool for deriving predictions, of definite or statistical character, as regards information obtainable under experimental conditions described in classical terms and specified by means of parameters entering into the algebraic or differential equations . . . These symbols themselves are not susceptible to pictorial [i.e. classical] interpretation' (Bohr [2], 314). The general philosophical background of this 'instrumentalist' view has been discussed by K. R. Popper in [21].

[28] For this point cf. n.10 of the present paper as well as the corresponding passages in the text.

[29] Cf. e.g. [3] as well as [4], 38ff, especially 40f.

[30] More detailed arguments may be found in chs. 2, 3 and 4 above.

REFERENCES

1. Bohm, D. *Quantum Mechanics* (Princeton, 1951).
2. Bohr, N. 'On the notion of Causality and Complementarity', *Dialectica*, 2 (1948), 312ff.
3. Carnap, R. 'Foundations of Logic and Mathematics' in *International Encyclopedia of Unified Science* (Chicago, 1939), 1/3, sec. 23.
4. Carnap, R. 'The Methodological Character of Theoretical Concepts' in *Minnesota Studies in the Philosophy of Science* (Minnesota, 1956), 1.
5. Dirac, P. A. M. *The Principles of Quantum Mechanics* (Oxford, 1947).
6. Einstein, A. 'Physik und Realität', *J. Frankl. Inst.*, 221 (1936), 313ff.
7. Einstein, A. 'Elementare Überlegungen zur Interpretation der Grundlagen der Quantenmechanik' in *Scientific Papers Presented to Max Born* (Edinburgh, 1953).
8. Farquhar-Landsberg, P. T. 'On the Quantum Statistical Ergodic and H-theorems', *Proc. R. Soc. Lond.*, A, 239 (1957), 134ff.
9. Feyerabend, P. K. 'Zur Quantentheorie der Messung', *Z. Phys.*, 148 (1957).
10. Furry, R. 'Note on the Quantum Mechanical Theory of Measurement', *Phys. Rev.*, 49 (1936), 393.
11. Groenewold, H. J. 'On the Principles of Elementary Quantum Mechanics', *Physica*, 12 (1946), 405ff.
12. Heisenberg, W. *Physikalische Principien der Quantentheorie* (Leipzig, 1944).
13. Heisenberg, W. 'The Development of the Interpretation of the Quantum Theory' in *Niels Bohr and the Development of Physics*, ed. W. Pauli (London, 1955), 12ff.
14. Jordan, P. 'On the Process of Measurement in Quantum Mechanics', *Phil. Sci.*, 16 (1949), 269ff.
15. Kemble, E. C. *Fundamental Principles of Quantum Mechanics* (New York, 1937).
16. Kümmel, L. 'Zur quantentheoretischen Begründung der klassischen Physik', *Nuovo Cimento*, 1:6 (1955), 1057ff.
17. Ludwig, G. 'Der Messprozess', *Z. Phys.*, 135 (1953), 483ff.
18. Neumann, J. von. *Mathematical Foundations of Quantum-Mechanics* (Princeton, 1955).
19. Popper, K. R. *Logik der Forschung* (Vienna, 1935).
20. Popper, K. R. 'The Propensity-Interpretation of the Calculus of Probability and the Quantum-Theory' in *Observation and Interpretation*, ed. S. Körner (London, 1957).
21. Popper, K. R. *Three Views of Human Knowledge* (London, 1957).
22. Schrödinger, E. 'Die neuere Lage in der Quantenmechanik', *Naturwissenschaften*, 23 (1935), 483ff.
23. Schrödinger, E. 'Are there Quantum Jumps?', *Br. J. Phil. Sci.*, 3 (1952), 109ff.
24. Slater, R. 'The Physical Meaning of Wave Mechanics', *J. Frankl. Inst.*, 207 (1929), 449ff.
25. Süssmann, G. 'A Quantum-Mechanical Analysis of the Measuring-Process' in *Observation and Interpretation*, ed. S. Körner (London, 1957).
26. ter Haar, D. 'The Foundations of Statistical Mechanics', *Rev. Mod. Phys.*, 27 (1955), 289ff.
27. Tolman, R. C. *The Principles of Statistical Mechanics* (Oxford, 1938).
28. Vleck, J. van. 'The Physical Meaning of Wave Mechanics', *J. Frankl. Inst.*, 207 (1929), 475ff.

14
Professor Bohm's philosophy of nature

1. This is a belated review of a highly interesting and thought provoking book.[1] Although dealing with some difficulties of a very specialized modern theory, the quantum theory, it should yet be of interest to the many non-physicists who want to know about the world we live in as well as about the ideas which are at present being developed for understanding this world. It is often assumed – and the basic philosophy of many contemporary physicists supports this assumption – that within the sciences speculation and ingenuity cannot play a very great role as physical theories are more or less uniquely determined by the facts. It is of course also assumed that our present knowledge about the microcosm is determined in exactly this way and is therefore irrevocable, at least in its main features. The book shows that this is not correct, it shows that today there exists a clash of ideas about some very fundamental things, that the imposing and perhaps a little terrifying picture of science as an unalterable and steadily increasing collection of facts is nothing but a myth, and that ingenuity and speculation play in physics as great a role as anywhere else. It also shows that even now it is possible to present difficult matters in an interesting and understandable way. It shows thereby that the separation, so often deplored, between the sciences and the humanities is due to a false picture, if not a caricature of science. It is this false picture which is attacked throughout the book. More especially, the book contains an explicit refutation of the idea that complementarity, and complementarity alone, solves all the ontological and conceptual problems of microphysics; that this solution possesses absolute validity; that the only thing left to the physicist of the future is to find, and to solve equations for the prediction of events which are otherwise well understood. In short, it contains a refutation of the idea that the physicist of the future is bound to be very similar to the more dogmatic of the medieval scholars with the sole exception that Bohr, and not Aristotle, will be his authority in matters metaphysical.

Secondly, the book presents, in qualitative terms, a new interpretation of some microphysical theories, and especially of the elementary quantum theory of Schrödinger and Heisenberg. It attempts to develop, again in

[1] David Bohm, *Causality and Chance in Modern Physics* (New York, 1957). Numbers in parentheses refer to pages of this book.

qualitative terms, a general picture of the universe which can give an account of statistical phenomena without assuming that they are irreducible. It discusses, on the basis of the picture presented, such fundamental problems of scientific method as the problem of induction, and the problem of the validity of empirical generalizations and of universal theories. Doing this without any discussion of 'ordinary language' or of language systems it (implicitly)[2] refutes another idea that it is very fashionable today, the idea that the only fruitful way of discussing more general problems of knowledge is either to analyse 'ordinary' language (whatever that may mean), or to construct formal systems and to investigate their properties.

Having stated in the above two paragraphs that I consider Bohm's book to be a major contribution to the contemporary philosophy of nature I must at once add that there are many things in it which I cannot accept and that more especially his discussion of the problem of induction seems to me to be highly unsatisfactory. Bohm's physical ideas are original, refreshing, and sorely needed in a time of complacency with respect to fundamentals. But the philosphical standpoint taken up with respect to both physics and cosmology is traditional, and perhaps even reactionary: it is a curious mixture of the methodological doctrine of inductivism and of ideas which may be found in various dialectical philosophies. This will become evident from a more detailed investigation of the book.

2. In order to enter into Bohm's theory, I shall first discuss the Copenhagen point of view. When it was first conceived this point of view constituted an interpretational feat of great importance. One realizes this when the historical situation is considered a little more closely. The early quantum theory of Bohr and Sommerfeld, although experimentally very successful, was yet regarded as unsatisfactory by many physicists. Its main fault was seen to lie in the fact that it combined classical and non-classical assumptions in a way that made a coherent interpretation impossible. For many physicists it was nothing more than a stepping stone on the way to a really satisfactory theory, i.e. to a theory which would give us not only correct predictions, but also some insight into the nature and the dynamics of miscroscopic entities. It is quite true that Bohr, Heisenberg and others worked along very different lines. Their main objective was not the construction of a new physical theory about a world that existed independently of measurement and observation; their main objective was rather the construction of a logical machinery for the utilization of those parts of classical physics which could still be said to lead to correct predictions. Quite obviously a theory of this latter type does not admit of a realistic interpretation: the classical signs it contains cannot be interpreted realistically as they are no longer universally applicable. And the non-classical signs it contains cannot be

[2] Cf. also the explicit discussion of the merits of conceptual analysis (156).

interpreted realistically as they are elements of the logical machinery used for the purpose of prediction, and possess no meaning apart from that usage. However that may be, the philosophical spirit behind the 'Korrespondenz-denken' was by no means shared by everybody. Now the most important thing is that Schrödinger's wave mechanics, which was conceived in an entirely different spirit, and which seemed to present the long awaited new and coherent account of the microscopic entities, encountered peculiar difficulties when the attempt was made to connect it with a universal interpretation of the kind that was applicable to the earlier theories. Any attempt to interpret wave mechanics as descriptive of entities which, although possessing new and surprising features, were still elements of an objective physical universe, any such attempt was found to lead to paradoxical consequences. It was Bohr's great merit that in this situation he developed an intuitive idea, the *idea of complementarity*, which, although incompatible with a straightforward realism, nevertheless gave the physicists a much needed intuitive aid for the handling of concrete problems.

According to this idea properties can be ascribed to a microscopic system only when it interacts with a suitable classical (i.e. macroscopic) piece of matter. Apart from the interaction the system possesses no dynamical properties at all.[3] It is also asserted that the totality of classical measuring instruments divides into pairs of kinds which are mutually incompatible in the following sense: if the system under investigation interacts with a measuring instrument which belongs to one of two mutually incompatible kinds, then all the properties defined by interaction with the other kind will be wholly undetermined.[4] And 'wholly undetermined' means that it would be meaningless to ascribe such a property to the system just as it would be meaningless to ascribe to a fluid a certain value on the Mohs scale of scratchability. It is clear that the *uncertainty relations* now indicate the domain of permissible applicability of classical functors (such as the functor 'position') rather than the mean deviations of their otherwise well-defined values in large *ensembles*.

The idea of complementarity can be interpreted in two different ways. It can be interpreted as an attempt to provide an intuitive picture for an existing theory, viz. wave mechanics, and as a heuristic principle guiding future research. This interpretation is undogmatic as it admits the possibility of alternatives, and even of preferable alternatives. A physicist who looks at complementarity in this way will regard it as an interesting fact about quantum theory that it is *compatible* with a relational point of view

[3] Dynamical properties are properties such as position, momentum and spin which can change during the course of a movement. Mass and charge are not dynamical properties in this sense.

[4] This totality comprises pieces of matter which have not been prepared by a physicist for the purpose of measurement, but which, by accident, as it were, satisfy some very general conditions not to be discussed here.

where interaction is a necessary condition of the meaningful applicability of terms which within classical physics (relativity included) are definable without such reference. He will also point out that there exist no satisfactory alternatives. But he will never go as far as to assert that such alternatives will never be found, or that they would be logically inconsistent, or that they would contradict the facts. But Bohr's idea of complementarity can also be interpreted in a different way. It can be interpreted as a basic philosophical principle which is incapable of refutation and to which any future theory *must* conform. Bohr himself most certainly took this stronger point of view.[5] 'Thus rather than consider the indeterminacy relations primarily as a deduction from quantum mechanics in its current form he postulates these relationships directly as a basic law of nature and assumes . . . that all other laws will have to be consistent with these relationships' (83, referring to Heisenberg). His assumption was 'that the basic properties of matter can *never* be understood rationally in terms of unique and unambiguous models' which implies that 'the use of complementary pairs of imprecisely defined concepts will be necessary for the detailed treatment of every domain that will ever be investigated' (94). It is true that some followers of the Copenhagen school have denied that this absolutism is part of complementarity. Thus in a discussion Rosenfeld has asserted that 'nobody thinks of attributing an absolute validity to the principles of quantum theory'.[6] But quite apart from the fact that he himself said in the lecture preceding this discussion that 'every feature' of the theory 'is forced upon us',[7] there is Bohr's explicit statement that 'it would be a misconception to believe that the difficulties of the atomic theory may be evaded by eventually replacing the concepts of classical physics by new conceptual forms'.[8]

3. This dogmatism with respect to fundamental principles is attacked and refuted in ch. 3 of Bohm's book, which contains an extremely lucid description of the development of the quantum theory and the various interpretations that have been suggested for it. It explains the reasonable elements of the point of view of Bohr and Heisenberg. This point of view is presented with a clarity that is sadly missing in many writers who support Bohr, and with an understanding and authority that reveals the former follower and expositor of Bohr's ideas.[9] The idea of its final and absolute validity is refuted by showing that all attempts to prove it (as indeed all attempts of a 'transcendental deduction' of physical principles) are circular. Thus, in

[5] However this was the result of a long development in the course of which many other interpretations were examined and refuted. For details cf. ch. 16.6.
[6] L. Rosenfeld, *Observation and Interpretation*, ed. S. Körner (London, 1957), 52.
[7] *Ibid.*, 41.
[8] N. Bohr, *Atomic Theory and the Description of Nature* (Cambridge, 1932), 16.
[9] Cf. Bohm, *Quantum Theory* (Princeton, 1951).

Heisenberg's 'proof' of the uncertainty principle (which is often used as an argument for its absolute validity) 'it was essential to use three properties; namely the quantization of energy and momentum in all interactions; the existence of these quanta; and the unpredictable and uncontrollable character of certain features of the individual quantum process. It is certainly true that these properties follow from the quantum theory' (94). However, in order to show the basic and irrefutable character of the uncertainty principle these features themselves would have to be demonstrated as basic and irrefutable. Quite obviously such a demonstration cannot be achieved by pointing to some theorems of *wave mechanics* (such as von Neumann's theorems) as this would only lead to the further question whether wave mechanics is valid in all domains of experimentation (95). Nor can it be achieved, as has been attempted by many inductivists, by utilizing the fact (if it is a fact) that either wave mechanics, or some part of it, is highly confirmed. In order to see this most clearly we need only realize that the assertion of the absolute validity of a physical principle implies the denial of any theory that contains its negation. More especially, the assertion of the absolute validity of the uncertainty principle implies the denial of any theory that ascribes to it only a limited validity in a restricted domain. But how could such a denial be justified by *experience* if the denied theory is so constructed that it gives the same predictions as the defended principle wherever the latter has been found to be confirmed by experience?[10]

It follows that neither experience nor mathematics can help if a decision is to be made between wave mechanics and an alternative theory which agrees with it in all those points where the latter has been found to be empirically successful. Now the idea of complementarity is well fitted to the structure of wave mechanics. As we cannot make any restrictions upon the structure of the empirically satisfactory alternatives of wave mechanics it also follows that its interpretation as a basic and irrefutable principle must be given up. Neither mathematics nor experience can be used to support such an interpretation. All this means, of course, that the position of complementarity is a metaphysical position,[11] which can be defended by arguments of plausibility only.

[10] Quantum mechanics is not the first theory that has been used to exclude alternatives. Using the fact that certain theorems of Newtonian mechanics contradicted the second law of thermodynamics, Ostwald and Mach argued that a mechanical account of heat was impossible, and that Newton's laws could not be universally valid. It turned out, however, that it was the second law that was not universally valid (fluctuations). Quite clearly the Ostwald–Mach argument suffered from the same deficiency as the more recent arguments of Born, Rosenfeld, and others. They argued: the second law is highly confirmed; classical mechanics contradicts the second law; hence classical mechanics is not universally valid. They overlooked (a) that confirmation does not imply truth; (b) that the mechanical theory of heat contradicted the second law in a domain in which it had not yet been tested, and in which it was therefore neither confirmed nor disconfirmed.

[11] I use here the word 'metaphysical' in the same sense in which it is used by the adherents of

4. So far only the (empirical and logical) *possibility* of alternative points of view has been shown. In ch. 4 of his book Bohm turns to the discussion of some alternatives that have actually been proposed in the literature and he also expounds some of his own ideas. I shall now give an outline of the epistemological background of all these alternatives.

One of the basic assumptions of the orthodox is that 'in our description of nature the purpose is . . . to trace down, as far as it is possible, relations between the manifold aspects of our experience'.[12] For them the facts of experience play the role of building stones out of which a theory may be constructed but which themselves neither can, nor should be modified. If we add to this the idea that 'only with the help of classical ideas is it possible to ascribe an unambiguous meaning to the results of observation'[13] (which means that the building stones referred to in the first quotation are classical states of affairs) we arrive at once at the result that a microscopic theory cannot be anything but a device for the prediction of a particular kind of fact, viz. of classical states of affairs. It is quite true that this point of view has led to some useful results (example: the dispersion formula of Ladenburg–Kramers; the first investigations of Heisenberg). It is also true that the quantum theory is the first theory of importance which to some extent satisfies the programme of Berkeley and Mach (classical states of affairs replacing the 'perceptions' of the former and the 'elements' of the latter). But it must not be forgotten that there is a whole tradition which is connected with the philosophical position of realism and which went along completely different lines.[14] In this tradition the facts of experience, whether or not they are now describable in terms of a universal theory (such as classical mechanics), are not regarded as unalterable building stones of knowledge; they are regarded as capable of analysis, of improvement, and it is even assumed that such an analysis and improvement is absolutely necessary. Indeed, the new theory of motion which was developed by Galileo and

---

the orthodox point of view, viz. in the sense of 'neither mathematical, nor empirical'. That the Copenhagen Interpretation is metaphysical in this sense has been asserted, in slightly different words, by Heisenberg who declared in 1930 (*Die physikalischen Grundlagen der Quantentheorie*, 15), that its adoption was a 'question of taste'. This he repeated in 1958 in the now more fashionable linguistic terminology (cf. *Physics and Philosophy* (New York, 1958), 29f). However at the very same place a highly objectionable criticism is found of Bohm's model of 1952. This model, it is asserted, 'cannot be refuted by experiment since [it] only repeat[s] the Copenhagen interpretation in a different language. From a strictly positivistic standpoint', Heisenberg continues, 'one may even say that we are here concerned not with counterproposals to the Copenhagen interpretation, but with its exact repetition in a different language'. Is it really the case that Bohm's counter-example against the assertion, made by von Neumann and others, that quantum theory does not allow for the addition of *untestable* hidden parameters (cf. J. von Neumann, *Mathematical Foundations of Quantum Mechanics* (Princeton, 1955), 326) is nothing but the 'exact repetition' of this assertion 'in a different language'?

[12] Bohr, *Atomic Theory and the Description of Nature*, 18.
[13] *Ibid.*, 17.   [14] For details see chs. 1, 8 and 11, as well as vol. 2, ch. 3.

Newton could not possibly be understood as a device for establishing 'relations between the manifold aspects of our experience', the simple reason being that, according to this very theory, observable motion would at best give us an approximation to its fundamental laws. Similarly the atomic theory of the late nineteenth century was not only not suggested, it was even contradicted by what was then regarded as an account of 'experience', viz. classical thermodynamics. This tradition proceeds from the very reasonable assumption that our ideas *as well as* our experiences may be erroneous and that the latter gives at most an approximate account of what is going on in reality. Bohm's own point of view is closely connected with this tradition. Having shown that all the attempts to prove the uniqueness of the Copenhagen Interpretation are invalid, he suggests 'to take the field and particle concepts of classical physics as starting points and to *modify* and enrich them in such a way that they are able to deal with the new combination of wave and particle properties that is implied in the quantum theory' (98, my italics). Such modified concepts, or even a completely new conceptual apparatus which does not any longer make use of classical ideas, will of course at first be 'extraphysical' (99) in the sense that it will not be accessible to test with the help of methods available *before* it was conceived. However, 'the history of scientific research is full of examples in which it was very fruitful indeed to assume that certain objects and elements might be real, long before any procedures were known that could permit them to be observed directly' (99). Assumptions of this kind then

> ultimately lead to new kinds of experiments and thus to the discovery of new facts. In the light of this historical experience [Bohm continues] positivism (i.e. the point of view expressed in the two above quotations) is seen to lead to a one sided point of view of the possible means of carrying out research. For while it recognizes the importance of the empirical data, positivism flies into the face of the historically demonstrable fact that the proposal of new concepts and theories having certain speculative aspects (e.g. the atomic theory) has quite frequently turned out to be as important in the long run as empirical discoveries have been (99).

In this way positivism 'constitutes a dogmatic restriction of the possible forms of future experience' which in the case of quantum mechanics leads to the belief 'that the success of probabilistic theories of the type of the current quantum mechanics indicates that in the next domain it is very likely that we shall be led to theories that are . . . even more probabilistic than those of the current quantum domain' (104).[15]

[15] A terminological remark: quantum physicists have sometimes refused to be called 'positivists' on account of the fact that they accepted the Copenhagen point of view. Thus in *Niels Bohr and the Development of Physics* (London, 1955), 22, Heisenberg asserts that 'the Copenhagen interpretation . . . is in no way positivistic. For whereas positivism is based upon the sensual perceptions of the observer . . . the Copenhagen interpretation regards things and processes which are describable in terms of classical concepts . . . as the founda-

5. More concretely, Bohm's ideas as presented in the book under review may be regarded as an adaptation, to the case of the quantum theory, of the situation described by the classical kinetic theory of matter. The kinetic theory was an attempt to give an explanation, in terms of the motion of small, and as yet unobserved, particles, of the behaviour of thermodynamic systems. According to this theory continuous improvement of the precision of measurements will lead to the following phenomena (we assume that we move outside the domain where relativistic effects become noticeable): as long as we are dealing with large systems the classical laws of motion (and the second law of thermodynamics) will be found to hold with absolute precision. However when experimenting with fairly small systems such as dust particles which are immersed into a surrounding medium, a completely new type of behaviour becomes apparent. These particles experience random displacements for which no explanation can be given in terms of the movements of bodies of a similar size. The laws describing this type of behaviour are not any longer the laws of classical mechanics. They are purely probabilistic and allow us to predict averages in large *ensembles* rather than individual processes. Within the framework of these laws no reason can be given for the occurrence of a particular movement of a particular particle. It can even be shown (107) that for particles under the conditions described above there exist laws which are formally identical with the uncertainty relationships, the diffusion of the embedding medium taking the place of Planck's constant h. But the situation changes again when we further improve the precision of our measurements or else use experiments of an altogether different type. We shall then find that the random behaviour of the dust particles is explainable in terms of a new set of causal laws referring to very small particles which are the ultimate constituents of the medium in which the dust particles are immersed. (In the case of the kinetic theory these new laws happen to coincide with the laws of classical mechanics from which we started. However, it is necessary to point out, in accordance with Bohm's more general ideas, that this need not always be the case.)

Speaking more generally one may now say that according to the kinetic theory there exist three different levels of experimentation which are characterized by three different sets of laws. There is the macroscopic level where the laws of classical mechanics hold exactly. More precise experiments show then that these laws are not universally valid, and thereby delimit the domain of their applicability. At the same time they lead to a new set of laws governing phenomena which are *qualitatively* different from the phenomena we meet on the macrolevel, as they involve randomness.

tion of any physical interpretation'. This is quite true. However this 'foundation' is again assumed to be 'given' in the sense that it cannot be further analysed or explained, an attitude which to a certain extent still justifies the term 'positivism'.

These new laws in their turn are not universally valid as they can be shown to be the result of the very complex, but again causal, behaviour of entities on a still deeper level.

It is Bohm's contention that the situation in the domain of the quantum phenomena is similar to the one just described. As opposed to the opinion of the majority of physicists he assumes that the probability laws of the present quantum theory are the result of the very complex interplay of entities on a deeper level, and are therefore neither ultimate nor irreducible. Ch. 4 contains a general discussion of various ways in which such a sub-quantum-mechanical level can be conceived. These considerations have been criticized by some members of the Copenhagen circle. One of the most frequent criticisms is that nobody has as yet succeeded in constructing a theory along these lines which can match the customary theory in predictive success. This criticism seems to proceed from the assumption that the existence of a certain theory and the absence of a theory, which is connected with a different 'ideology' as it were, may be regarded as an implicit criticism of the latter. However the fact that this pragmatic criticism can also be directed against the dynamical investigations of Galileo and Kepler (the successful theory being in this case Aristotle's theory of motion) should be sufficient to make its proponents a little more cautious about its force. A second criticism points out that the present theories, and the philosophical structure connected with them, are firmly based upon experience. This criticism has already been dismissed earlier in the present review. Indeed, we have seen that the customary point of view about microphysics cannot produce any empirical or logical argument against a procedure such as Bohm's. And assertions such as 'it is idle to "hope" that the cure of our troubles will come from underpinning quantum theory with some deterministic substratum' can at most be regarded as affirmations of faith.[16]

6. I leave now the physics of the book and turn to a discussion of the cosmology and methodology developed in it. Both these fields are dealt with on the basis of a generalization of the situation described by the kinetic

---

[16] Rosenfeld, in *Observation and Interpretation*, 44. In his review of the present book in the *Manchester Guardian*, Rosenfeld accused Bohm of contradicting the 'exigencies of sound scientific method' and he described the followers of Niels Bohr (and presumably also himself) as possessing the 'uncommitted, commonsense attitude of the true scientist'. Now first of all an attitude can hardly be called 'uncommitted' if it appeals to the principle that experience alone can be the judge of our theories, and at the same time is singled out neither by experience, nor by mathematics. Secondly the history of science has given ample evidence for the fact that it is 'sound scientific method' not to take experience at its face value, even if it should be expressed in very complicated (classical) terms, but to try to explain it as the result of processes which are not immediately accessible to observation. It is strange indeed to see that Rosenfeld describes as 'uncommitted' the attitude of those who because of their observationalistic bias distort both history and scientific method.

theory. The cosmological generalization, as I understand it, is as follows: the world contains infinitely many levels. Each level is characterized by a set of laws which may be causal, or probabilistic, or both. The validity of these laws need not extend beyond the level to which they belong. When a certain level is left qualitatively new processes appear which have to be described by a new set of laws. Bohm recognizes that sometimes these new laws may be general enough to allow for the derivation of the more specific laws of the preceding level (example: special relativity – general relativity; cf. 141). However he points out – and this must be regarded as a highly important contribution to cosmology – that such a reduction need not always be possible. Let us assume, for example, that the level L_1 of causal laws possesses a substratum L_2 of probability laws which are the outcome of the causal interplay of entities of a level L_3 which in its turn possesses a probabilistic substratum L_4, and so on. Now the fact that the laws of L_2 can be explained by reference to complicated causal mechanisms on L_3 shows that they cannot be entirely random. On the other hand the laws of L_3 are not absolutely causal either as they are limited by the fluctuations which appear upon L_4. A complete explanation of the laws of L_1 (or of any set of causal laws or of probability laws) would therefore have to take into account an infinity of laws and levels. Clearly, then, an explanation of the laws of L_1 in terms of finite sequence of substrata cannot be regarded as a *reduction* of L_1 to these substrata. Each level, and each set of laws possesses a surplus over and above any finite set of more general laws. It is only if we take all the mutually irreducible properties and laws together that we may hope to get a complete account of one particular level. This is the way in which Bohm makes physical sense of the idea of emergence and the irreducibility of qualities. At the same time it is suggested, at least by the cosmological model we are discussing at the present moment, that qualities may be reducible after all if only appropriate mathematical instruments are found for the handling of infinites of relatively self-contained experimental domains. The model also suggests a new interpretation of the difficult problems of probability, randomness and statistical independence. In this interpretation neither the idea of a deterministic law, nor the idea of randomness is given absolute preference (20f). The laws of nature, whether they appear in the form of causal laws, or in the form of probability laws, are regarded as a Hegelian synthesis, as it were, of the idea of absolute determination (the thesis), and of absolute randomness (the antithesis). This way of describing Bohm's procedure is by no means a mere verbal trick, for it is Bohm's conviction that in all fields the alternative use of opposite sets of concepts is to be preferred to the exclusive utilization of only one of them.

7. However, the model which we have just described and which plays an important role in Bohm's analysis of probability is not the one he uses in his

discussion of scientific method. He is 'not even supposing that the general pattern of levels that has been so widely found in nature thus far must necessarily continue without limit'. He admits the possibility that 'even the pattern of levels itself will eventually fade out and be replaced by something quite different' (139). The structure of levels, he asserts, is only one way in which the *qualitative infinity of nature* may represent itself to the experimenter. This qualitative infinity of nature is one of the basic postulates of Bohm's cosmology. He incessantly insists upon the 'inexhaustible *depth* in the properties and qualities that matter' (138) which is such that no finite system of laws and categories can ever express it adequately. Every thing and every process has infinitely many sides to it which are such that at any stage of scientific development they will only approximately be expressed by the laws and the concepts then in use. That such an approximate representation is at all possible is due to the further fact that there exists 'some degree of autonomy and stability' in the mode of being of the things around us (139). For example, 'we may say that [a] real fluid is enormously richer in qualities and properties than is our macroscopic concept of it. It is richer, however, in just such a way that these additional characteristics may, in a wide variety of cases, be ignored in the macrodomain' (155). In spite of the fact that in every real fluid an infinite variety of processes is going on which are not covered by our macroscopic description of it, these processes just so counterbalance each other that relative stability is achieved upon the macrolevel, and the macroscopic description is in this way made applicable within its proper domain of validity. In short: the world is infinite as regards the properties and processes which are present in every part of it. But these properties are arranged in complexes of relative stability which may then be described with the help of scientific theories employing a finite number of concepts only. Every such description is true within a certain domain of validity. On the other hand the presence of further properties which are not covered by the description, and which slightly influence the elements of the complex implies that 'associated with any given law there must be errors that are essential and objective features of that law resulting from the multitude of diverse factors that the law in question must neglect. Thus each law inevitably has its errors, and these are just as necessary a part of its true significance as are those of its consequences that are correct' (166). It is important to repeat that for Bohm the errors referred to in the above quotation are not purely subjective phenomena; they possess an objective counterpart in the way in which the interplay between the elements of the relatively stable complexes as well as the qualities that have been left out delimits the validity of the laws describing the behaviour of the complexes. 'It is clear from the above discussion', Bohm continues (166), 'that scientific research does not, and cannot lead to a knowledge of nature that is completely free from error.'

8. The application to *scientific method* is now quite straightforward. Nature is such that no law can ever be universally valid. Hence, it is sound scientific method to restrict the laws we find to a certain domain (135). It is unsound method to apply them outside this domain. And never should we be so bold as to proclaim a certain law as universally valid, i.e. as valid in all domains of experimentation, and under all possible conditions. On the other hand, if we are careful enough in our pronouncements about the applicability of a scientific theory, and if we always restrict it to its proper domain, we do not run the risk of being refuted by new discoveries. For 'a new theory to which the discovery of . . . errors will eventually give rise, does not invalidate the old theories. Rather . . . it corrects the older theories in the domain in which they are inadequate and, in doing so, it helps to define the conditions under which they are valid' (31). Only a philosophical idea, and not sound scientific method can lead to the attempt to apply a theory to every possible domain. Thus the assumption 'that all the various levels, all qualitative changes, and all chance fluctuations will, eventually be reducible completely . . . to effects of some fixed . . . scheme of purely quantitative laws . . . is . . . essentially philosophical in character' (62). More especially the assumption that Newton's laws are universally valid 'has implications not necessarily following from the science of mechanics itself, but rather from the *unlimited* extrapolation of this science . . . Such an extrapolation is evidently . . . not founded . . . on what is known scientifically. Instead, it is in a large measure a consequence of a *philosophical* point of view . . .' (37). It is this methodological doctrine which I find highly questionable and which I shall attempt to criticize in the following last part of my review.

9. First of all, how does Bohm justify his two basic cosmological principles, viz. the principle of the infinity of nature and the principle that there exist complexes which are relatively stable over a certain period of time and which therefore allow for the description, in terms of finite sets of laws and concepts, of parts of nature? The principle of the infinity of nature he tries to justify partly by reference to experience which shows us a great variety of qualities; partly by reference to the history of science which shows that every set of laws has at some time been found to be valid in a restricted domain only; and partly by reference to the 'basic spirit of scientific method itself, which requires that *every* feature be subjected to continuous probing and testing' (132). The principle of the existence of complexes of relative independence and stability is again justified by reference to experience; but it is also justified by some kind of 'transcendental' reasoning according to which in a world of a different structure the concept of a thing would not be applicable and science would be impossible (139f). Now if we look at these arguments we find that they are all unsatisfactory. To start with, Bohm's

methodological rules which have been stated above would forbid us to draw consequences from experience and to apply them universally. Yet this is just what is done in the first argument. The appeal to the history of science cannot be accepted either. For it could also have been used by the Aristotelians *against* the assumption that human knowledge gave at most an approximate account of what went on in nature.[17] Thirdly the transcendental argument is not of the slightest use as long as we do not know whether our theories express knowledge or whether they are not merely well-fabricated dreams. But knowing this would presuppose knowledge of exactly those states of affairs whose existence is to be proved with the help of the argument. And finally the methodological argument is of no help either as it might well be the case that all the tests we carry out with respect to a certain theory lead to its corroboration and thereby to the corroboration of the idea that the world possesses a finite number of basic properties after all. We see, then, that Bohm's two basic principles are not supported by the arguments he uses in their favour. They are not even empirical, or scientific in Bohm's own sense (cf. 166), as he is not prepared to admit that they may be valid in a certain domain only and give way to some kind of mechanicism in all the remaining domains of experimentation. They represent an *absolute truth* which is not capable of improvement by taking into account errors (169f). Yet they are cosmological principles, i.e. principles describing the basic structure of our world. This, then, is my first criticism: that there is not the slightest reason for not treating the most general cosmological principles, such as the principle of the infinity of nature on a par with less general laws. There is not the slightest reason for denying them the status of all the other laws, that is, as provisional.

10. However, it seems to me that this criticism does not yet go to the heart of the matter. For it leaves out one of the most important arguments that Bohm could adduce in favour of the absolute character of his two principles. I did not find the argument in the book, but I trust that it may be constructed along the following lines. Consider a law that is valid in a certain domain only. When this law is properly stated we shall soon discover its limitations. We are able to do so because there exists another domain which is not covered by the law, and whose presence is responsible for the errors it possesses. The conditioned validity of the law and its approximative character are thus wholly dependent upon the objective existence of such other domains. It would then seem to follow that for lack of domains outside the domain of its applicability a statement about 'the infinite totality of matter in the process of becoming' (170) must be

---

[17] The Copenhagen Interpretation of quantum mechanics which Bohr adopted after various unsuccessful attempts at a 'deeper' explanation (cf. ch. 16), reintroduces 'Aristotelian' features, but with excellent arguments.

unconditionally and absolutely valid. It is this argument which will be the starting point of my second criticism.

It is assumed in this argument that the provisional and approximative character of a scientific law is *wholly* due to the objective limitations of the stability of the entities, or of the domain it describes. We must correct the law not because we had a wrong idea about the properties of the things described. We must correct it because these properties themselves are the relatively stable result of a very complicated interplay of an infinity of processes, and because they are therefore subject to slight changes and transformations. *But if we keep well within the domain of application of the law, then we cannot possibly be mistaken.*

This last principle has the following very interesting corollary: every description of nature that has ever been uttered is true within its domain, and conversely, it exhibits the existence of a domain to which it properly applies. There does not exist any description that is wholly mistaken and without a corresponding reality. Or, to express it differently, when describing our surroundings *we always speak the truth* (relative truth, that is), *and we are also always in contact with some part of reality.* Now this corollary has so little *prima facie* plausibility that I must defend it before trying to show its shortcomings. 'Is it really the case', one may easily feel oneself inclined to object, 'that the savage who believes in, and claims to have observed, the actions of ghosts, tribal spirits and the like, is talking about entities which have some kind of existence in a restricted domain?' To this objection the retort may well be that a savage could not have described, or interpreted what he saw as indicating the existence of a ghost, if there had not been a justification for doing so. After all, he does not, *and cannot,* make arbitrary judgements in matters which may be of importance to his well being, and even to his life. Neither for him, nor for us would it be possible

> to choose the natural laws holding within a given degree of approximation, and in a particular set of conditions at will . . . This does not mean that we cannot, in general, make our own choices as to what we will, or will not do. But unless these choices are guided by concepts that correctly reflect the necessary relationships that exist in nature, the consequences of our actions will not in general be what we choose, but rather something different [165].

In short, every theory of the universe, whether mythological or scientific in content, possesses some degree of truth, as the choice of a false theory would lead to undesirable consequences and would therefore be at once abandoned. *Nature itself forces man to speak the truth,* and it also forces him to speak in such a way that his theories have objective reference.

This, then, is the epistemology behind Bohm's belief that every theory, however absurd it may seem at first sight, has some kind of truth in it and correctly mirrors what exists in the universe: the lack of success of a theory which is downright wrong and does not describe anything whatever is a corrective which after a very short time forces us to abandon it (if we were ever foolish enough to put it forth). Knowledge is a natural process which leads to a mirroring, in the head of man, of the properties of the universe. The mirror-image may be distorted at the edges. But first of all this distortion is due to a similar objective distortion of the processes in the world. And secondly this distortion does not reach into the centre of the mirror which perfectly represents the situation at a certain level.

I do not believe that this account of our knowledge is a correct one. The simplest reason I can give for this contention of mine is that I believe man to be a little more whimsical and capricious than is assumed in the above picture of him. For in this picture it is assumed that *as a matter of fact* we recognize our mistakes, take them into account, and learn from them how to behave better. It is assumed that this process works like a well lubricated machine so that in the end whatever has been said contains some truth in it. (I suspect that a consistent elaboration of this epistemology will finally lead to the result that errors – subjective errors, that is – are never made: quite obviously Hegel's notorious 'Alles Vernünftige ist wirklich' is here lurking in the background.) But only a little knowledge of history will show that this assumption is factually false for at least two reasons. First, because there are enough examples of men, or of whole groups, who are not prepared to admit that they have been mistaken. And secondly because even death may not be a sufficient reason for changing ideas which have led to it. Quite the contrary we often find, even in our own times, that failure of an ill-conceived undertaking, and death resulting from it, are both regarded as values, and we also find the corresponding assumption that fate will sometimes deal roughly with its protegés. Furthermore, to turn to more theoretical considerations, is it not well known that refuting instances can with some ingenuity always be turned into confirming instances and that there exist elaborate theories which perform this transformation nearly automatically? Quite clearly such theories cannot be said to be in contact with reality and this in spite of their sophistication and in spite of the many fascinating statements they contain. From all this we have to conclude that *nature can never force us to admit that we have been mistaken.* Nor can it force us to recognize our mistakes. A mistake will be recognized as such only if first the conscious *decision* has been made not to make use of *ad hoc* hypotheses and to eliminate theories which do not allow of falsification. It is true that as a matter of historical fact this decision has been made by nearly all great scientists (although the present quantum theory seems to present an exception to this rule). What is of importance here is that they never were, and never could be forced to proceed in that way, either by nature or by society.

11. To sum up: at the back of Bohm's theory of knowledge there is the idea that facts and decisions both obey the same kind of laws, i.e. the laws of the

material world in which we live. It is the idea that the development of moral codes, or of the laws which govern the non-moral behaviour of the members of a society, or the development of knowledge, is nothing but an aspect of the development of this material universe. This idea implies that neither the moral behaviour, nor the social behaviour, nor even the status of our knowledge can be changed on the basis of an explicit decision. It is quite impossible to entertain a point of view which has no reference to any facts whatever. And it is equally impossible to introduce a new moral system unless it is somewhat related to situations already existent. This doctrine of *naturalism* can be given various forms.[18] It exists in a form which allows for the accommodation of the most revolutionary changes by simply asserting that these changes had already been prepared by the development either of the material universe or of society. In this form the doctrine is nothing but a verbal manoeuvre. Another form of the doctrine decrees that some existing pieces of knowledge, or of morals, are unchangeable, because a change would amount to nothing less than a change of the unalterable course of events and of the laws which govern the universe. In this form the doctrine has very often been held by the defenders of the status quo. The simple logical point that decisions are never derivable from facts should show that in all its forms naturalism is based upon a logical fallacy.[19] Bohm's own doctrine, although related to the doctrine of naturalism, is more detailed and less radical. He seems to admit that at times ideas may be invented which have very little to do with the facts. What he contends is, however, that these ideas will very soon be eliminated by a kind of natural selection which works either against those who hold them (they die), or against the ideas themselves (they are given up). That is, Bohm allows for deviations, but at the same time he assumes the existence of a corrective mechanism which quickly eliminates pipe-dreams and falsehoods. I want to show that although the doctrine in this form allows us to say that we sometimes speak the truth, it nevertheless does not give us any indication whatever as to which particular point of view expresses the truth. This we see when we ask the following important question: how long does it take this mechanism to eliminate a false hypothesis? Most certainly the length of time will depend upon the frequency with which the theory is tested, upon the decisiveness of the tests, as well as upon the intention, on the side of the scientist, to take refutations seriously. Laziness and *ad hoc* manoeuvres may extend the periods of correction indefinitely. And the scientist, or whoever else is defending a certain point of view, need not perish in the course of events as he may well be careful enough to avoid tests which endanger his personal safety (there are numerous examples of this kind in the so-called 'primitive'

[18] For an excellent discussion of this doctrine, its history, and its shortcomings see K. R. Popper, *The Open Society and It's Enemies* (Princeton, 1954). ch. 5. For a criticism see the addition to this chapter, below.   [19] This 'point' has little argumentative force. Cf. the addition below.

societies). Furthermore, who says that we shall at once stumble upon a refuting instance? Perhaps human lifetimes could be extended indefinitely if we had, and acted upon, the correct ideas about the nature of life. But if this is so then Bohm's idea of the self-correcting character of knowledge does not help us at all to distinguish truth from falsehood. For all we know *all* our ideas may be quite thoroughly mistaken.

If this is the case, and if it is further admitted that we are able to discover our errors when trying to apply the ideas we possess (provided of course, we have first *decided* to give them a form in which they are testable, and we have also decided to take refutations seriously) then the only path open to us is that we must attempt relentlessly to falsify our theories. As we do not know which part of them is true, in what domain they are true, and whether they are true at all, we must attempt the falsification under all possible conditions. Testing them under all possible conditions means assuming *first* that they are *universally valid* and *then* trying to find out the limitations of this assumption. It is this fact that we never know to what extent our theories are correct which makes us first apply them universally. If we use a theory in this way we by no means assume, as Bohm seems to think (cf. his criticism of mechanism, discussed above) that the theory *will be found to be correct* in all domains. The universal application of a theory means rather that *we are prepared to collect refuting instances from all domains*. The reason why I cannot accept Bohm's methodology of caution, and why I prefer to it the methodology of falsification as it has been developed by Popper, is therefore that the methodology of caution assumes the existence of things we know for certain, whereas I believe on the basis of the above consideration that this is much too optimistic a view of the status of our knowledge.

These, if I understand the book correctly, are the criticisms which I think must be made. But let me at once repeat that I do not therefore think the book to be less valuable. Quite the contrary, it is the repeated discussion and criticism of various points of view which leads to an advance of knowledge, and not the repetition of plain statements in which nobody can find any fault. To have in this way contributed to the theory of knowledge, and also to have shown the unity of (physical, philosophical, etc.) knowledge is the great merit of the book.[20]

[20] *Added 1980.* My criticism in sections 10 and 11 is a paradigm case of abstract reasoning, and shows all its weaknesses. The argument assumes (a) that methodological stipulations can be introduced and enforced independently of what goes on in the world and (b) that developing them is just a matter of imagination, not of real relations that might hinder or further the work of the theoretician. In a word – it is assumed that the domain of knowledge-construction and knowledge-modification is unencumbered by the laws of nature, it is a special domain in which the laws of nature are suspended, but whose products can transform nature. One only needs to formulate this assumption to see its absurdity. Yet philosophers of science who praise methodological rules because they circumvent some recondite logical problems (such as 'Hume's problem') and regard them as adequate on those grounds alone make this absurd assumption.

15

Reichenbach's interpretation of quantum mechanics

In section 3 of his paper 'Three-valued logic',[1] Hilary Putnam deals with
Reichenbach's attempt to interpret quantum mechanics on the basis of
three-valued logic, and he uses some arguments of his own in order to show
that this attempt is 'a move in the direction of simplifying the whole system
of laws' (171). I believe that the Reichenbach–Putnam procedure cannot be
defended and that it leads to undesirable consequences. There are the
reasons for my belief:

1.THREE-VALUED LOGIC AND CONTACT-ACTION

Putnam asserts (a) that 'the laws of quantum mechanics . . . are logically
incompatible with' the principle of contact-action 'if ordinary two-valued
logic is used'; and (b) that 'adopting a three-valued logic permits one to
preserve both the laws of quantum mechanics and the principle that no
causal signal travels with infinite speed'. Assuming for a moment that (a)
and (b) give a correct statement of Reichenbach's position (which they do
not – see section 3) and that (a) is true we can at once say that adopting
the procedure suggested in (b) would violate one of the most fundamental
principles of scientific methodology, namely, the principle to take refutations seriously. The statement that there is no velocity greater than the
velocity of light is a well-corroborated statement of physics. If, as is asserted
in (a), quantum mechanics implies the negation of that statement, we
should consider it as refuted and look for a better theory. This is what has in
fact happened. Ever since the invention of elementary quantum mechanics
(which is not relativistically invariant) physicists have tried to design a
two-valued relativistic theory. These attempts, although by no means completely successful, have yet led to some promising results such as, for
example, Dirac's theory of the electron and the prediction of the existence of
the positron.

Now consider the alternative suggestion in (b). This alternative removes
the need to modify either quantum mechanics or the principle of contact-
action as it devises a language in which the statement that both are
incompatible cannot be asserted. It thereby presents a defective theory

---

[1] In *Philosophical Papers* (Cambridge, 1975), I, 166ff.

(quantum mechanics) in such a way that its defects (it is not Lorentz-invariant) do not become apparent and that no need is felt to look for a better theory.

It is evident that this sly procedure is only one (the most 'modern' one) of the many devices which have been invented for the purpose of saving an incorrect theory in the face of refuting evidence and that, consistently applied, it must lead to the arrest of scientific progress and to stagnation. In a private communication Putnam has evoked the example of non-Euclidian geometry as a case where it was suggested to change the formal structure of physical theories and he has said that it is analogous to the present case. But this comparison is altogether misleading. The application of non-Euclidean geometry to physics led to fruitful new theories; it suggested new experiments and enabled physicists to explain phenomena which so far had defied any attempt at explanation (the advance of the perihelion of Mercury is one of them). Nothing of that kind results from the application of a three-valued logic to quantum mechanics. On the contrary, important problems (such as how to relativize elementary quantum mechanics) are covered up, objectionable theories (elementary quantum mechanics) are preserved, fruitful lines of research (attempts to find a general relativistic theory of micro-objects) are blocked. Hence, no physicist in his right mind would adopt procedure (b).

2. EXHAUSTIVE INTERPRETATIONS AND THEIR ANOMALIES

Reichenbach's main problem is the interpretation of the unobservables of quantum mechanics. In this connection he considers what he calls 'exhaustive interpretations'. At least two non-synonymous explanations are given for interpretations of that kind. According to the first explanation an exhaustive interpretation is an interpretation which 'includes a complete description of interphenomena', i.e. of quantum-mechanical entities (*PF*, 33).[2] An exhaustive interpretation in this sense does not employ any special assumption about the nature of the things to be interpreted. The only conditions to be satisfied are that the interpretation be consistent as well as compatible with the theory used. According to another explanation, an exhaustive interpretation is an interpretation which 'attributes definite values to the unobservables' (PA, 342; cf. *PF*, 139). An exhaustive interpretation in this more specific sense (silently) employs an assumption (we shall call it assumption C) which may be expressed as follows: (a) Divide the class of all the properties which the entities in question may possess *at*

---

[2] The following abbreviations will be used: *P F* for *Philosophical Foundations of Quantum Mechanics* (Berkeley, 1946); P A for 'The Principle of Anomaly in Quantum Mechanics', *Dialectica* 7/8 (1948) 337ff; F L for 'Les fondements logiques de la théorie des quanta' in *Applications Scientifiques de la Logique Mathématique* (Paris, 1954), 103ff.

some time into subclasses comprising only those properties which exlude each other. These subclasses will be called the *categories* belonging to the entities in question. Then each entity possesses *always* one property out of each category. (b) The categories to be used are the classical categories.[3] Applied to the case of an electron, C asserts that the electron always possesses a well-defined position and a well-defined momentum.

It is evident that an exhaustive interpretation of the first kind (an E_1) i.e. an attempt to state what the nature of quantum-mechanical entities is, need not be an exhaustive interpretation of the second kind (an E_2), i.e. it need not be an attempt to represent quantum-mechanical systems as things which always possess some property out of each classical category relevant to them. An E_1 need not even comply with assumption C(a): It is not the case that water has always a well-defined surface tension (it possesses a surface tension only if it is in its fluid state); nor is it the case that it has always a well-defined value on the Mohs scale (it possesses such a value only if it is in its solid state). Yet one can explain what kind of entity water is.

Reichenbach shows that all E_2 lead to causal anomalies.[4] Those anomalies are not unusual *physical processes* although Reichenbach's wording some-

---

[3] This omits spin.

[4] This is expressed by Reichenbach's *Principle of Anomaly* (which he assumes to be independent of the uncertainty principle *PF*, 44). It is worthwhile considering the transformations this principle undergoes in Reichenbach's book. It is introduced as saying that 'the class of descriptions of interphenomena contains no normal system' (*PF*, 33) which means, when decoded, that the laws for quantum-mechanical objects cannot be *formulated* in such a way that they coincide with the laws governing the behaviour of observable objects, viz. the classical laws (cf. *PF* 19). As it stands the principle is obviously refuted by the fact that formulations of quantum mechanics and of classical physics exist which are identical. We may, however, interpret the principle as saying that the *laws* (not their formulations) for quantum-mechanical objects are not the same as the laws governing the behaviour of observed objects or, to use Reichenbach's terminology (which is supposed to express 'the quantum mechanical analogue of the distinction between observed and unobserved objects' (*PF*, 21)) that the laws of interphenomena are not the same as the laws of phenomena. In this case the truth of the principle follows from the definitions of 'phenomenon' and 'interphenomenon' provided by Reichenbach which say that the 'phenomena are determinate in the same sense as the unobserved objects of classical physics' (*PF*, 21) whereas the introduction of 'inter-phenomena can only be given within the frame of quantum mechanical laws' (*PF*, 21). For according to these definitions the principle of anomaly asserts that the laws of classical physics are different from the laws of quantum mechanics, which is of course true but does not justify the introduction of the principle as an independent assumption (see the beginning of this note). However, this is not the sense in which the principle is used at other places of the book where it is meant to say that 'every exhaustive interpretation' (in the second sense) 'leads to causal anomalies' (*PF*, 136). Having introduced this latter sense of the principle and having announced that it will be proved later on the basis of the principles of quantum mechanics, Reichenbach swiftly returns to the first interpretation (in which, as we have seen, the principle follows trivially from the definitions given for its two main terms together with the fact that classical physics is not quantum mechanics) and derives from it that the idea of the uniformity of nature (same laws for observables and unobservables) must be given up (*PF*, 39). These are only some of the confusions found in a book which demands that 'the philosophy of physics should be as neat and clear as physics itself' (*PF*, vii).

times suggests that they are.[5] Assume, for example, that we try to interpret the behaviour of electrons by waves. As soon as an electron is localized (at P) the wave collapses (into a narrow bundle around P). This sudden collapse cannot be understood on the basis of the wave equation which means that electrons are not (classical) waves.

On the other hand, consider the particle picture. If we want to explain interference (in the two-slit experiment) on the basis of the particle picture we must assume that the particle can 'know' what happens at distant points (cf. the discussion in 7 of PF). This 'knowledge' cannot be provided by any physical means (e.g. by a signal travelling with infinite speed) since (a) there is no independent evidence of the existence of such signals (hence the hypothesis that they exist would be an *ad hoc* hypothesis) and since (b) in the case of the existence of such signals the wave picture (which does not assume their existence) would lead to incorrect results even in those situations where it has been found to be correct. One may, of course, say that the wave picture provides us with a *description* of the dependences, existing between the state of a particle and some distant event (such as the event 'opening of the second slit') – but this amounts to saying that the particle picture is incorrect. Result: the so-called 'anomalies' are nothing but facts which show that quantum mechanics, interpreted in accordance with an E_2, leads to incorrect predictions. And the 'principle of anomaly' (in its second interpretation; see n.4) must be read as saying that *for any theory which consists of the mathematical formalism of quantum mechanics together with some E_2 there exist refuting instances.*

Reichenbach discusses four methods to solve this difficulty: Method 1 suggests that we should 'become accustomed' to the anomalies (PF, 37), i.e. it expects us not to be worried by the fact that the interpretation used turns quantum mechanics into a false theory. Method 2 advises us to use a certain interpretation only for describing those parts of the world where it works and to switch over to another interpretation as soon as a difficulty arises. The only difference between this method and method 1 is that the former uses alternatively two or more anomalous interpretations where the latter uses only one. It also leads to a renunciation of the idea that nature is

---

[5] Reichenbach realizes that his 'anomalies' are not simply physical phenomena which exist in addition to the phenomena implied by some E_2; he calls them 'pseudoanomalies' which are of a 'ghostlike character. . . . They can always be banished from the part of the world in which we happen to be interested although they cannot be banished from the world as a whole' (PF, 40). This means that they are not physical phenomena (physical phenomena cannot be 'banished' from the part of the world in which they happen to occur) but are due to a deficiency of the picture chosen (which, of course, can be explained away). More especially the existence of anomalies is not the same as the existence of signals with velocities greater than the speed of light as may be seen from the fact that Reichenbach counts among the anomalies of the theory of the pilot wave that 'this wave field possesses no energy' (PF, 32) and that he refers to the 'anomaly connected with potential barriers' (PF, 165) which simply consists in the violation of the principle of conservation of energy.

uniform in the sense that the same laws apply to both observables and unobservables (*PF*, 39, bottom paragraph). Method 3 suggests that we stop interpreting altogether and that we regard the statements of the theory as cognitively meaningless instruments of prediction (*PF*, 40). Method 4, which is the one adopted by Reichenbach, suggests that we change the laws of logic in such a way that the statements which show the inadequacy of one of the chosen interpretations 'can never be asserted as true' (*PF*, 42). Reichenbach seems to assume that the principle of anomaly forces us to adopt one of those four methods.

This wild conclusion is completely unjustified. The principle of anomaly applies to E_2 only and shows that they are untenable. There is no reason to assume that they should be tenable being based as they are upon the (classical) principle C. Classical theory is incorrect. Therefore, it is to be expected that also the more general notions of classical thought such as are incorporated in C (which is not even correct in all classical cases – see above) will turn out to be true only in a restricted number of cases. It is only when one does not realize that this assumption is part and parcel of classical thinking (rather than an *a priori* principle to be satisfied by any interpretation, whether E_1 or E_2) that one will be inclined to sense a breakdown of realism, or logic, or of the simple idea that theories are not only instruments of prediction but also descriptions of the world. Reichenbach is one of those thinkers who are prepared to give up realism and even classical logic because they cannot adjust themselves to the fact that a familiar and well-understood theory has turned out to be false (that not all interpretations are E_2).[6]

But the methods suggested by Reichenbach can be criticized also independently of these more general remarks. The criticism of 1 and 2 is obvious. Method 4 was criticized already in section 2. Some further criticism will be developed in section 4.

3. ANOMALIES AND THE PRINCIPLE OF CONTACT-ACTION

The above discussion (especially n.5) shows that the difficulty of the E_2 cannot be described by saying that '*the laws of quantum mechanics* . . . are logically incompatible with' contact-action (cf. the beginning of the first paragraph of section 1). The reason is first that those difficulties arise only if we use the laws of quantum mechanics *together with* assumption C (which is not a law of quantum mechanics). Second, it would even be incorrect to assume that the conjunction of the laws of quantum mechanics with C is incompatible with contact-action. The principle of contact-action applies to fields which can be used for the transmission of signals. Neither the collapse of

---

[6] It appears that the principle of complementarity owes its existence to the same reluctance to part with classical ideas.

a wave (in the wave interpretation) nor the telepathic information conveyed to particles (in the particle interpretation) can be used in this way. Neither of these phenomena contradicts contact-action (as should be seen from the discussion in section 2 and especially from n.5). It follows that the above description of Reichenbach's point of departure (due to Putnam) is incorrect.

4. THE POSITION OF LAWS IN THE SUGGESTED INTERPRETATION

A criterion of adequacy of the interpretation by means of a three-valued logic is this: (a) every statement expressing an 'anomaly' should have the truth-value 'indeterminate' (PF, 42); (b) every law of quantum mechanics should have either the truth-value 'true' or the truth-value 'false', but never the truth-value 'indeterminate' (PF, 160; F L, 105). It turns out (PF, 158f) that in the special system of three-valued logic used, formulas can be constructed which satisfy (b). The question arises whether every law of quantum mechanics satisfies (b).

Consider for that purpose the law of conservation of energy and assume that it is formulated as saying that the sum of the potential energy and the kinetic energy (both taken in their classical sense) is a constant. Now according to (a),[7] either the statement that the first part of the sum has a definite value is indeterminate, or the statement that the second part has a definite value is indeterminate, or both statements are indeterminate, whence it follows that the statements of the conservation of energy itself will always have the value 'indeterminate'. The same results if we use the statement in the form in which it appears in quantum mechanics. In this form the statement asserts that the sum of various operators, not all of them commuting, will disappear. According to (a) the statement that an operator has a certain value is indeterminate unless the operator is diagonal. As the only statements admitted to E_2 are statements to the effect that an operator *has* a certain value it follows again that the law of conservation of energy can only possess the truth-value 'indeterminate'.

The last argument admits of generalization: Every quantum-mechanical statement containing non-commuting operators can only possess the truth-value 'indeterminate'. This implies that *the commutation rules* which range among the basic laws of quantum mechanics *as well as the equations of motion* (consider them in their Heisenberg form) *will be indeterminate* and hence 'neither verifiable nor falsifiable' (PF, 42). We have to conclude that Reichenbach's interpretation does not satisfy his own criterion of adequacy since it violates (b).[8]

---

[7] In applications to concrete cases, such as the one under review, Reichenbach uses the stronger condition that statements about unobserved values should be indeterminate (PF, 145). The following arguments will be based upon this stronger condition.

[8] Reichenbach admits that in his interpretation 'the principle of the conservation of energy is

5. THE COPENHAGEN INTERPRETATION

Reichenbach considers his interpretation superior to what he thinks is the Copenhagen Interpretation. According to him this interpretation admits statements about (classically describable) phenomena only and calls meaningless all statements about unmeasured entities (PF, 40). Although the view is supported by many more or less vague pronouncements made by members of the Copenhagen circle, it is yet somewhat misleading. The correct account of the matter seems to be somewhat like this: In his earlier writings Bohr ascribed the change of the state of a system due to measurement to the *interaction* between the measuring device and the system measured. However, later on,[9] he made a distinction between *physical changes* ('mechanical disturbances') of the state of a system which are caused by physical fields of force, and changes of 'the very conditions which define the possible types of predictions regarding the future behaviour of the system'.[10] The fact that as an example of conditions of this latter kind Bohr mentions the reference systems introduced by the theory of relativity seems to indicate that what is meant here is a *logical property* of the state function and hence of any statement which ascribes a certain value to some variable of a quantum-mechanical system. On measurement the state S of a system s changes, not only because forces are acting upon the system, but also because it is a *relation between the system and a certain kind of physical preparation*. The analogues of classical properties are defined for a restricted class of states only, i.e. properties of a (classically explained) category are applicable to a system only if the system has been prepared in a certain way. If this interpretation is carried through consistently, that is, if it is separated from the instrumentalist philosophy which is an altogether independent (though never clearly separated) element of the Copenhagen view, it may be used as an E_1, (since it contradicts C). In this interpretation statements

eliminated . . . from the domain of true statements' (PF, 166). Six pages earlier he asserts that 'it will be the leading idea' (of the interpretation used) 'to put into the true–false class those statements which we call quantum mechanical *laws*' (original italics). From this I can only conclude either that on p. 166 Reichenbach has already forgotten what he said on p. 160, or that for him the principle of conservation of energy (and, we shall have to add, the quantum conditions as well as the equations of motion) are not quantum mechanical *laws*. It seems that the latter is the case; for Reichenbach mentions as a case where (b) is satisfied the '*law* of complementarity' (F L, 105; PF, 159; my italics), which has certainly never been listed as a physical law. In 'Three-Valued Logic' Putnam asserts that in the case of quantum mechanics 'the suggestion of using a three-valued logic makes sense . . . as a move in the direction of *simplifying*' (my italics) 'the whole system of laws'. This statement is of course true, but it is true in an unexpected sense. Quantum mechanics is 'simplified' indeed as all important laws are 'eliminated . . . from the domain of true statements'.

[9] Cf. especially 'Can the Quantum Mechanical Description of Reality be regarded as Complete', *Phys. Rev.* 48 (1935), 696ff, and here especially the last paragraph, as well as *Albert Einstein, Philosopher-Scientist*, ed. P. A. Schilpp (Evanston, Ill., 1949), especially 231ff.

[10] *Albert Einstein*, 234.

about 'interphenomena' are meaningful (hence it is an E_1) but statements such as 'the position of the electron at time t is x' ('P $(s, t) = x$') are occasionally meaningless (hence it is not an E_2). It is mainly for the latter reason that Reichenbach finds the Copenhagen Interpretation unsatisfactory. In the next section we shall consider some of his (and Putnam's) arguments which are intended to show that their interpretation is better than the Copenhagen Interpretation.

6. ARGUMENTS AGAINST IT CONSIDERED

Three arguments are used in FL to show that 'three-valued logic, and it alone, provides an adequate interpretation of quantum physics'. They seem to be the only arguments which Reichenbach has at his disposal. The first argument may be formulated thus: The statement that 'P $(s, t) = x$' is meaningful only if certain conditions are realized amounts to saying that a statement is meaningless at the time t if no observer is testing it at time t. 'The whole domain of unknown truth would thus be eliminated from physical language' (PA, 347). Putnam uses a similar argument: the Copenhagen Interpretation is unsatisfactory since it allows us to call a statement meaningful only if we actually *look* at some measuring apparatus used for testing the statement.

Two points must be distinguished in this argument, namely (a) the (correct) point that predictions belonging to classical categories can be applied to a physical system only if certain conditions are fulfilled; and (b) the (incorrect) assertion that these conditions include an observation by a conscious observer. To deal with (b) first it must be pointed out that actual observation is by no means necessary in order to enable one to say that a certain predicate (like 'position') applies to a physical system. If the system is in such (physically definable) conditions that its state may be represented as a superposition of spatially well-defined bundles with negligible interference between them, then we may say that it possesses some position. If on the other hand the physical conditions are such that narrow trains with approximately the same frequency are fairly well isolated, then we may say that the system possesses some (perhaps unknown) momentum. Hence, whether or not a system possesses momentum or position depends on the existence of physical conditions (which *may* have been realized by a measuring instrument: every measuring instrument is devised to provide a separation of wave trains such that predicates of a given classical category become applicable to the system) and not on the presence of an observer. But is the fact that *such* a dependence exists in itself unsatisfactory (point (a))? Reichenbach's and Putnam's answer to this question is that sentences like 's has the position x at time t' 'admittedly have a very clear cognitive use; hence it is unnatural to regard them as "meaningless" ' (Putnam, 102;

cf. also F L, 105). This answer smells dangerously of apriorism. It also overlooks that in our search for better theories we frequently discover that situations we thought would obtain universally do in fact exist only under special conditions, which implies that the properties of these situations are applicable in those conditions only. Within the theory this new dependence is then expressed by introducing a relation for something which was so far described by a property. The theory of relativity is a familiar example of this procedure which in the above argument is described as 'unnatural'.[11] Is it perhaps more 'natural' to stick to the notions of an overthrown theory at the expense of epistemology (transition to idealism) and even of classical logic?

Reichenbach's second argument runs as follows: If it is meaningless to say, in the case of the two-slit experiment, that an electron has passed through, say slit 1, then it is also meaningless to say that it passed through one of the two slits. Yet we would like to assert the latter statement. In Reichenbach's interpretation such an assertion is possible since the disjunctive statement 'the electron passed through slit one or it passed through slit two' may be true (PF, 41, 163f; F L, 104). Hence this interpretation is preferable to the usual interpretation. Here it must be pointed out that the statement that every electron which arrives at the photographic plate has passed either through slit 1 or through slit 2 is of course excluded by quantum mechanics since it would imply that there is no interrelation between the situation at slit 1 and the situation at slit 2. But is does not follow, as Reichenbach seems to assume (F L, 104) that on that account we are unable to say that what has arrived at the photographic plate has passed through the slits and not through the wall in between; for it is not true that only particles can pass through slits and be intercepted by walls. The correct description consists in saying that within a certain interval of time (to be determined by the latitude of knowledge of energy) interfering parts of the electron passed through the (not simply connected) opening 'slit-1-plus-slit-2'.

---

[1] Reichenbach and Putnam express in the 'formal mode of speech' a type of argument which has frequently been used by 'traditional' philosophers against the conceptions introduced through new theories: it was regarded as 'unnatural' to let simultaneity depend on the coordinate system chosen (and to assume that '$Sim(xy)$' is not well formed and hence, meaningless); yet it had to be admitted that special relativity was more successful than pre-relativistic physics. In order to solve this difficulty traditional philosophers usually adopted what we have called method 3 (section 2 above; cf. Philipp Frank, *Relativity, a Richer Truth* (Boston, 1952)). That is, they regarded relativity proper as a set of cognitively meaningless sentences which nevertheless could be used as parts (cogwheels, so to speak) of a good prediction machine. However objectionable this method may be, traditional philosophers took contradictions seriously and tried to remove them. It was left to Reichenbach (who argued against 'speculative philosophy which must appear outmoded in the age of empiricism', PF, vii) to provide the above approach with two further methods, viz. the 'method' to call contradictions 'anomalies' and 'to become accustomed to them' (his methods 1 and 2), and the 'method' of dropping two-valued logic.

Reichenbach's third argument consists in pointing out (FL, 105) that in his interpetation all the laws of quantum mechanics are statements which are either true or false, but never indeterminate (whereas within the Copenhagen Interpretation they may be meaningless). We have shown above that this statement is incorrect.

We may summarize this section by saying that none of Reichenbach's arguments in favour of his own position and against the Copenhagen Interpretation are tenable.

7. FORMALIZATION

The arguments of the above sections have mainly been formulated in the 'material mode of speech' to use an apt expression of Carnap's, i.e. they have been formulated as arguments about the properties of quantum-mechanical systems. These arguments may also be expressed in the formal mode of speech, i.e. as arguments 'concerning the structure of the language in which this world can be described' (PF, 177) and, more especially, as arguments concerning the structure of the language of quantum mechanics. Both kinds of arguments may be found in Reichenbach's book. It is to Putnam's merit that he separated them more clearly by describing the problem at issue as *the problem to find an adequate formalization of quantum mechanics*.

Yet one must realize that by using the formal mode of speech the problem at issue has not been removed from the domain of physical argument. Assume that we use a formalization in which (a) the logic is two-valued and (b) for every s some atomic sentences are of the form '$P(s, t) = x$.' Any theory which has been formalized in this way is committed to the assertion that a system with a single degree of freedom has always a well defined position, i.e. it is committed to the particle interpretation. Similar remarks apply to other types of E_2. Hence, all our arguments against E_2 can be repeated against the corresponding formalizations and they show that a theory formulated in accordance with some such rule as (a) and (b) above will lead to 'anomalies'. Quantum mechanics before any formalization does not lead into anomalies. Hence, the formalizations considered are inadequate (here I have used the principle that T' is an adequate formalization of T only if there is no empirical statement which follows from T and does not follow from T' and vice versa). And Reichenbach's interpretation may now be described as an interpretation which shows that even an inadequately formalized quantum mechanics can be made compatible with some very distressing facts, if a three-valued logic is used. However, it would obviously be more 'natural' to use an adequate formalization.

Against this argument Putnam has asserted (in a private communication) that we cannot say what 'follows' from a theory unless we are already using some formalization, and that what I have done above amounts to nothing but a comparison of the results of one formalization with the results of another formalization. Now even even if this were the case, I would still maintain that my arguments are good arguments and show that my (alleged) formalization is preferable to Reichenbach's and Putnam's. But a brief consideration of the arguments used in section 2 will, I think, convince everybody that the point at issue here does not presuppose any specific formalization (although it may be possible to present it formally in a more satisfactory way).

I must conclude, then, that in spite of Putnam's arguments, in his paper and also in private discussion, I still feel that Reichenbach's suggestions cannot be regarded as a step toward a better understanding of the 'logic' (in a not strictly formal sense) of quantum mechanics.

16
Niels Bohr's world view*

1. INTRODUCTION

In his essay 'Quantum Mechanics Without "the Observer"' (in [20], 7–44), Popper criticizes the Copenhagen Interpretation and suggests construing the quantum theory as a 'generalization of classical statistical mechanics' ([20], 16). The uncertainty relations, he says, set 'limits to the statistical dispersion . . . of the results of *sequences* of experiments' ([20], 20) and not to what can be said about individual systems. Some strange features of orthodox microphysics are due either to a misinterpretation of probability, involving some 'very simple mistakes' ([20], 42); or else they are a straightforward consequence of the fact that the quantum theory is a statistical theory. For example, 'the reduction of the wave packet . . . has nothing to do with quantum theory: it is a trivial feature of probability theory' ([20], 37). The idea of probability that is presupposed in all these arguments, however, is Popper's propensity interpretation. Adopting this interpretation suffices for making clear what the quantum theory really is: a 'generalization of classical statistical mechanics' ([20], 16). This is Popper's position as I understand it.

This position, it seems to me, is attractive enough as long as one does not carry the analysis too far. But it breaks down when confronted with *all* the facts which Bohr wanted to take into account. Altogether Popper's criticism and his alternative proposals once more reveal the strong points of the Copenhagen philosophy, and especially of the ideas of Niels Bohr.

Today it is very difficult to find a satisfactory account of these ideas.[1] Both the opponents and the followers of Bohr have changed and to some extent distorted his views.[2] And their identification with the so-called

---

* An earlier version of the paper was criticized by Joseph Agassi and J. W. N. Watkins, Dr Bub and Imre Lakatos, as well as by Mr Musgrave and an anonymous referee. I have made use of some of their suggestions. This paper is a belated after effect of a discussion with C. F. von Weiszäcker in autumn 1965. For support of research I am again indebted to the National Science Foundation.

[1] This is no longer entirely correct especially in view of Dr Meyer-Abich's marvellous book [76], which should be made available in an English translation.
[2] As regards his followers we have Bohr's own testimony. As an example, see Bohr's criticism of Heisenberg as reported in section 7 below. Heisenberg, of course, went his own way ([82], 15; [87], cf. also nn.79 and 87) but his views are often confounded with those of Bohr.

'Copenhagen Interpretation', which is a *fait accompli* as far as textbooks and philosophical discussions are concerned, has almost put an end to the possibility of a fair examination. For the 'Copenhagen point of view', as it is understood by some critics, and, one must admit, by some of its adherents,[3] is not a single idea but a mixed bag of interesting conjectures, dogmatic declarations, and philosophical absurdities. Some versions which are discussed at great length in the literature are not even a real thing but an arbitrary construction on the part of certain philosophers and scientists who in their eagerness to prove Bohr wrong have collected *prima facie* absurd statements wherever they could find them,[4] without regard for context, or for idiosyncrasy of expression.[5] Is it worthwhile to solemnly criticize spectres of this kind? I do not think so. But it is worthwhile to examine a philosophy that has led to concrete physical suggestions,[6] including criticisms of premature declarations of success,[7] that is largely responsible for the discovery and the proper understanding of one of the most fascinating contemporary theories and that entails also some very interesting ideas concerning the relation between subject and object, concept and fact, and knowledge and observation. I shall therefore use Bohr rather than the 'Copenhagen school' as a measure of the adequacy of Popper's criticism and of his positive suggestions.

2. PROPENSITY: A PART OF COMPLEMENTARITY

I shall start my presentation by exhibiting an interesting parallelism between the propensity interpretation and complementarity. This parallelism is rather striking for it makes Popper stand much closer to Bohr whom he attacks than to Einstein whom he defends.[8]

Probabilities, according to Popper ([20], 32ff), are objective properties of experimental arrangements which depend both on the system examined

---

[3] See e.g. Feynman's version as reported in n.62 below.

[4] Jocular asides and popular simplifications included.

[5] E.g. Bohr's use of the term 'subject', or 'observation'. Cf. below, section 6, n.81.

[6] 'In fact, there was rarely in the history of physics a comprehensive theory which owed so much to one principle as quantum mechanics owed to Bohr's correspondence principle.' (M. Jammer [56], 118).

[7] See Bohr's criticism of the quantum theory of Schwarzschild, Epstein and Sommerfeld as reported in section 6 below.

[8] This is not the only instance where the positive view of an author who attacks Bohr turns out to be rather close to the view which Bohr actually holds. Thus Professor Margenau first characterizes Bohr as a 'formalist' and a 'positivist' and ascribes to him the view that *particles have position in space and time under all circumstances* but atomic nature is so constituted that we often cannot know them' ([74], 8 original italics); having done this he introduces a 'third view', his own, which is indistinguishable from Bohr's (from Bohr's real view and not from his view as described by Margenau) except for some fancy terminology ('possessed' and 'latent' variables). The same remark applies to some of Margenau's pupils such as McKnight [77] and Park. Cf. also n.13 for similar comments on Landé's work.

(e.g. a die) and on the physical conditions under which it is being examined. Thus the probability of a loaded die showing a six will change when we remove it from the surface of the earth, and it will change again when we bring it into a strong magnetic field (assuming that the loading has been done with a piece of metal). This joint dependence of the outcome of a probability experiment on system and conditions suggests not separating the two, even in words, but regarding them as different features of a single indivisible block: die-plus-surroundings. The statement $p[a, e] = \alpha$ can correspondingly be interpreted in two ways, the one still reflecting the older prevalence of the system that 'has' the probabilities, the other making allowance for the fact that system and surroundings enter the situation on an equal basis, and that the probabilities belong neither to the one, nor to the other, but to both ('Propensities', says Popper, 'are properties of neither particles, nor photons, nor pennies. They are properties of repeatable experimental arrangements' ([20], 38 – italics omitted; cf. also [20], 40)).

As an example of the first way, we may say that a stands for 'this individual die is showing a six' and that e characterizes the experimental set-up. Or, as the second way, we may say that a stands for block: 'die-plus-surroundings showing a six' while e stands for the block 'die-plus-surroundings'. (We are here always speaking of a particular individual die and of particular individual surroundings. However, it is not difficult to change the interpretation and to speak of dice of a certain kind in surroundings of a certain kind.) If I interpret Popper's intentions correctly, then the second formulation would definitely be preferable to the first as it makes it clear that α is predicated of the relation in which the die stands to the surroundings and not of the die taken by itself.

Once these matters have been settled it is evident that as long as we consider probabilities a particular system cannot be separated from its surroundings, not even in thought. The probability of a die that has been shaken in a randomizer cannot therefore be explained as the result of an interaction between the die and the randomizer, or of a disturbance of the die by the randomizer. Any such account assumes that die and randomizer are separate systems which may affect each other in various ways. And while this is a perfectly adequate assumption in dynamics it must not be made in the case of probabilities which apply to the total experimental set-up.

We can go a step further. Assume we stop throwing the die in a random fashion and start examining its dynamical behaviour, such as the trajectory its centre of gravity will describe when it is put into the planetary system. Extrapolating the propensity interpretation into dynamics, we may then say that we have ceased to discuss the 'block' die-cum-randomizer, and have started discussing the block die-cum-planetary system (in which the probabilities happen to be very close to one). Seen in this way (which should be compared with Exner's interpretation of mechanics as explained in the famous fourth chapter of his *Vorlesungen ueber die Physikalischen Grundlagen der Naturwissenschaften*) Newtonian mechanics is not a theory that describes the behaviour of physical systems under all possible circumstances and therefore also in randomizers. It is rather a theory which applies to specific 'blocks' such as material particles in planetary systems, in double stars, on frictionless planes and so on, but which is entirely irrelevant as far as the behaviour of other blocks is concerned.

The reader who is familiar with complementarity will at once recognize the great similarity with the view just described. Ever since 1935 (cf. especially [9] and the 'Discussions with Einstein' in [34]) and in slightly different terms even long before,[9] Bohr has emphasized the need to refer the

---

[9] See especially [6], 118, as well as [8], *passim*. The expression 'experimental arrangement' in the sense that is relevant here occurs for the first time in the introduction to [8] which was written in 1929: 'Now what gives to the quantum theoretical description its peculiar characteristic is just this, that in order to evade the quantum of action we must use *separate experimental arrangements* to obtain accurate measurements of different quantities . . . and further that these experimental results cannot be supplemented by repeated measurements' ([8], 17 – my italics).

The idea behind the expression occurs much earlier and may be said to form the background of Bohr's very first paper. At this early stage the emphasis is upon the 'abstract' ([6], 57) character of classical concepts such as the 'concept of radiation in free space as well as [the concept of] isolated material particles' ([6], 56f) which derives from the assumption, implicit in their meaning, 'that the phenomena . . . may be observed without disturbing them appreciably' ([6], 53). As a consequence classical concepts may be used only in situations where the disturbance is negligible: 'one is obliged . . . to always keep in mind the[ir] *domain of application*' ([6], 118) – which is precisely the restriction expressed at a later stage by the reference to experimental arrangements.

Bohr *originally* believed that *all* quantum theoretical considerations have approximate or, as he often expressed himself, 'formal' validity only. Thus the existence of stationary states and the possibility of calculating them with the help of ordinary mechanics and electrodynamics 'must not be asserted with a precision that is greater than the validity of [Hamilton's equations] whose approximate character follows already from the omission of the radiative forces of classical electromagnetic theory' ([6], 127; 119, 129, 130, 153ff as well as the criticism of Epstein's more dogmatic position on p. 128, n.1). Cf. also the difference between Munich and Copenhagen as reported below in section 6. 'This lack of sharpness in the description of the movement of the electrons within an atom entails an uncertainty in the definition of stationary states whose recognition may on occasions be of essential importance' ([6], 152: the quantum theoretical uncertainties already announce themselves!). *Later on* it was discovered, mainly as a result of the experiments carried out in an effort to test the theory of Bohr, Kramers and Slater [13], that certain classical laws remained valid on the microlevel also. It now became possible to give a more precise account of the restricted application of classical terms. In this account one started from the idea, implicit in the principle of correspondence, that classical terms are freely applicable in the 'interpretation of experimental evidence' ([8], 56) arising in the 'special problems' ([8], 72) which are defined by an appropriate 'experimental device' ([8], 66). There arose then the demand to refer all statistical results to specific experimental arrangements and to 'constantly keep the possibilities of definition' (i.e. the concepts permitted – cf. also n.85) 'as well as of observation' (i.e. the experimental conditions) before the mind' ([8], 73). This development was furthered by the arrival of wave mechanics (cf. n.87). The appropriate terminology (elimination of such phrases as 'disturbing the object by measurement' – cf. the next footnote) was introduced in 1935; and it was consolidated in 1948, in Bohr's contribution to the Einstein volume.

statistical results of the quantum theory to experimental conditions and to explain some of their changes not by a causal influence, but by a change of these conditions. It is in this spirit that he criticizes phrases such as 'disturbing the phenomena by observation', etc.,[10] which suggest that the reduction of the wave packet is due to an *interaction* between system and experimental conditions, both being regarded as separate entities bound together by a strong force.[11] (In the theory of relativity which also ascribes properties to the total situation rather than to the individual system such a 'mechanistic' interpretation is provided by the point of view of Lorentz. See below, section 6.[12]) He quite explicitly reminds us that one must distinguish between 'a mechanical disturbance of the system under investigation' and 'an influence *on the very conditions* which define the possible types of [statistical] predictions regarding the future behaviour of the system'; and he compares this latter influence with 'the dependence on the reference

[10] Such phrases abound in Heisenberg. Thus we read ([51], 495): 'The most important distinction between quantum theory and the classical theories consists in the fact that when observing a physical magnitude we must take into consideration the *disturbance* which the experiment effects in the system measured. It is quite generally the case that an observation *changes* the physical behaviour of the system. The so far unsolvable paradoxes of the quantum theory all arose from neglecting this *disturbance* which is necessarily connected with every observation; according to Bohr they could be simply explained by a careful discussion of the *interaction* between the measuring apparatus and the system to be measured.' (My italics; they emphasize the critical terms which Bohr wanted to eliminate. Cf. also the next footnote.) We find these phrases also in Bohr [8], 5, 11, 15, 54, 57, 68, 93, 115; [10], 315), even as late as 1961: the indeterminacy discovered by Heisenberg's 'appears not only as an immediate consequence of the commutation relations ... but also directly reflects the *interaction* between the system under observation and the tools of measurement'. But Bohr immediately adds, in accordance with the caveat formulated in the next footnote, that 'the full recognition of the last crucial point involves ... the question of the scope of unambiguous application of classical physical concepts' ([98], 26 – my italics. Cf. also the discussion of the uncertainty relations in section 7 below.)

[11] 'In this connexion I warned especially against phrases, often found in the physical literature, such as "disturbing of phenomena by observation" or "creating physical attributes to atomic objects by measurement." Such phrases which may serve to remind of the apparent paradoxes, in quantum theory, are at the same time apt to cause confusion, since words like "phenomena" and "observations", just as "attribute" and "measurement" are used in a way hardly compatible with common language and practical definition.'

[12] 'Notwithstanding all differences between the physical problems which have given rise to the relativity theory and quantum theory, respectively, a comparison of purely logical aspects of relativistic and complementary argumentation reveals striking similarities as regards the renunciation of the absolute significance of conventional physical attributes of objects. Also the neglect of the atomic constitution of the measuring instruments themselves, in the account of actual experience, is equally characteristic of the application of relativity and quantum theory' [12], 64. The last sentence alludes to the fact that Einstein was looking for laws which characterize physical systems in a general way, and quite independently of their physical constitution. Cf. [61], esp. 510 and 515, concerning 'theories of principle'. One should notice the great similarity, emerging from this essay, between Einstein's procedure and that of Bohr. 'I no longer ask whether these quanta really exist', Einstein writes to Michele Besso in 1911 ([61],514) 'nor am I trying any longer to construct them, because I now know that my brain is incapable of accomplishing such a thing. But I am searching through the consequences [of the quantum hypothesis] as carefully as possible, *in order to learn the domain of applicability of this concept*' (my italics).

system, in relativity theory, of all readings of scales and clocks' ([9], 704).

The very same dependence of statistical results on experimental conditions is emphasized, about ten years later, by Popper who writes: 'We may say that *the experiment as a whole* determines a certain probability distribution' ([20], 35) which will change or not '*relative to what we are going to regard as a repetition of our experiment*; . . . in other words, [probabilities] are relative to what experiments are, or are not regarded as *relevant to our statistical test*' ([20], 36 – original italics).

'The experiment as a whole' – that is precisely what Bohr means to express by his notion of a *phenomenon* which is supposed to 'refer to the observations obtained under specific circumstances *including an account of the whole experimental arrangement*' ([12], 64, 50, and *passim*).

Comparing quotations like these we see that complementarity and the propensity interpretation coincide – as far as probabilities are concerned.[13] In both cases probabilities are properties of 'blocks', be they now called 'experimental arrangements', or 'phenomena'. They are not properties, not even tendencies, of individual physical systems.

3. MEASUREMENT: CLASSICAL LIMIT

I shall digress for a moment in order to mention an interesting consequence of the block-interpretation of probability which is quite clear in Bohr but which may also be developed in Popper. It has been said in section 2 that it is possible to extend propensity to purely dynamical theories such as Newton's mechanics. Such an extension means relating the mechanical behaviour of physical objects outside randomizers to special conditions also, and turning every statement of Newton's mechanics into a relational statement involving these conditions. As far as I am aware, such an extension does not appear in Popper's writings. I do not even know whether he would be prepared to take the step (his insistence on trajectories in probability contexts ([20], 25) is perhaps an indication that he would not). However, his assertion that 'statistical conclusions cannot be obtained

---

[13] In his essay 'Measurement and Quantum States' ([75], section 4) Margenau proposes a very similar interpretation of probability. He points out that quantum theoretical probabilities are irreducible in the sense that (excluding mixtures) they cannot be changed by knowledge, or by evidence, but only by a 'refinement or narrowing of the *physical conditions* to which the probabilities refer' ([75], 11 – my italics). This use of probability, he says, is possible in the classical case also: we can ascribe to a die a certain fixed probability to show a six 'which has reference to conditions inherent in the die' and does not change with the discovery that the die actually shows a six ([75], 10f) – an account that can easily be extended so as to include the physical conditions in the neighbourhood. Landé, in his beautiful unification of the quantum theory [71] has paid special attention to the dependence of probabilities on experimental arrangements, 'meters', or 'filters', as he calls them. One wonders what prompts him (and Rosenfeld) to believe that he is in opposition to Bohr: cf. [11], 133 as well as Landé's reply in [72], 132–4.

without statistical premises' ([20], 17) which has always formed a very essential part of his interpretation of probability (cf. [83], section 70) would in this way obtain much needed support.[14] A derivation of this kind would now be impossible not because the dynamical premises taken by themselves are too poor, as it were, for yielding statistical results; it would be impossible because they involve experimental conditions different from those presupposed in the statistical case. The reverse is of course also true: dynamical consequences cannot be derived from statistical premises either, however closely the formulae of the two domains and the numerical values obtained with their help may be made to approach each other. For what would be needed for such a derivation is not just an identification of formulae and of numerical values, but an identification of different experimental conditions – and this cannot be done. More especially, it is now impossible in principle to derive classical physics (classical mechanics) from the quantum theory.

This last assertion plays an important part in Bohr's philosophy where it is pointed out that classical mechanics and quantum theory are 'caricatures', depicting two separate, non-overlapping parts of the world, so that it would be in vain to try to reduce the one to the other.[15] Remarks like these can be found quite early, in connection with the *principle of correspondence*. Thus Bohr warns us in 1922 ([6], 144) 'that the asymptotic connexion' between the quantum theory and classical physics 'as it is assumed in the principle of correspondence ... does not at all entail a gradual disappearance of the difference between the quantum theoretical treatment of radiation phenomena and the ideas of classical electrodynamics; all that is asserted is an asymptotic agreement of the numerical statistical results'. In other words, the principle of correspondence asserts an agreement of numbers, not of concepts.[16] Nor is this numerical agreement assumed to be complete. Indeed, the simile of the 'caricatures' suggests that there are problems which are adequately dealt with by classical electrodynamics but

[14] Concerning the need for such support, cf. the last two sections of my essay 'In Defence of Classical Physics' ([68]).

[15] 'Bohr has expressed himself in discussions somewhat as follows: classical physics and the quantum theory, taken as descriptions of nature, are both caricatures; they allow us, so to speak, to asymptotically represent actual events in two extreme regions of phenomena.' H. A. Kramers [63], 559.

[16] According to Bohr this agreement of numbers has even a certain disadvantage, for it obscures (verschleiert) 'the difference in principle between the laws which govern the actual mechanism of microprocesses and the continuous laws of the classical point of view' [6], 129; see also [8], 85, 87f. Bohr has therefore repeatedly emphasized that 'the principle of correspondence must be regarded as a purely quantum-theoretical law which cannot in any way diminish the contrast between the postulates [i.e. the postulate of the existence of stationary states and the transition postulate] and electromagnetic theory' ([6], 142 n. and *passim*; see also [7], 159). The difficulties arising from a neglect of this situation have been exhibited very clearly by the late N. R. Hanson ([47], chapter 6; see also my comments on Hanson in [39], esp. 251).

which fall outside the competence of the quantum theory. Such problems exhibit 'a total failure' (*'ein voelliges Versagen'*) of the postulates of this theory ([6], 156). They show that the quantum theory, taken by itself, is too weak to give us classical physics – or at least it was too weak in 1922 (the year of the above quotation), before the arrival of the new quantum mechanics of Schrödinger and Heisenberg. According to Bohr these new theories 'can be regarded as a precise formulation of the tendencies embodied in the correspondence principle' ([91], 49).[17] One may therefore conjecture that they will still show essential lacunae which must be filled from the outside, with the help of classical considerations. It seems that this intuition of Bohr's and the vivid image of the 'caricatures', mentioned above, must indeed be regarded as essentially correct. Or, to speak more cautiously, the opposite position according to which the quantum theory is self-contained and 'reduces to classical mechanics as a special case' (Margenau [23], 349[18]) encounters difficulties which are serious enough to regard it as refuted. We shall now briefly look at this matter.

Consider the following two questions.

(i) Is it true that all problems which are successfully dealt with by (non-relativistic) classical physics (classical mechanics) can also be treated by the elementary quantum theory without help from additional assumptions which (a) do, (b) do not involve classical ideas?

(ii) Is it true that the quantum theory does not make any assertion that disagrees with assertions made by classical physics (classical mechanics) in its own proper domain, and not contradicted by experience? (The last qualification is added in order to exclude such cases as superfluidity, superconductivity, the existence of stable material objects, visual fluctuations under weak illumination, and so on.)

A negative answer to (i) would force us to *supplement* the quantum theory with principles obtained from inspecting the macrodomain and expressed, in the case of (i) (a), in classical terms.

A negative answer to (ii) would force us to *eliminate* certain consequences of the quantum theory – and this again on the basis of known macroscopic facts.

---

[17] The reference is here to Heisenberg's early investigations which are indeed a direct outcome of the correspondence principle and especially of Kramers' dispersion theory. However, in his late papers Bohr makes it clear that the same is true of the fully developed elementary theory of Heisenberg and Schrödinger: 'In fact, wave mechanics, just as the matrix theory, on this view represents a symbolic transcription of the problem of motion of classical mechanics adapted to the requirements of the quantum theory' ([8], 74). Even the various theories of elementary particles which were developed at the time are assumed by him to 'fulfil' the correspondence requirements' ([8], 90).

[18] As an argument Margenau refers to the pseudoclassical behaviour of special wave-packets associated with macroscopic mass points. But the existence of such cases which are well behaved from a classical point of view does not of course guarantee the non-existence of other consequences of the quantum theory which are not so well behaved.

It seems that neither (i) nor (ii) admits of an affirmative answer.

As regards (i) we can start by pointing out that a complete subsumption of the macrolevel under the laws and the concepts of the quantum theory assumes the existence of quantum theoretical descriptions for every macroscopic situation, or else for every situation that can be described in classical terms. For example, there must be Hermitian operators corresponding to the Eulerian angles and the temporal changes of the Eulerian angles of a rigid body so that the essential theorems of rigid body mechanics can at least be stated in quantum-mechanical terms (we are not as yet talking of a derivation of these theorems from the quantum theory). Also – and this point is often overlooked in the discussion of problems such as the one just introduced – the quantum theory of measurement must be extended to macroscopic situations and it must be explained how the complementary features of micro-objects manage to disappear on the macrolevel. This programme runs into trouble at the very beginning. The problems which arise are connected partly with the quantum theory of measurement proper where 'for some observables, in fact for the majority of them (such as xyp_z) nobody seriously believes that a measuring apparatus exists' ([101], 14 – the reference is to microscopic magnitudes). And as regards macroscopic magnitudes we encounter difficulties of definition already in the simplest cases, such as in the case, mentioned above, of the Eulerian angles of a rigid body.[19] The attempt at a *general characterization* of macroscopic variables has more recently been taken up by various authors such as Ludwig [15][20] and the members of the Italian school ([21], [27], [28]), but this very attempt has revealed the need for new principles which one has then tried to discover with the help of an 'inverted principle of correspondence which, starting from the quantum theory allows us to guess at the limits of very large systems, the main features of macrophysics acting as guides in our guesswork' ([15], 160). Now while such additional principles (which have been compared with the exclusion principle ([15], 159)) might be accepted as legitimate additions which do not transcend the framework of the quantum theory, this can no longer be said about the additions and modifications required in connection with the second question.

[19] Schrödinger [90] and [91] has shown the difficulties of defining the angle between two mutually inclined surfaces of a crystal.
[20] 'Geloeste und Ungeloeste Probleme des Messprozesses in der Quanten mechanik': this essay contains an excellent survey of the problems which one encounters when trying to link the quantum theory with the macrolevel. Cf. also Ludwig's essay in [21] for a more technical presentation. The drawing below is taken from [15], 159 and agrees with the spirit expressed in the simile of the 'caricatures'. Cf. n.15 of the present paper.

As regards this second question it suffices to point out that the quantum theory is a linear theory whereas classical mechanics (with such idealized exceptions as the theory of small vibrations) is not. Now all those approximations which are based on quantum theoretical hypotheses about the macrolevel preserve linearity so that deviations are bound to occur.[21] Drastic examples of such deviations have been given by Schrödinger ([89], 812) and by Einstein [92]. *And exactly these drastic deviations are produced by quantum-mechanical measurements*:[22] a measurement of a magnitude in a system that is not in one of its eigenstates separates these eigenstates (e.g. by spreading them out in space, as is done in the Stern–Gerlach experiment) but it does not destroy the interference terms (this is true even if the original state of both the system and the measuring apparatus should happen to be a mixture rather than a pure state ([101], 11)). Thus the state-function of a single electron that has passed a device showing some kind of periodicity and now starts interacting with the molecules on a photographic plate is broken up into many tiny packets with complex phase relations between them. And as long as there are these phase relations the electron cannot be said to be in any particular packet, but must be described as being 'in all of them at once' ([29], para. 3).[23] But then the quantum theory as we have described it so far (which is not yet the quantum theory of Bohr and Heisenberg) does not allow us to say that the electron has interacted with a particular molecule and has left its mark. And this indefiniteness will spread the more objects participate in the interaction, and it may even reach the mind of the conscious observer making it impossible for him to say that he has received definite information, no matter how certain he himself may feel about it.[24]

At this point one might be tempted to say that the difficulty is purely imaginary. One might wish to emphasize that while there are these phase relations between different wave packets, even on the macroscopic level, and while it may be difficult to deal with them mathematically, the logic of the situation is quite simple: owing to their complexity, and owing also to

---

[21] Rosen [85] gives specific examples and discusses modifications of the quantum theory which might prevent these deviations from appearing. Hill ([42], 430ff) discusses in detail some of the problems which arise in the attempt to reduce classical mechanics to the quantum theory.

[22] This is discussed in detail in [101].

[23] Of course we do not here regard the electron as a perennial particle (as Popper does [20], 27). For arguments against such an interpretation, cf. section 5 of the present paper and the discussion of the uncertainty relations in section 7.

[24] For these points, see ch. 13. Professor Wigner, in [101], has objected to extending the argument to the subjective impressions of the observer which to him contain no element of uncertainty whatever. But we must of course also know that the impressions have something to do with the case (the electron) – and here certainty can no longer be guaranteed. Besides, Wigner's point supports what we want to establish: that the application of the quantum theory to the macrolevel (sensations included) is impossible without additional assumptions.

the crudity of our observations on the macrolevel (looking, etc.) they will forever escape detection.[25] But as was pointed out by Hilary Putnam in conversations in 1957,[26] and as Bohm and Bub have explained more recently,[27] the removal of the interference terms does not yet return us to the classical level. This removal changes an assembly of interfering wave packets which are jointly occupied by the electron into an assembly of isolated wave packets which, however, are still jointly occupied by the electron. And as the classical level is reached only when we are allowed to assign the electron to a single wave packet we need a further transition which cannot in any way be regarded as an approximation.[28] It was a discovery of the first order to show that interpreting the terminal assembly – the assembly that emerges in our natural surroundings and can be described in the usual macroscopic terms – as a collective in the sense of classical statistics is a consistent procedure which leads to correct predictions.[29] It was this discovery which allowed physicists to combine wave mechanics with Born's interpretation and which was adopted by the founders of the quantum theory, and especially by Niels Bohr. Let us repeat the assumptions which are involved in it:

(1) Interference terms are disregarded – but the approximation is excellent and the differences undiscoverable on the macrolevel.[30]

(2) Macro-objects are assumed to have well-defined properties which inhere in them independently of measurement and which do not spread.

(3) These properties are identified with the properties ascribed to the objects by classical physics (such as the position of a pointer, or of a mark on a photographic plate, and so on).

(1)–(3), which together constitute the notorious reduction of the wave packet, are collapsed into a single step by the Born Interpretation and by more popular presentations. Thus for example Landau and Lifshitz write ([69], 22): 'The classical nature of the apparatus means that . . . the

---

[25] This was asserted in [62], 122. For mathematical details cf. Jauch [57]. The same idea seems to inspire the work of the Italian school. For an analysis of the latter which despite its brevity contains many extremely useful observations see [58].

[26] Cf. the brief report in 'Problems of Microphysics' [22], n.203 and text. A report of the conversations can be obtained from the files of the Minnesota Center for the Philosophy of Science.

[27] 'When [the wave packet of the electron] arrives at the film it begins to interact with the wave functions of the atoms of the film. In the interaction it is 'broken up' into many very small packets which cease to interfere coherently. What is still unexplained, however, is that only one of these packets contains the electron' ([4], 459).

[28] The situation is further aggravated by the fact that a mixture does not uniquely determine its elements unless the weights are all different ([80], 175f; cf. also ch. 13.4).

[29] For part of the mathematics see von Neumann's proof that the 'cut' between observer and observed object can be moved back and forth in an arbitrary manner ([80], 225ff).

[30] Ludwig ([15], 157) writes: 'To test the quantum theory in the case of macro-objects is . . . impossible *in principle* once the objects are big enough.' Ludwig here refers to the fact that a measuring apparatus testing the fine quantum-mechanical properties of a large object would exceed the size of the universe.

reading of the apparatus . . . has some definite value. This enables us to say that the state of the system apparatus–electron after the measurement will in actual fact be described, not by the entire sum $\Sigma\, A_n(q)\,\Phi_n(\xi)$ where q is the coordinate of the electron and ξ the apparatus coordinate, but only by the one term which corresponds to the 'reading' q_n of the apparatus, $A_n(q)\,\Phi_n(\xi)$.'

It is extremely important to see that (2) and (3) are not necessary for turning the formalism of the quantum theory into a physical theory.[31] We can connect the formalism with 'the facts' without using either assumption. As a matter of fact we do connect it in this manner at the beginning of a measurement when we represent both the object and the apparatus by wave functions without paying attention to the demands of the classical level (use of plane waves in scattering theory as well as in the above thought experiment with electrons). But if the theory resulting from such a use of the formalism is to satisfy the demands of the correspondence principle then its predictions must be drastically modified. It is in this connection that assumptions (1)–(3) become relevant. (1) is of course harmless. It is an approximation of exactly the same kind as we use in classical physics. (2) and (3) are very different. They eliminate not some minute and perhaps negligible quantitative consequences of the principle of superposition, but some quite noticeable qualitative effects. It is only after the application of (2) and (3) that we obtain agreement with the principle of correspondence and can guarantee the universal (macroscopic) validity of the Born Interpretation.[32] *At the same time we must deny the universal validity of the superposition principle and must admit that it is but a (very useful) instrument of prediction.*[33] 'In fact', says Bohr quite explicitly ([8], p. 75), 'wave mechanics, just as the matrix theory, on this view represents a symbolic[34] transcription of the problem of motion of classical mechanics adapted to the requirements of the quantum theory and only to be interpreted by an explicit use of the quantum postulate', i.e. 'with the aid of the concept of free particles' ([8], 79). We see that the relation of quantum theory to classical physics is indeed very different from that of a comprehensive point of view to a special case of it, using as it does essential parts of the classical framework in the prediction of phenomena.[35]

---

[31] This paragraph was added to deal with an objection raised in conversation by Dr Heinz Post.

[32] Von Neumann's proof of the arbitrariness of the 'cut' between observer and observed object ([80], 225ff) shows that (2) and (3) suffice and that (1) can be eliminated.

[33] This background of quantum-mechanical instrumentalism is nowhere mentioned in Popper's attack on the Copenhagen Interpretation. Quite the contrary, he asserts ([84], 99f) that instrumentalism 'has won the battle without another shot being fired . . .' Cf. section 6 below. Cf. also ch. 11.

[34] Bohr's term 'symbolic' or 'formal' corresponds to Popper's 'instrumentalistic'. Cf. section 6 of the present paper.

[35] An excellent summary of the above results is contained in [2], ch. 23. 'At first sight', writes

It remains to say a few words about Popper's contention that 'the reduction of the wave packet ... has nothing to do with the quantum theory: it is a trivial feature of probability theory' ([20], 37, 34). The 'feature' which Popper has in mind is of course the dependence of probabilities on experimental conditions which entails that they change abruptly when the conditions change. Now the fact that the Born Interpretation is part of the quantum theory entails that this feature is part of the quantum theory also (this we already found in section 2). But while Popper, having before himself the ready-made result of the labours of the Copenhagen school, can regard the situation as a 'trivial feature of probability theory',[36] Bohr has shown how difficult it is to have both the quantum theory and a propensity interpretation of its predictions, and how many modifications are needed in order to achieve a combination of this kind (one of the modifications which Popper does not seem properly to appreciate is the need to regard the quantum theory as an instrument for the prediction of classical results rather than as a universally valid description of the world). Now the 'reduction of the wave packet', which leads from the unmodified quantum theory to the classical level with its propensities, contains all these modifications, including the approximations which are needed to eliminate the unwanted effects of a realistic interpretation of the principle of

Bohm ([2], 624) 'one would ...be tempted to conclude that classical theory is merely a limiting form of the quantum theory . . . [But it emerges] that quantum theory in its present form actually presupposes the correctness of classical concepts... [These concepts] cannot be regarded as limiting forms of quantum concepts, but must instead be combined with quantum concepts in such a way that, in a complete description, each complements the other.' 'It is in connexion with the interpretation of the wave function that classical and quantum theories meet. For the physical interpretation of the wave function is always in terms of the probability that when a system interacts with a suitable measuring apparatus, it will develop a definite value of the variable that is being measured. But . . . the last stages of a measuring apparatus are always classically describable' (this collapses the three stages of our analysis). 'In fact it is only at the classical level that definite results for an experiment can be obtained in the form of distinct events... This means that without an appeal to a classical level quantum theory would have no meaning.' (This is not entirely correct, as we have seen. It would have a meaning, but would make incorrect predictions.) 'We conclude then that quantum theory presupposes the classical level and the general correctness of classical concepts in describing this level; it does not deduce classical concepts as limiting cases of quantum concepts (as, for example, one deduces Newtonian mechanics as a limiting case of special relativity' ([2], 625; original italics; the parenthetical remark at the end occurs in a footnote). 'The necessity for presupposing a classical level and the appropriate classical concepts implies that the large scale behaviour of a system is not completely expressible in terms of concepts that are appropriate at the small scale level. . . . As we go from small scale to large scale level, new (classical) properties . . . appear which cannot be deduced from the quantum description in terms of the wave function alone. . . . These new properties manifest themselves . . . in the appearance of definite objects and events which cannot exist at the quantum level' ([2], 626). What Bohm does not mention in his summary is that the transition to the classical level involves not only additions (definite objects) but also eliminations (interference terms). The latter are however dealt with in ch. 22 of his book which develops a quantum theory of measurement.

[36] Different views are called by him 'great quantum muddle[s]' ([20], 19) or 'very simple mistakes' ([20], 42).

superposition (sections 1–3 above).[37] It is therefore much more complex than the simple 'reduction' of propensities for which it makes room. Of course, there is no observable difference between the two reductions and an instrumentalist will feel no compunction in identifying them. But such an identification does little to advance our understanding of the quantum theory.

To sum up the last two sections: we have seen that propensity is part of complementarity. We have also indicated how the quantum theory must be modified if it is to admit propensities. This has led us into the quantum theory of measurement and forced us to consider some rather technical arguments. These arguments – and with this we continue the considerations of section 2 – can be presented in a different way by appealing, not to the mathematical structure of the quantum theory but to some quite simple physical facts. Such a procedure is also more in accordance with the ideas of Bohr who 'feared that the formal mathematical structure could obscure the physical core of the problem and [who] . . . was convinced that a complete physical explanation should absolutely precede the mathematical formulation' (Heisenberg, [87], 98). It is of course the idea of complementarity which is supposed to give us this 'complete physical explanation'. We now turn to those elements of complementarity which go beyond the propensity interpretation.

4. THE RELATIONAL CHARACTER OF QUANTUM-MECHANICAL STATES

The propensity interpretation as explained in section 2 takes probabilities out of the individual physical system and attributes them to the experimental arrangement. Complementarity does the same; but in addition it takes position, momentum, and all the other dynamical variables out of the individual physical system and attributes them to the experimental arrangement. The only properties which remain in the object and which can be attributed to it independently of the situation are mass, charge, baryon-number and similar 'non-dynamical' characteristics. We shall say that *complementarity asserts the relational character not only of probability, but of all dynamical magnitudes.*

One of the consequences of this procedure is that 'phenomena' cannot be 'subdivided' ([2], 50), which just means that dynamical magnitudes cannot be separated from the conditions of their application: 'this crucial point . . . implies the impossibility of any sharp separation between the behaviour of atomic objects and the interactions with the measuring instrument which serve to define the conditions under which the phenomena appear' ([12], 39ff).[38] Another consequence is that there arises an 'ambiguity in ascribing

---

[37] 'Realistic' in Popper's sense. See his 'Three Views Concerning Human Knowledge' in [84].
[38] Note the objective character of the explanation which does not in any way refer to observers, or to consciousness. In earlier papers one hears occasionally of the 'impossibility of

customary physical attributes to atomic objects' ([12], 51) – for these attributes no longer apply to the object *per se* but to the whole experimental arrangement, different assertions being made in different contexts.

Complementarity also offers a rough quantitative estimate of the extent to which experimental arrangements allowing for the application of certain notions (such as position) still permit the application of other notions (such as momentum). I am of course referring to the *uncertainty principle* which describes objective relations between states of systems in different experimental arrangements. It is this feature of complementarity, this relativization of dynamical magnitudes in addition to probabilities which Popper wants to reject. I shall now discuss the issue, using Popper's classical examples as a starting point.

5. TRAJECTORIES IN CLASSICAL PHYSICS AND IN THE QUANTUM THEORY

According to Popper the discussion of classical devices such as pinboards and dice in beakers reveals features of probability which so far have been regarded as belonging to the quantum theory proper. In cases like these one can inquire about the probabilities and their change on transition from one arrangement to the next; and one can also inquire about the behaviour of each individual system that passes through the device. For example, one can inquire about the behaviour of an individual ball that enters the pinboard, hits various pins, and finally reaches the bottom. Popper, if I understand him correctly, would want us to believe that problems of the second kind do not arise in the quantum theory which, according to him, is 'essentially of a statistical or probabilistic character' ([20], 17). And this is precisely the point at issue. It is here that the difference between Bohr and Popper arises. Moreover, even purely statistical theories may involve elements obeying non-statistical laws of their own. Classical statistical mechanics certainly is an example. We are then faced with the problem of whether the elements of the ensembles used continue to obey these laws when embedded into sufficiently complex surroundings; or whether they violate these laws[39]; or whether the question, perhaps, ceases to be meaningful. We are assured by Popper that he does not accept the last alternative: particles always have well-defined trajectories ([20], 27 – this seems to exclude the

distinguishing . . . between physical phenomena and their observation' ([8], 15), but it is quite clear that the term 'observation' does not involve subjective elements but merely refers to the physical 'agencies' by which [the phenomenon] is observed' ([18], 11). For the notion of a phenomenon see again the explanation given in section 2.

[39] One interpretation of the ergodic theorem is that it establishes the consistency between the *a priori* probabilities of statistical mechanics (including whatever assumptions of statistical independence are made) and the Hamiltonian equations.

extrapolation mentioned at the beginning of section 3 as well as Landé's position of 1955: [71], 16ff, 26). We can therefore ask the question: how do they behave? It is in the treatment of this question that the difference between Bohr and Popper arises.

To see this, imagine a classical pinboard and paths which are influenced only by a direct encounter with obstacles such as pins but are inertial otherwise. We assume, then, that all outside forces have been removed. We also assume that there is only one ball (one 'particle') in the arrangement at one time.[40] Now consider all those paths which come sufficiently close to a particular point P' of the board which may, or may not, contain a pin. Let their class be A. Changing the situation at P' (removing a pin if one happens to be there; drilling a little hole; inserting a pin; adding some other kind of obstacle; etc.) will influence the paths of kind A only and will not affect any other path.

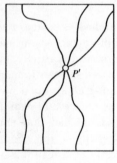

Fig. 1

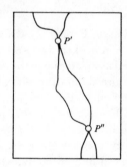

Fig. 2

Now consider those paths of A which go through P'' also (fig. 2). Call the class of these paths the paths of kind B, where $B \subset A$. Again changing the situation at either P' or P'' will change the paths of kind B only, and will not influence any other path.

Finally, we consider the path(s) which go(es) through a (dense) set S of points on the pinboard. By elimination we may say that this path is influenced only by events in its close neighbourhood and is quite independent of what happens to the remainder of the board (fig. 3). We may cut off this remainder, and the path will still be the same, provided the appropriate initial conditions have been realized and provided also the above mentioned assumptions are satisfied. Result: each path is passed through as if the only physical events were those occurring in the close neighbourhood of its points. All this is of course pretty trivial, but it is in trivial matters such as these that the worst mistakes are made.

[40] It seems that this is the case which Popper has in mind in his discussion in [20], 33ff.

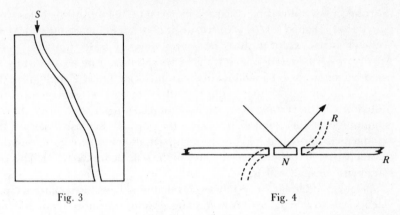

Fig. 3   Fig. 4

Take another example (which is supposed to satisfy the same conditions): a ball hits an extended surface and is deflected by it. Its behaviour again depends on the close neighbourhood of the point of impact, i.e., of the region N (fig. 4), and is quite unaffected by the remainder R of the surface. As far as its trajectory is concerned this remainder might as well not exist. Remember that we are all the time dealing with the behaviour of individual particles. We are not discussing probabilities and we have not yet asked how a change of the experimental arrangement will affect these probabilities.

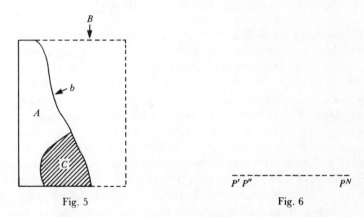

Fig. 5   Fig. 6

We shall now draw some conclusions from the fact that the experimental arrangements discussed by Popper produce classes of trajectories of the kind just described.

The most important result is the following: the totality of trajectories that can be realized in a given experimental set-up E contains all the trajectories which occur when only part of E is actually present. This result we shall call **A**. Thus the trajectories that can be realized on board B (fig. 5) contain all the trajectories which can be realized in part A of B as a proper subclass (the boundary b is supposed to be 'open', i.e. it does not influence the behaviour of the elements except in those rare cases where it passes through a pin). Again (fig. 6) the totality of all trajectories resulting from a ball hitting an extended surface will be the sum of the trajectories hitting P' only, plus the trajectories hitting P'' only, plus – and so on. As a consequence, the probability of a certain event in the total arrangement is the sum of probabilities of this event in any one of the mutually exclusive parts of the arrangement, provided, of course, each part allows for the formation of a complete path (C in fig. 5 is not a 'part' in the sense assumed here). This last statement we shall call **B**.

Moreover, adding to a certain experimental arrangement (for example, extending A to B in fig. 5, or N to R' in fig. 4) adds trajectories, but does not change the trajectories already present. For example, it does not force them to crowd together, or to redistribute themselves with lesser density. Nor are conditions created which make it impossible for certain trajectories to occur. This statement we shall call **C**.

Each of the three statements, **A**, **B** and **C**, is derived under the assumption (which we shall call assumption **T**) that the particle, the ball, the die move along a well-defined path that is influenced by the situation in its close neighbourhood only. **T** (together with some simple definitions introducing probabilities) entails **A**, **B**, as well as **C**.

Now one knows very well, and it is also emphasized by Popper ([20], 40–last paragraph), that the quantum theory is inconsistent with **A**, **B** and **C**. Probabilities are no longer additive and events which occur when only part of the experimental arrangement is active (for example, when only one slit is open in the two-slit experiment) do not occur when we use the total arrangement (both slits are open). From the way in which **A**, **B** and **C** were obtained we conclude that **T** cannot be correct either. It is this situation which, through a series of conjectures and refutations has finally led to the Copenhagen Interpretation. Note that it concerns not only probabilities, but also the individual case and its properties (**T** is about the individual case). It will be of advantage to elucidate the matter once more in a different way. This will make it even clearer that we are faced here with a real difficulty (for the assumption of particles moving along well-defined trajectories) and not just with a 'trivial misunderstanding'.

Assume that R is a diffraction grid and N one particular narrow part of it. We direct our attention to all the particles which hit N and N only and are then deflected. We call these particles particles of kind A. Following the above train of reasoning we can point out that the behaviour of the particles of kind A (the behaviour of each individual particle of kind A) is determined by N only (this is assumption **T**). R has no influence on it. Hence, it makes no difference, for the behaviour of the particles of kind A, if we use R, or R', or if we altogether omit the remainder of the grid. Whatever this remainder may 'know' ([20], 33–last but one paragraph) is irrelevant for the behaviour of the particles of kind A. Each single particle behaves as if there existed absolutely nothing outside its own path. But then the class of all particles reflected by the grid must contain all particles of kind A (this is consequence **C**, and **B**). Yet such is not the case. The pattern of the total grid contains minima in places that can be reached by the particles of kind A so that adding the remainder *does* seem to have an effect, contrary to the dynamical assumptions we have made about the individual particle. This is one of the problems which the Copenhagen Interpretations wants to resolve.

Incidentally, the transition, made here, from statistical laws to individual events (and vice versa) is quite harmless and involves no 'great quantum muddle' ([20], 19): in the qualitative form which we have adopted here the statistical quantum law already denies the existence of certain individual events (no particles at the minimum) whereas the dynamical considerations entail that such events must occur. And even if we do not consider the minima (or deny the existence of absolute minima)[41] we must still admit that there is a redistribution of paths, contrary to what is said in **B** and **C**.

One does not see how Popper's insistence on a purely statistical approach can get around this difficulty. He pleads with us not to be surprised when a change of experimental conditions also changes the probabilities. This, he says, is 'a trivial feature of probability theory' ([20], 37; cf. p. 35, last paragraph). Quite correct – but irrelevant. For what surprises us (and what led to the Copenhagen Interpretation) is not the fact that there is *some change*; what surprises us is the *kind of change* encountered: trajectories which from a classical standpoint are perfectly feasible are suddenly forbidden and are not entered by any particle. It is in order to explain these curious occurrences that the Copenhagen Interpretation was gradually built up. Let us therefore look at them a little more closely.

How can one possibly explain the shifting, the redistribution of trajectories, and the complete disappearance of certain paths? (That there is such a shifting and redistribution can be seen at once by adopting Popper's method ([20], 26) for the construction of the paths.) Two possibilities suggest themselves.[42]

One might want to point out that the assumptions formulated at the beginning of the argument (inertial paths between obstacles; no outside forces) are not valid on the microlevel. Such explanations and their difficulties will be discussed only briefly in the present essay (see the discussion of

---

[41] As was suggested to me by Professor Landé.
[42] If we try to stay close to Popper's ideas, that is. Otherwise one would have to consider hidden variable theories, non-classical logics, and further hypotheses.

the paper of Einstein and Ehrenfest below). It is worth mentioning that they in effect introduce an 'interaction' in those very same areas which in the wave picture are occupied by the wave front without providing the appropriate energies of interaction. Strictly speaking there is therefore no interaction at all, but only a correlation via a phase: the 'interaction' of the parts of extended arrangements is given by the wave theory and not by considering particle trajectories.[43] Popper considers neither this solution nor, for that matter, any other dynamical theory. All he does is emphasize again and again the relational character of the probabilities.

We may therefore assume – and this is a second possible explanation of the interference patterns – that Popper takes the redistribution of paths to be a primitive phenomenon, not in need of further analysis (except in the framework of the propensity theory): each particular experimental arrangement is associated with its own distributions of paths (or, more generally, of physical events) and transitions between different distributions are 'trivial feature[s] of probability theory' ([20], 37), not to be further explained. This intriguing hypothesis has much in common with an earlier conjecture of Bohr, Kramers and Slater[44] – and just like the conjecture it is refuted by the experiments of Bothe and Geiger ([17], [18]) and Compton and Simon [25].[45] These experiments show that energy, momentum and other dynamical variables are conserved not only on the average, so that one could postulate a redistribution without asking for some dynamical cause, *but in each single interaction*. 'Purely statistical' redistributions are therefore out. Each single change of path must be accounted for. But this turns us right back to what was above called the first possibility and to all the difficulties it encounters.

As an example of such difficulties I shall now briefly report the content of

---

[43] One can translate the wave language into classical interaction language by introducing a 'quantum potential' ([3], 166f). This is merely a formal trick, however, for there exists no possibility of testing independently the properties of this potential (such tests may become possible when the formalism of the quantum theory is changed in certain ways).

[44] See [13]. The paper can be interpreted as dealing with a special case of the condition-dependence of probabilities which Popper discusses, viz. with the way in which the emission and absorption of quanta by atoms depend on the material conditions in the neighbourhood. These conditions create a 'virtual field' that interferes (in accordance with the usual laws of interference via a phase) with the virtual field of any addition one might wish to make to them. Speaking in a somewhat picturesque fashion one might say that the theory lets the conditions interfere with each other. It is obvious that **A**, **B** and **C** now cease to hold. They cease to hold, not because of additional forces of the kind excluded at the beginning of section 5, but because the material conditions are no longer additive. It is this kind of interference which Landé seems to have in mind in his discussion of Duane's suggestion of 1923 ([71], 74) and to which Popper alludes when he says (without further comment) that 'the board as a whole "knows" ' ([20], 33). The theory entails the statistical validity of the conservation principles: 'The occurrence of photo electrons with the energy h means that we must either give up our old view that light comes in waves, or abandon the doctrine of the conservation of energy' ([24], 509).

[45] Cf. also the excellent survey by Compton [24].

a paper by Einstein and Ehrenfest on the Stern–Gerlach experiment, [34]. The paper was written in the earlier period of the quantum theory when a coherent interpretation was not yet available and when numerous experiments were discussed in the hope of finding such an interpretation. It quite clearly shows the difficulties arising from every attempt to give, not a causal explanation, but merely a detailed description of the behaviour of individual systems.

The Stern–Gerlach experiment on the one hand gives 'strong support to the idea of stationary states'.[46] On the other hand the situation does not seem to permit a detailed account of the manner in which an individual silver atom (a) acquires its orientation; (b) enters the beam in which it is found after passage through the magnetic field. Einstein and Ehrenfest, who discuss (a) only, mention the following difficulties:

(i) We assume that the atoms obey the laws of classical electrodynamics throughout their path. Then the magnetic field will cause them to rotate around the field lines, preserving the angle of inclination (Larmor precession). In the oven from which the atoms emerge the directions are distributed in a random manner. They should then exhibit a random distribution also after passing through the magnetic field. This is not the case.

(ii) There is a process of quantization which begins as soon as the atoms enter the magnetic field. In this case the atoms will absorb or emit radiation just as they do in any quantum transition and they will gradually move into their new position. The time required for such a transition is of order of magnitude $> 10^9$ seconds. The atoms, however, need only 10^{-4} seconds to traverse the magnetic field (this corresponds to the time differences calculated and observed in the photoelectric effect).

(iii) Quantization occurs automatically without involving any interactions and the atoms are always in a well defined quantum state. This assumption which seems to be closest to Popper's point of view where each experimental arrangement effects an instantaneous redistribution of trajectories (and, we may conjecture, of directions and positions) evidently violates the conservation principles: it is assumed that the atoms follow even the weakest field 'without inertia', i.e. without the delay due to their mechanical structure and the interaction with the field. Einstein and Ehrenfest criticize (iii) for exactly this reason, as involving 'a violation of the equations of motion' ([34], 33). Since then (1922) the case for rejection has been strengthened by the experiments of Compton and Simon and Bothe and Geiger which demonstrate some conservation laws directly, and for each single interaction.

At this point it is perhaps appropriate to inform the reader that Popper

---

[46] Bohr, 'Discussions with Einstein' (where the paper by Einstein and Ehrenfest is briefly mentioned) in [12], 37.

has always viewed these and similar experiments with a good deal of suspicion and that he has doubted that they were properly interpreted when taken to establish the conservation laws for each single case. He has been much more reluctant than Bohr to make the step from the philosophy of Bohr–Kramers–Slater, for which he still seems to have considerable sympathy, to the new quantum mechanics of Bohr and Heisenberg. Whatever one's opinion on these matters should be, one must admit that the accepted interpretation of the Bothe–Geiger and Compton–Simon experiments is a possible interpretation and that it is also a reasonable interpretation. Moreover, it seems to be the most natural interpretation of these experiments which were designed precisely for the purpose of testing whether the conservation laws were satisfied in each single case, or on the average only.[47] The idea of complementarity is therefore not just the result of having pursued a mistaken programme to the bitter end as Popper would want us to believe ([20], 16). Bohr, after all, did consider a purely statistical theory. He did consider such a theory despite the fact that it was not in line with his own point of view[48] (it was suggested by Einstein's treatment of the radiation problem [32] which Bohr always regarded as somewhat extreme[49]), but then he found that such a theory could not be upheld. True, Bohr was originally interested in a generalization, not of classical statistical mechanics, but of classical particle mechanics. This part of Popper's account is unobjectionable ([20], 16). However, Bohr was also open minded enough to generalize Einstein's considerations and to develop and publish a purely statistical particle theory with waves acting as probability fields only. It was only after the refutation of this theory that he returned to his earlier philosophy – and this time with very good reasons. This important episode is not mentioned anywhere in Popper's paper – a very unfor-

[47] Cf. the advertisement [17] by Bothe and Geiger as well as the introduction and the summary to [18]. Popper's opinions as reported above were expressed by him in discussions (1958).

[48] It seems that it was the combined use of the correspondence principle and of the facts known about interactions (photoelectric effect; recoil; Compton effect) that made Bohr extend Einstein's ideas of 1916 (published 1917: [32]) into a purely statistical theory whose 'formal' (i.e. instrumentalistic) character was explicitly emphasized by him ([13], 790). The correspondence principle is interpreted as demanding retention of 'the laws of propagation of radiation in free space' ([13], 793) as well as dependence of 'transition[s] in a given atom . . . on the initial stationary state of this atom itself and on the state of the atoms with which it is in communication through the virtual radiation field, but not on the occurrence of transition processes in the latter atoms' ([13], 791). The 'but not' refers to Slater's original suggestion [13] according to which each transition produces quanta moving 'for example' along the direction of Poynting's vector and causing transitions in other atoms. It is clear that such a theory violates correspondence, as the frequency of radiation need no longer approach the orbital frequency of the vibrating electron that corresponds to it in the classical model. For the difference between Slater's original hypothesis and Bohr–Kramers–Slater cf. also [100], 11–15.

[49] For an excellent and detailed account of the background and the repercussions of Bohr–Kramers–Slater see section 4.3 of [56].

tunate omission that makes the idea of complementarity appear much more dogmatic than it actually is.

To return to the paper of Einstein and Ehrenfest. The fourth suggestion discussed by these authors is that:

(iv) The atoms are not always prefectly quantized, but transitions into states which are perfectly quantized do occur, and with speeds much greater than those discussed in (ii). This conjecture restricts quantization to charged systems as only such systems can interact with the field. A consideration of the specific heat of diamond and of H_2 shows that such a restriction is refuted by the evidence.

The authors conclude that some apparently quite natural attempts to explain the reorientation observed in the experiment of Stern and Gerlach are 'quite unsatisfactory' ([34], 34). 'Bohr's idea, however, viz. that [during interaction] there is not sharp quantization, remained unexamined.'[50] It is now time to comment on this basic hypothesis of Bohr's and to develop its philosophical background. For a formulation that brings out its objective character the reader is invited to consult again sections 2 and 4.

6. A SKETCH OF BOHR'S POINT OF VIEW

'Bohr was primarily a philosopher, not a physicist, but he understood that natural philosophy . . . carries weight only if its every detail can be subjected to the . . . test of experiment' (Heisenberg in [87], 95).[51] As a result his approach differed from that of the school-philosophers whom he regarded with a somewhat 'sceptical attitude, to say the least' ([87], 129) and whose lack of interest in 'the important viewpoint which had emerged during the development of atomic physics' he noticed with regret ([87], 183). But it also differed, and to a considerable degree, from the spirit of what T. S. Kuhn has called a 'normal science' [65]. Looking at Bohr's method of research we see that technical problems, however remote, are always related to a philosophical point of view; they are never treated as 'tiny puzzles' [64] whose solution is valuable in itself, even if one has not the faintest idea what it means, and where it leads: 'For me', Bohr writes to Sommerfeld in 1922 ([87], 71) '[the quantum theory] is not a matter for petty didactic details, but a serious attempt to reach . . . an inner coherence.'[52] Emphasis is put on matters of principle ([87], 36) and minor

---

[50] [34], 34. My translation.

[51] 'He would hardly ever end a discussion without conceding, with a disarming smile, that he himself occupied the position of a "superdilettante"' (J. Kalckar, [87], 234). 'The first thing Bohr said to me was that it would only then be profitable to work with him if I understood that he was a dilettante' ([87], 218). 'Bohr's strength lay in his formidable intuition . . . not at all in his erudition' ([87], 218).

[52] This applies to all of Bohr's work, including the very detailed research of 1913 and after. As an example we mention, Bohr's account of the minute difference (0.04 per cent) observed by

discrepancies, or 'puzzles' in the sense of Kuhn, instead of being de-emphasized, and assimilated to the older paradigm, are turned into fun-damental difficulties by looking at them from a new direction, and by testing their background 'in its furthest consequences by exaggeration' ([87], 329).[53] A noteworthy example of Bohr's 'non-normal' and rather metaphysical approach to physics is his scepticism in the face of the success of his own atomic model.

From the very beginning 'Sommerfeld . . . was inclined to take the early theories more literally than Bohr considered justified' ([87], 72). His attitude was reinforced by the success of Delauny's perturbation theoretic methods in the treatment of problems of atomic mechanics. Utilizing these powerful and beautiful methods Schwarzschild and Epstein had arrived at an unambiguous formulation of the quantum conditions and they had also succeeded in giving an excellent account of the Stark effect in hydrogen. So good was the agreement that Epstein could declare: 'We believe that the reported results prove the correctness of Bohr's atomic model with such striking evidence that even our conservative colleagues cannot deny its cogency. It seems that the potentialities of the quantum theory as applied to this model are almost miraculous and far from being exhausted' ([36], 150). 'Until a few years ago', wrote Sommerfeld ([99], 674), 'one could believe that the Hamilton–Jacobi method in mechanics was of no use for physics and that it could only serve the needs of mathematics and of astronomical perturbation theory. . . However since the work of Schwarz-schild and Epstein . . . [it] seems to be ideally suited just for the most important problems of physical mechanics.' Thus according to a fairly prevalent view the formal apparatus for the calculation of stationary states had finally been found: one adds a 'fourth Keplerian law', the quantum

A. Fowler ([45], 95) in the spectrum of ionized helium. He shows ([46], 231f) that the discrepancy can be explained when one replaces the mass of the nucleus by its reduced mass. This investigation and similar investigations between 1913 and 1924 seem to be normal science *kat'exochen*, were it not for the fact that they are guided by a philosophical principle (the exceptional role of classical concepts, an idea that is close to Kantianism) which is occasionally used to debunk and to philosophically dissolve successes in physics proper (see the text immediately below). Also Bohr has no compunction whatever in turning minor discrepancies into difficulties of principle wherever he believes that a change of point of view is necessary. His whole early work was designed to increase rather than to alleviate the conflict between certain experimental facts and 'the admirably coherent group of concep-tions which have been rightly termed the classical theory of electrodynamics' ([5], 19). Philosophical considerations of this kind are entirely unknown to normal science.
[53] These quotations should be compared with Einstein's lack of interest in 'confirmation by little effects' ([54], 242) and with his tendency to 'raise [a] difficulty' (the reference is to the failure to discover a motion through the ether) 'into a principle' of a new theory (the theory of relativity) instead of regarding it as a 'puzzle' (or as a 'cloud', cf. [60]) of the classical electron theory (the quotation is from Einstein's 'Zur Elektrodynamik Bewegter Koerper', quoted from [1], 26. Bohr admired Einstein 'precisely for the way in which he had laid stress on the epistemological aspects of classical physics and, at an early stage, of quantum theory also; ([87], 131).

conditions, to the laws of classical mechanics and calculates the resulting problem by the Hamilton–Jacobi method.

'Bohr's view of [the situation] was much more sceptical than that of many other physicists' ([87], 95). For example, he emphasized again and again that one was still moving within the framework of classical thought, that the elements of this framework were being used in a non-classical manner and that one was therefore obliged 'always to remember the domain of application of the theory, especially at the present state of science' ([6], 118). He pointed out that Hamilton's equations could for this reason yield the stationary states 'to a high degree of approximation only', which approximation was determined by 'the extent to which the reactions of the electromagnetic field can be neglected' ([6], 119; cf. also the criticism of Epstein in [6], 128, n.1. One sees that the criticism is epistemological, not physical in the traditional sense of the word). Sommerfeld had overlooked precisely this restriction asserting that 'the "aether" and the atom can be regarded as independent of each other, being connected only during the process of emission' ([99], 314; cf. however the footnote on that page which formulates, and accepts, Bohr's objection). Other writers such as Max Born ([16], 283) who recognized the 'approximate' and 'formal' nature of the microcelestial mechanics of Sommerfeld, Schwarzschild and Epstein restricted this judgement to the calculation of intensities, giving a separate treatment to the stationary states ([16], 18, 67). Bohr not only upheld the approximate character of all microphysics, stationary states included, but he also stressed the 'formal' character of the new laws, that is the fact that they aided in the prediction of phenomena, but did not yet provide any understanding (see below). His intention in those early days seems to have been the elaboration of a predictive scheme that could aid in the discovery of a physical theory in the full sense of the word, including an account even of the mechanism of the transitions ([13], 785, 788). The difference between Bohr and Sommerfeld is best described as the difference between a mathematical physicist who is content with formally satisfactory and factually adequate equations and a philosopher who looks beyond success and who realizes the need for a sense of perspective, even in the face of the most surprising confirmations.

It must have been very hard in the early days, with all the patent success of the rather primitive atom model, to shake off unfounded optimism regarding the fruitfulness of the orbit conception and to accept the necessity of a description renouncing in principle such pictures, but based merely on the correspondence principle. This is vividly shown in a letter Bohr sent to Oseen in January 1926, very shortly after the new theory [of Schrödinger and Heisenberg] was put forward . . . 'We are slowly progressing, I hope, but in every result lurks the temptation to stray from the right path. This is so true in atomic theory that at the present stage of the development of the quantum theory we can hardly say whether it was good luck or bad luck that the properties of the Kepler motion could be brought into such simple connection with the hydrogen spectrum, as was believed possible at the time.[54] If this connection had merely had that asymptotic character which one might expect from the correspondence principle, then we should not have been tempted to apply mechanics as crudely as we believed possible for some time. On the other hand it was just these mechanical considerations that were helpful in building up the analysis of the optical phenomena which gradually led to quantum mechanics'[55] (L. Rosenfeld and E. Rüdinger in [87], 73).

Bohr's realization of the need to retain a sense of perspective, even in the face of the most surprising successes (and failures!) is clearly seen in the style of his papers which are heavily seasoned with historical material,[56] and are best characterized as preliminary summaries[57] surveying the

---

[54] As will be explained below it is exactly this lesson which Bohr applied (in 1961) in his criticism of certain attitudes in the quantum theory of fields, pointing out that it is necessary to ask 'whether the whole mathematical approach has not been twisted in some particular direction' ([98], 257).

[55] Cf. also F. Hund, 'Goettingen, Copenhagen, Leipzig im Rueckblick' ([15], 13): 'In Munich one used more concrete formulations and was therefore more easily understood; one had been successful in the systematization of spectra and in the use of the vector model. In Copenhagen, however, one believed that an adequate language for the new [phenomena] had not yet been found, one was reticent in the face of too definite formulations, one expressed oneself more cautiously and more in general terms, and was therefore much more difficult to understand.' One should also consult Sommerfeld's biographical sketch in the Appendix to [66], 144 and 146.

[56] 'I must admit', writes A. Pais concerning a paper of Bohr's in whose preparation he participated, 'that in the early stages of the collaboration I did not follow Bohr's line of thinking a good deal of the time and was, in fact, often quite bewildered, I failed to see the relevance of such remarks as that Schrödinger was completely shocked in 1925 when he was told of the probability interpretation of quantum mechanics, or a reference to some objection by Einstein in 1928, which apparently had no bearing whatever on the subject at hand. But it did not take very long before the fog started to lift. I began to grasp not only the thread of Bohr's arguments but also their purpose. Just as in many sports a player goes through warming up exercises before entering the arena, so Bohr would relive the struggles which it took before the content of quantum mechanics was understood and accepted. I can say that in Bohr's mind this struggle started all over every single day. This, I am convinced, was Bohr's inexhaustible source of identity' ([87], 218f). That we are dealing here with more than a psychological process can be seen from Pais' remark that to start with he, Pais, like all 'men of [his] generation [had] received quantum mechanics served up ready made. . . . Through steady exposure to Bohr's "daily struggle" and his ever repeated emphasis on the "epistemological lessons which quantum mechanics has taught us", to use a favourite phrase of his, my understanding deepened not only of the history of physics, but of physics itself. In fact, the many hours which Bohr spent talking to me about complementarity, have had a liberating effect on every aspect of my thinking' ([87], 219). It will be noticed that the role of history in this process is very similar to the role it plays in the *Phänomenologie des Geistes*, a parallel which is emphasized by Bohr's remark 'that since those distant days when human beings began to use words like "here" and "there" and "before" and "now", there had been no further advance in epistemology until Einstein's theory of relativity' ([87], 236).

[57] 'he would never try to outline any finished picture, but would patiently go through all the phases of the development of a problem, starting from some apparent paradox and gradually leading to its elucidation. In fact, he never regarded achieved results in any other light than as starting points for further exploration. In speculating about the prospects of some line of

past,[58] giving an account of 'the present state of knowledge'[59] and making general suggestions, in the spirit of the correspondence principle, on the possible course of future research. (In this respect Bohr's papers differ considerably from Einstein's early papers which spring to light like Pallas Athene from the forehead of Zeus.[60]) Such an approach which tempers physics with history and defuses its success by a philosophical criticism,[61] has

investigation, he would dismiss the usual considerations of simplicity, elegance or even consistency with the remark that such qualities can only be properly judged *after* the event . . .' ([87], 117).

[58] 'Bis hierher ist das Bewusstsein gekommen' Hegel concludes his survey of history, and of the 'principal moments of the form in which the principle of freedom has [so far] realised itself' ([49], 562) and he has often been criticized for regarding the Prussian State as the pinnacle of history when all he did was to point out that historically seen this was the last, the most recent, manifestation of freedom. For a brief analysis see [59], 259f.

[59] 'When I was requested', writes Bohr ([8], 5f), 'to write a paper for the Year Book 1929 of the University of Copenhagen, I first intended to give, in the simplest form, an account of the new points of view brought about by the quantum theory, starting from an analysis of the elementary concepts on which our description of nature is founded. However my occupation with other duties did not leave me sufficient time to complete such an account, the difficulty of which arose, not least, from *the continuous development of the points of view in question.* Sensing this difficulty, I gave up the idea of preparing a new exposition and was led to consider using instead a translation into Danish, made for this occasion, of some articles which, during recent years, I have published in foreign journals as contributions to the discussions of the problems of the quantum theory. These articles belong to a series of lectures and papers in which, from time to time, I have attempted to give a coherent survey of the state of the atomic theory *at the moment.* Some previous articles of this series form in some respects a background for the three articles which are reproduced here. . . They are [all] intimately connected with each other, in that they all discuss *the latest phase in the development of the atomic theory. . .*' Considering the italicized passages (which are but a small sample of numerous similar passages in the book and in all his articles) the reader may wonder how the myth of Bohr's dogmatism could possible have arisen. Cf. also the new introduction (1961) to the book from which the above passage is taken which asserts that 'the old articles, which are here reprinted . . . contain utterances which may be formulated in a more precise manner' utilizing 'a more adequate terminology'. (This refers to the change of terminology described in n.10 and text.)

[60] Cf. the judgement of R. S. Shankland [94], 48, n.3: 'The whole paper [of 1905, on the special theory of relativity] is rather strange in the respect that Einstein reveals very little about what he knows to be experimentally verified and that he makes no specific reference to the work of others. *The paper in fact represents an enigma* in that it is very difficult to see how much is an inference from experimental results (or a theoretical formulation of them) of which Einstein had knowledge.' The same may be said about the two other papers of 1905 which contain many puzzling features (thus the role of Wien's law is what is commonly called the paper on the photoelectric effect becomes clear only in 1909 when Einstein publishes his results on the fluctuations of the electromagnetic field ([31], 185ff). However, Einstein has also written papers of a very different kind and he has criticized Hilbert for remaining silent about the method by means of which he reached his results. 'Hilbert's presentation' he writes to Ehrenfest (Postcard of May 24, 1916, quoted from Carl Seelig [93], 276) 'does not please me at all. It is unnecessarily specialized . . ., unnecessarily complicated, not honest (Gaussian) in the way in which it has been built up (pretense of superhuman powers by concealment of the method used).' Professor G. Holton who has examined the peculiarities of the 1905 paper on relativity in some detail has published some very interesting conjectures concerning the predecessors of the style and method used by Einstein. Cf. [54].

[61] 'Acknowledging, as he did, no boundaries in the range of rational investigation . . .' ([87], 236).

occasionally annoyed the experts, and some smart alecks of the younger generation[62] have insisted that their problems could be solved without 'having to go all the way back'.[63] A broader view cannot encourage such optimism, however, for the hypertrophy of the mathematical framework raises the question 'whether one has got hold of all the points and whether the whole mathematical approach has not been twisted in some particular direction' ([98], 257).

This last remark and the similarity it introduces between the most recent developments and an equally hypertrophical episode of the early quantum theory (the quantum theory of Sommerfeld and Epstein) reveals another characteristic aspect of Bohr's philosophy: 'I noticed', writes Heisenberg ([87], 98), 'that mathematical clarity had in itself no virtue for Bohr. He feared that the formal mathematical structure would obscure the physical core of the problem, and in any case he was convinced that a complete physical explanation should absolutely precede the mathematical formulation.'[64] It was in this spirit that Bohr observed,

> in one of his last conversations, . . . that the reason why no progress was being made in the theory of the transformations of matter occurring at very high energy [was] that we have not so far found among these processes any one exhibiting a sufficiently violent contradiction with what could be expected from current ideas to give us a clear and unambiguous indication of how we have to modify these ideas [[87], 118].

And it was for the very same reason that he tried to base the interpretation

---

[62] Who 'received quantum mechanics served up ready made' [87], 219).
One sign of this ready-made character is the way in which rules of interpretation are formulated today. e.g. in [43], section 3/4. It is here stated that the probability of the alternate occurence of events which are in principle distinguishable is the sum of the squares of the corresponding amplitudes, while the probability of the alternate occurrence of events which are in principle indistinguishable is the square of the sum of these amplitudes. This is of course a correct rule – but it leaves unexplained the status of the superposition principle, of Schrödinger's equation, of the uncertainty relations (the situation is here exactly the same as it is in Popper's 'interpretation'). The loss in clarity and in accuracy becomes very obvious when we look at Feynman's 'Philosophical Implications', sections 2–6. There is no trace here of the series of conjectures and refutations that led to the idea of complementarity and of the evidence that can be adduced in its support. Rather, we meet an odd assortment of slogans, combined with a dash of operationalism and some of Heisenberg's more popular ideas.

[63] Feynman ([98], 256), discussion remark, commenting on a preceding discussion remark of Bohr's. The remark of Bohr quoted in the text below is part of his reply to Feynman's criticism.

[64] Bohr would on occasions be critical of von Neumann's approach which, according to him, did not solve problems, but created imaginary difficulties (such as the problem where the 'cut' between the observer and the things observed should be placed). During a seminar in Askov in 1949 I also had the impression that he definitely preferred his own qualitative remarks to the machinery of the famous 'von Neumann Proof'. Nor was he too fond of the rising axiomania emphasizing, as he did, that 'every sentence I say must be understood not as an affirmation, but as a question' (quoted from Jammer ([56], 175)).

of the quantum theory not so much on the properties of the formal mathematical scheme (as did Heisenberg – cf. ([87], 98), and ([82], 15)) as on
some simple qualitative considerations or 'pictures' as he called them. Such
pictures which may be regarded as concise and intuitively convincing
summaries of the results of past research exhibit his astounding skill in
'obtaining qualitative or semi-quantitative results without detailed calculation' ([87], 110).[65]

There arose then what one might call two different versions of the
uncertainty relations, one emerging from the qualitative discussions just
alluded to and embodying the lessons of twenty years of physical research
(as well as a whole treasure of valuable classical results), the other being
obtained through a somewhat complicated derivation from the formalism
(Born's interpretation included). For Bohr it was the former version that
was really important as it both tested and gave content to the mathematical
formulae (it retained its importance even after the acceptance of the Born
interpretation which as we have seen (section 3) is but a shorthand account
of the rather complex transitions connecting the formalism with reality).
The value of qualitative considerations of this kind in addition to an
inspection of the symmetries of the formalism lies in the fact that the formal
features of a theory do not always adequately mirror the physical
situation.[66] Thus Bohr was able to show that the intrinsic magnetic moment
of the electron and angular momentum in the usual sense are related to
measurement in very different ways though they appear in an exactly
analogous way in the theory.[67] Similarly, the classical examination of the
physical content of quantum electrodynamics in [14] revealed 'how many
extreme abstractions have to be made in order to describe [the] measurement' of the observables of the theory ([101], 18), thereby again emphasizing the difference between the symmetries of the formalism and the

---

[65] 'This is no way to make physics – without any guidance by an intuitive physical principle'
said Einsten to Weyl (Carl Seelig, 274. Cf. however his letter of March 18, 1918, expressing
his enthusiasm about *Raum, Zeit, Materie* quoted on p. 277). Einstein's non-formal explanation of the principle of equivalence and of its consequences is very similar to many of Bohr's
qualitative considerations.

[66] This becomes clear at once when one considers how many different formalisms can be used
to express the facts of classical celestial mechanics and how careful one must therefore be in
the attempt to use the features of some particular formalism as a basis for ontological
inference. 'The variety of ways, in part based on entirely different sets of concepts, in which
we can express the fundamental laws for gravitating matter, all leading to identical physical
predictions, should caution us not to put undue stress on the supposed implications of a
particular formulation of a theory, *even if other formulations might not be available at a given time.*'
([48], 963; my italics.)

[67] For detailed arguments concerning this surprising feature ('I never felt quite at ease about
his argument that the spin cannot be observed by classical means' ([87], 111)) see [97],
217ff and the discussion, with comments by Bohr, ([97], 276ff). Pauli shows, among
other things, that the attempt to measure the magnetic moments of free electrons by a
Stern–Gerlach experiment cannot succeed as the wave nature of the electron obliterates the
very effects one is inquiring into.

symmetries of the facts it tries to represent.[68] The question, however, how, and in what terms, 'actual physical situations' are to be discussed now leads us one step further into Bohr's philosophy.

One can present this philosophy in various ways. One can give a general outline. One can discuss the specific shape it was given within the framework of the older quantum theory. And one can also discuss the form it assumed in the context of the idea of complementarity. The central problem is in all cases the relation between subject and object or, to use a phrase that occurs already at a very early stage, the problem of an observer who is both actor and spectator in the world that surrounds him ([8], 119).[69] It seems that Bohr from the very beginning adopted a certain idea

[68] It is in this spirit that Bohr raised the question (in 1961) 'how far the technical mathematical devices used in the account of . . . phenomena [at very high energies] take into consideration all aspects of the situation' ([98], 254). Cf. also Feynman's answer and Bohr's retort as reported above. His criticism of the argument of Einstein, Podolsky and Rosen proceeded from the very same reluctance to take complicated consequences of an opaque formalism at their face value. (In the case of Einstein, Podolsky and Rosen there were other factors to be considered also, such as the authors' 'Criterion of Reality', see ch. 17.)

Bohr's habit of always distinguishing between formalisms and their physical content, and his attempts to make this content as clear as possible in a qualitative way, have their exact parallel in Einstein's insistence that 'the main thing is the content, not the mathematics. For mathematics can be used to prove everything' (Report, by Dr Hans Tanner, on Einstein's lectures in Zürich, quoted from Carl Seelig [93], 174). A participant of Einstein's evening lectures in Zürich in 1911 writes: 'He presented the most difficult issues and developments without any notes but he never failed to start with a purely qualitative analysis of the physical meaning and content of a problem . . . in order to make the train of thought as clear as possible. The mathematical solution was discussed only after the physical side had been made crystal clear, and he usually introduced it with the words "now we shall write down the x-es and y-s" ' ('nun wollen wir x-en' – report by W. Janitzky in Zug, quoted from Seelig, 238). This background of Bohr's version of the uncertainty principle and of his use of 'pictures' is thoroughly misrepresented by Popper. See below, end of section 8.

[69] This formulation occurs quite early and it is plausible to assume that Bohr obtained it from [78] where we read: 'Certainly I have seen before thought put on paper; but since I have come distinctly to perceive the contradiction implied in such an action, I feel completely incapable of forming a single written sentence. And although experience has shown in-numerable times that it can be done, I torture myself to solve the unaccountable puzzle, how one can think, talk, or write. You see, my friend, a movement presupposes a direction. The mind cannot proceed without moving along a certain line; but before following this line, it must already have thought it. Therefore one has already thought every thought before one thinks it. Thus every thought, which seems the work of a minute, presupposes an eternity. This could almost drive me to madness. – How could then any thought arise since it must have existed before it is produced? When you write a sentence, you must have it in your head before you write it; but before you have it in your head, you must have thought, otherwise how could you know that a sentence can be produced? And before you think it, you must have had an idea of it, otherwise how could it have occurred to you to think it. And so it goes on to infinity, and this infinity is enclosed in an instant. – Bless me, said Fritz, while you are proving that thoughts cannot move, yours are proceeding briskly forth. – That is just the knot, replied the student. This increases the hopeless mix-up, which no mortal can ever sort out. The insight into the impossibility of thinking contains itself an impossibility, the recognition of which again implies an inexplicable contradiction. . . Thus on many occasions man divides himself into two persons, one of whom tries to fool the other, while a third one who in fact is the same as the other two, is filled with wonder at this confusion. In short, thinking becomes dramatic and quietly acts the most complicated plots with itself, *and the*

concerning this problem but that the elusiveness of psychological matters prevented him from arriving at a satisfactory solution. His research in physics then showed him both a similar problem and, because of the definiteness of the circumstances in which it arose, a definite, 'hard' answer.[70] The answer allowed him to return, from a much more advanced point of view, to the problems that had originally aroused his curiosity.[71] It is reported that the resulting convergence of the psychological and the physical problem caught him somewhat by surprise,[72] as he had not expected to rediscover the psychological situation in the 'hard facts' of physics. We shall now give a very brief account of Bohr's philosophical ideas in the form in which they appeared in the early quantum theory. The main purpose of this sketch is to dispel certain myths which have recently arisen as the result, not of a study of Bohr's papers, but of a series of free associations, conceived in the presence of secondary material.

Two points must be emphasized. The first is that throughout the period of the older quantum theory Bohr underlined the 'formal' character of the new ideas which arose in connection with the principle of correspondence: these ideas, we can read repeatedly in his earlier papers, lead to some surprising predictive successes, and they also establish some order in the steadily accumulating empirical material. However, they do not form a theory in the ordinary, or 'classical' sense of the word, that is, they do not contain a coherent description of new and objective features. Nor can we say that they explain the relevant experimental facts. All this is emphasized

---

spectator again and again becomes actor.' (Quoted from Ruth Moore [79], 15f, my italics.) Møller's tale was well known to the members of the Copenhagen school and '[e]very one of those who came into close contact with Bohr at the institute, as soon as he showed himself sufficiently proficient in the Danish language, was acquainted with the little book: it was part of his initiation. Bohr would point to those scenes in which the licentiate loses the count of his many egos, or dissert on the impossibility of formulating a thought, and from these fanciful antinomies he would lead his interlocutor – along a path Poul Martin Møller never dreamt of – to the heart of the problem of unambiguous communication in science whose earnestness he then dramatically emphasized' [87], 121.

[70] J. Kalckar ([87], 231) reports Bohr's remark: 'People do not understand that complementarity is something hard and concrete.'

[71] 'If however', L. Rosenfeld continues his report (see end of n.69) 'we had only psychical experience to rely upon, it would be hard to make progress in the analysis of this remarkable epistemological problem. It is therefore fortunate that in atomic physics our attention has been forced upon a relation . . . of a very much simpler character, for which we could find a much more precise, indeed a mathematical formulation. What we have said above makes it clear that it is not by chance that this momentous advance in the theory of knowledge was made just by Bohr. Very early he had occasion to realize the relevance of his views on the ambiguity of language even in the realm of physics; he recognized that the concept of light presented such an ambiguity, inasmuch as it referred to aspects of the phenomena – light waves and light quanta – which were in that relationship of mutual exclusiveness he later called complementarity' ([87], 121). As regards the idea of ambiguity, cf. also the quotations in section 4 above.

[72] Professor Aage Petersen and Dr Klaus Meyer-Abich, private communication. For what follows the reader is also advised to consult Dr Meyer-Abich's account [76].

by Bohr himself and not only once, but in almost every paper he writes. He also emphasizes that the dependence on classical ideas that is implied by the use of the correspondence principle characterizes 'the present state of science', or 'the present state of our knowledge', and not the essence of physics. It is even regarded as somewhat undesirable. In these early years Bohr's aim seems to have been the gradual construction of an effective instrument of prediction which he hoped would act as a stimulus and as an aid towards the invention of a new atomic theory in the full sense of the word. A few quotations will support this assertion.

In an address delivered before the Physical Society in Copenhagen on December 20, 1913,[73] Bohr declared as his then objective not 'to propose an explanation of the spectral laws', but rather 'to indicate a way in which it appears possible to bring the spectral laws into close connection with other properties of the elements, which appear to be equally inexplicable on the basis of the present state of science' ([56], 87). He was well aware of the fact that 'the mixture of Planck's ideas with the old mechanics ma[d]e it very difficult to form a physical idea of what is the basis of it,'[74] and he himself repeatedly emphasized this lack of a theoretical foundation ([87], 60). The reason he gave for this peculiar and unsatisfactory situation was that new concepts were not yet available and that the classical concepts, though patently inadequate, still produced numerous correct predictions provided one restricts their application, is always aware of the approximations made, and refrains, for the time being, from all ontological inference. 'At the present state of physics', writes Bohr in 1922 ([6], 117) 'we must base each description of nature on an application of those concepts which are introduced by and defined in classical physics'; as a consequence 'one is obliged' (again 'at the present state of physics' 'to always keep in mind the domain of application of the theory' ([6], 118). Even the stationary states and the corresponding quantum numbers can be defined only to the extent to which 'the radiative reaction can be neglected' ([6], 119 – exactly the point that was overlooked by Sommerfeld, as we have seen above). 'This lack of sharpness in the description of the movement of the electron within an atom entails an uncertainty in the definition of stationary states whose recognition may on occasions be of essential importance' ([6], 152; the quantum-theoretical uncertainties already announce themselves).[75] It is for these reasons that the methods used 'for the determination of the stationary states' can be of a 'formal' nature only ([6], 131); they give us numbers; but they do not allow us to say what particular objective processes are responsible for the appearance of these numbers. Bohr is quite prepared to admit

---

[73] 'Which he himself . . . considered one of his best and clearest lectures' ([87], 60).

[74] Rutherford, letter to Bohr of March 20, 1913; quoted from Ruth Moore [79], 59.

[75] This is 'Bohr's idea' as described by Einstein and Ehrenfest in their paper ([34], 34) and mentioned at the end of section 5 above. Cf. also n.9.

the possibility of a deeper understanding which would depend, among other things, on 'assumptions concerning the nature of radiative processes' ([6], 155); but he points out, as a matter of historical fact, that such assumptions are not yet available. A new theory of the microlevel, a new theory of the nature of microprocesses, such a new theory has not yet been found. What has been found is a somewhat abstruse combination of classical concepts which is a purely 'formal' instrument of prediction (in addition Bohr indicates how he hopes to advance by 'emphasizing [the] conflict' between the new conceptual scheme and 'the admirably coherent group of conceptions which have been rightly called the classical theory of electrodynamics' ([56], 88)).[76]

The joint paper of Bohr, Kramers and Slater expresses this attitude even more clearly. In the attempt to explain the Compton effect the authors use the idea of virtual oscillators moving at a velocity different from the velocity of the illuminated electron and they emphasize the 'formal' character of this idea ([13], 799), i.e. the fact that one is dealing here not with an event that takes place in the real outer world, but rather with a predictive device for the proper arrangement of classical concepts such as 'the wave concept of radiation' ([13], 785; one wonders how Popper's realism is going to work in this case). The very first page of the paper contains the remark that 'at the present state of science [!] it does not seem possible to avoid the formal character of the quantum theory'.[77]

Compare this extremely condensed account with the following quotation from Popper's widely read 'Three Views of Human Knowledge' ([84], 99f):

> Today the view of physical science founded by Osiander, Cardinal Bellarmino, and Bishop Berkeley [who according to Popper declare either all of physics, or particular physical theories to be instruments of prediction only] has won the battle without another shot being fired. Without any further debate about the philosophical issue, without producing any new argument the *instrumentalistic view* . . . has become the accepted dogma. It may well now be called the 'official view' . . . and it has become part of the current teaching of physics.

Considering this quotation we can at once accept the assertion that instrumentalism is a well-entrenched belief of many contemporary physicists and that it is accepted by them as a matter of course, and 'without debate'. Trying to explain this phenomenon we can point to the complete lack of historical perspective on the part of the younger generation who 'received quantum mechanics served up ready made' ([87], 219) and who

---

[76] For a detailed discussion of this feature, see the first chapter of [76].

[77] Cf. also Bohr's comments on the 'formalistic' character of the photon theory of light ([6], 156ff – the whole third chapter of this paper is entitled 'On the Formal Nature of Quantum Theory') and Einstein's considerations of 1916 ([13], 788f).

'do not know the debates that settled [important] issues . . . for the fathers of the quantum theory' ([66], vi). But must we not say that the remarks of Popper and of some more recent critics of the Copenhagen Interpretation exhibit the very same ignorance or neglect of background knowledge? The realism–instrumentalism issue, after all, was one of the most central and most hotly debated topics of the older quantum theory and it was discussed to such an extent that its repercussions should have reached the ears even of a philosopher. Every paper of Bohr's emphasizes that so far an instrument of prediction is all one can have and that this shortcoming is due to the absence of unrefuted hypotheses about the nature of atomic processes. How does this fit with Popper's 'without any further debate'? It is true that Bohr eventually arrived at the position that 'the whole purpose of the formalism of the quantum theory is to derive expectations for observations obtained under given experimental conditions' ([98], 27), but this was the result of a series of refutations and discoveries which seemed to show that considerations of 'nature' (and the word here does not indicate essentialistic aberrations!) had been removed very far indeed.[78] Now it is of course Popper's privilege to disregard such refutations and to continue believing in the correctness of his own microphilosophy. But it is somewhat unjust to describe those who took the refutations seriously as philosophical dogmatists who never realized that there was an issue, and it is also somewhat optimistic, under such circumstances, to think that one can teach them a lesson.

We now come to the second point, Popper's accusation of subjectivism. Bohr is interested in the interaction between the subject and the object and he also emphasizes the similarity between physics and psychology in which latter science 'we are continually reminded of the *difficulty of distinguishing between subject and object*' ([8], 15 – original italics). But the 'subject' in physics is for him not the consciousness of the observer but 'the agency' used for observation, that is the material measuring instrument (including the body of the observer, and his sense organs); and the 'boundary' disappears, in physics, not from between the consciousness of the observer and 'the world'; it disappears from between 'the [atomic] phenomena [and] the [material] agencies of observation' ([8], 54).[79] There is therefore no 'ghost' to be exorcized from quantum mechanics.[80]

---

[78] Cf. sections 3, 5 and 7 of the present essay.

[79] According to Bunge ([20], 4) Heisenberg 'has recently admitted' that the talk about the observer must be extended to include the measuring instrument. Bunge also insinuates that this 'Turn of the Tide' is due to the fact that one now at last starts listening to some of those philosophers and physicists (Bunge?) who have tried to influence the Copenhagen school from the outside. I am sorry to say that this self-satisfied insinuation only reveals Bunge's ignorance of Bohr and of the actual development of ideas within the Copenhagen circle. *The objective notion of observation was there from the very beginning.* Cf. n.87.

[80] 'This is an attempt', Popper starts his paper ([20], 7) 'to exorcise the ghost called "consciousness", or the "observer" from the quantum mechanics. . .'

It remains for us to connect what has been said so far with the hypothesis of the relational character of the quantum-mechanical states which was formulated in section 4 above.

First of all it should now be clear that this hypothesis is a physical hypothesis and not the consequence of some inarticulate philosophical faith. It is to be admitted that Bohr also had philosophical reasons to expect a hypothesis of this kind to be true. But we have seen that this philosophy did not prevent him from exploring alternatives; and we have also seen that it was the refutation of these alternatives taken together with difficulties such as those mentioned in section 5 which led him back to his original ideas and convinced him, almost to his surprise, of the correctness of his more philosophical outlook. Secondly, the hypothesis does not introduce any subjective element but concerns the physical situation only (cf. the definition of a 'phenomenon' in section 2 where the similarity to Popper's own ideas is emphasized). The hypothesis has certain similarities with Kantianism, assuming that nature (and the word is here used in almost the same sense as in Kant) depends on our categories and forms of perception. The difference is that the physical interactions involved in any act of cognition are taken into account and that consequences for epistemology are drawn from such physical considerations (Bohr's view is therefore a step in the direction of materialism, as Fock [44] has pointed out).[81] But wider philosophical principles such as these are now beside the point. What is essential is that the hypothesis of the relational character of all dynamical states is a physical hypothesis as it is an attempt to account for a long series of interesting conjectures and drastic refutations. The reader may also convince himself, by a study of Bohr's writings, that it is possible to find a correlation between experimental conditions and magnitudes that avoids the difficulties discussed in section 5 as well as the almost innumerable further difficulties which arose during the period of the older quantum theory. Finally, it is evident that the 'observer' and his 'knowledge' nowhere enter the scene. All that is asserted is that there are objective conditions which do not allow for the application of certain magnitudes (and of the theories relating these magnitudes to each other) irrespective of whether these objective conditions are now used for improving our knowledge or not.[82] In order to make this feature of the hypothesis as clear as possible I shall now illustrate it, as did Bohr ([9] and [88]), with the help of the very similar situation in relativity.

For the theory of relativity, just as the quantum theory, contains non-

---

[81] Of course, Bohr often uses subjectivistic terms. But Popper himself has reminded us that words are not too important and that one should always try to discover the ideas they are intended to express. Fock [44] gives a very clear account of Bohr's ideas, comments on the misleading manner in which they are expressed ([44], 192) and distinguishes them from their 'false positivistic interpretation' ([44], 210).

[82] For an elaboration of this point see section 2 of my essay in [22].

statistical terms which express relations between an object and a reference frame. We can even introduce the very same notation we introduced in the case of propensity (section 2: $M[O, R] = \alpha$ means now (for example) that the mass of O in R is α or, using the 'second way' of speaking (section 2) that the mass number of the block O-in-R is α.

Just as in the case of probability, abrupt but objective changes occur when we replace one set of experimental conditions by another. An object may be heavy in one frame, light in another, i.e. $M[O, R']$ may be $\gg M[O, R'']$ and the value changes from one to the other depending on whether we consider the one or the other reference system. (This shows again that the 'reduction of the wave packet' cannot be a 'trivial feature of the probability theory' ([20], 37). For, if we use the hypothesis of the relational character of state descriptions, then individual changes analogous to those just described in the case of relativity will occur in addition to whatever happens to the probabilities.) But again, it would be quite mistaken to try explaining the change in a 'mechanistic' fashion, as the result of an interaction between O and R (as is attempted in the electron theory of H. A. Lorentz); and we must again recall Bohr's warning not to confuse 'a mechanical disturbance of the system under consideration' with an 'influence on the very conditions' of the experiment. Only this warning, which was used in section 2 in connection with probabilities, now applies to the individual case also.

There are much simpler examples of objective relational magnitudes which elucidate another feature of Bohr's hypothesis in that they are applicable only as long as certain conditions are first satisfied. Thus the concept of hardness as defined by the Mohs scale ceases to be applicable when the temperature becomes too high and the same is true of surface tension at low temperatures. There is no need to continue this list which shows quite clearly the existence of non-probabilisitic concepts which characterize experimental set-ups, are applicable only in certain physical conditions, and change abruptly when the conditions change. Bohr assumes that position, momentum, etc. are concepts of exactly this kind and he specifies the conditions under which they are applicable, and to what degree of precision.

7. THE UNCERTAINTY RELATIONS

With this section we now at last resume the discussion that was started in section 5 but interrupted in order to clarify certain mistakes, due to ignorance, concerning Bohr's general philosophy. We consider a very simple version of a pin board, namely a board that contains no pins at all. We turn the board around through an angle of 90° and have a slit of width Δx (fig. 7). We consider trajectories t of single particles behaving exactly as postulated in assumption **T** (and under the restrictions formulated at the beginning of section 5). Assuming that all particles move in such trajectories we obtain a Gaussian distribution (this is consequence **C**). Interference, on the other hand, leads to a redistribution of trajectories. Now (and here we simply repeat what was already said in section 5) we (i) cannot find any forces which could account for such a redistribution, and attempts to conceive such forces lead to trouble (this is illustrated by the second and fourth point

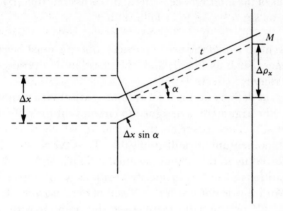

Fig. 7

of Einstein and Ehrenfest, [34]). On the other hand, (ii) the conservation laws are still valid in each individual case (this follows from the experiments of Bothe–Geiger and Compton–Simon). (ii) shows that a dynamical account is needed of the redistribution, at least as long as we continue using trajectories. (i) shows that such an account is impossible. The way out of the dilemma suggested by Bohr consists in assuming (A) that *particles do not always have trajectories* ([8], 81).[83] There are conditions which forbid us to ascribe to the particle a well-defined position together with a well-defined momentum.[84] To this assumption Bohr adds the hypothesis (B) that the wave theory which uses extended entities connected by a phase instead of particles and their trajectories becomes valid to the extent to which the particle 'picture' ceases to be valid, and vice versa. (A) and (B), taken

[83] The quotations which emphasize that assumption (A) removes any 'possibility of a contra-diction with the law of conservation' refer to 'knowledge' in a way which on superficial reading might raise the question of subjectivism. But all that is meant is that, as there are no trajectories, there cannot be any knowledge of them either (Heisenberg, of course, occa-sionally reverses the argument). Cf. n.83.

[84] 'From these results', writes Bohr ([8], 34) referring to the interference phenomena, the Compton effect, the experiments of Bothe and Geiger and of Compton and Simon, 'it seems to follow that, in the general problem of the quantum theory, one is faced not with a modification of the mechanical and electrodynamical theories describable in terms of the usual physical concepts, but with an essential failure of the pictures in space and time [such as trajectories] on which the description of natural phenomena has hitherto been based.'

together with de Broglie's relation $\lambda = h/p$ (which may be regarded simply as an empirical law) now permit a semi-quantitative estimate of the limits of notions such as positions, momentum, trajectories, etc.: the problem of the redistribution of paths disappears as soon as we deny that there is any well-defined path between the maximum and the first minimum M (this is assumption (A)). The first minimum is characterized by the angle α such that $\Delta x \sin \alpha = \lambda$ (λ the wavelength of the classical wave used for calculating the details of the interference pattern: this is assumption (B)). Now $\lambda = h/p_x$. Hence, as $\sin \alpha = \Delta p_x / p_x$ it follows that

$$\Delta x \Delta p_x > h \qquad\qquad (1)$$

gives the extent to which the use of particle concepts must be restricted in an arrangement such as Fig. 7, and, therefore, in any arrangement that permits interference effects.

Δx and Δp_x as defined in this brief argument are obviously not statistical magnitudes (the argument is designed to circumvent the difficulties of the individual case). They describe the extent to which concepts such as position and momentum are still applicable. Thus $\Delta x = \infty$ entails that the conditions forbid us to talk about position at all while $\Delta x = 1$ cm entails that the conditions forbid us to specify positions with a precision greater than 1 cm. And this is not just a restriction of our *knowledge*. It is not first asserted that we occasionally *cannot know* the position with a precision <1 cm and then concluded, by some kind of positivistic reasoning, that more precise statements are meaningless. Quite the contrary, it is asserted that under certain conditions positions with a precision <1 cm cannot be known because, once these conditions are realized, *there is no such feature in the world* (a block of ice may have a certain hardness on the Mohs scale; but when it melts, and turns into water, then its hardness simply ceases to exist).[85] In Bohr's view the application of the uncertainty relations to the past is not 'a matter of personal taste' ('eine Geschmackssache') as Heisenberg has said ([52], 20) but a direct consequence of hypotheses (A) and (B) which determine what can and what cannot be asserted under certain conditions.[86] Bohr has also been critical of other statements of

[85] Bohr usually expresses such matters in what Carnap has called the 'formal mode of speech', i.e. he speaks about the limits of definability rather than limits of applicability, or existence. Thus he says ([8], 56) that 'radiation in free space as well as isolated material particles are abstractions, their properties on the quantum theory being *definable* and observable only through their interaction with other systems', which simply means that notions such as wavelength and position are applicable only when certain physical conditions are first specified, that there are no such things as wavelengths in the world unless there are also these conditions, and that the pre-quantum notion of wavelength, and of position (just as the pre-relativity notion of length) did not take this relational property into account. For more on the interaction terminology cf. n.10, 11, and text.

[86] 'Indeed', says Bohr ([8], 66), 'the position of an individual particle at any two given moments can be measured with any desired degree of accuracy; but if, from such measurements, we would calculate the velocity of the individual in the ordinary way, it must be clearly understood that we are dealing with an abstraction from which no unambiguous

Heisenberg.[87] Thus, for example, Heisenberg wanted to ascribe the uncertainty of momentum resulting from a position measurement to the Compton effect; but Bohr again pointed out that it is due not to such an interaction which would still permit us at least to think of precise positions and momenta (a point that has been emphasized by Popper), but rather 'to the uncertainty [arising] in an arrangement that can be used for an accurate measurement of position' ([87], 88).[88] It is to be admitted that Bohr himself quite often uses subjectivistic terms, speaking of a restriction of 'knowledge', and so on.[89] However, once we go beyond this cloak of words

information concerning the *previous or future* behaviour of the individual can be obtained' (my italics).

[87] It is told (H. J. Grönewold, private communication) that Bohr regarded Heisenberg's publication of his [52] as premature and thought that a more careful and less subjectivistic presentation was needed. For Bohr's early criticism of Heisenberg see the *Nachtrag bei der Korrektur* to Heisenberg's important [50]. It is this criticism rather than any philosophical sermon by professional realists which Heisenberg seems to remember in his more recent writings. Cf. above, n.79.
Heisenberg has described the difference between Bohr and himself in the following way: 'It was very instructive for me to see how Bohr continued to try to advance through the physical interpretation of the formulae and thus to reach a decision, while it was much more natural for me to use a formal mathematical view'. 'Bohr considered Schrödinger's investigations to be very important for two reasons. For the first, it strengthened our trust in the correctness of the mathematical formalism, which could now be called, with equal justice, quantum mechanics or wave mechanics. Secondly, the question arose as to whether one should seek a physical interpretation of the formalism along quite other lines than those hitherto considered by the Copenhagen group. Bohr realized at once that it was here that we would find the solution to those fundamental problems with which he had struggled incessantly since 1913, and in the light of the newly won knowledge he concentrated all his thought on a critical test of those arguments which had led him to ideas such as stationary states and quantum transitions. . . I myself was not really willing to concede Schrödinger's theory a part in the interpretation of quantum theory. I considered it rather an extremely useful tool for solving the mathematical problems of quantum mechanics, but not more. Bohr, on the other hand, seemed inclined to place the wave-particle dualism among the basic assumptions of the theory' ([87], 100ff). For Bohr's insistence on wave features which occasionally led him astray see n. 48 and the literature there. Cf. also ([100], 11–14).
[88] 'In connection with the measurement of the position of a particle, one might, for example, ask whether the momentum transmitted by the scattering could not be determined by means of the conservation theorem from a measurement of the change of momentum of the microscope – including light source and photographic plate – during the process of observation. A closer investigation shows, however, that such a measurement is impossible, if at the same time one wants to know the position of the microscope with sufficient accuracy. In fact, it follows from the experiences which have found expression in the wave theory of matter that the position of the centre of gravity of a body and its total momentum can only be defined within the limits of reciprocal accuracy given by relation (2) – [(1) in the present paper]' ([8], 67). We see here very well how hypotheses (A) and (B are work together. Statistical redistribution is out because of the 'conservation theorem'. Each single path must be accounted for. But this is impossible because of 'the experiences which have found expression in the wave theory of matter'. It is therefore necessary to abandon trajectories 'within the limits of reciprocal accuracy given by relation[(1)]' and to 'constantly keep the possibilities of definition as well as of observation before the mind' ([8], 73: for the terminology of these remarks see n.85). Cf. also n.9 above.
[89] Cf. above, n.81. As an example we quote the following passage: 'We have recently experienced such a revision in the rise of the theory of relativity which, by a profound analysis

we find that the discussion is not about lack of knowledge, but about objective conditions of the individual case. Having presented the explanation, and the arguments in favour of the hypothesis of the relational character of quantum states, we shall now turn to Popper's objections against it.

8. REFUTATIONS OF TWO OBJECTIONS

The first objection, which already occurs in the *Logic of Scientific Discovery*, proceeds from Born's statistical rules. 'It has till now not been taken sufficiently into account', writes Popper ([83], 227), 'that to the mathematical derivation of the Heisenberg formulae there must correspond, precisely, a derivation of the *interpretation* of these fundamental equations.' And he points out that, given Born's rules and nothing else, we must interpret Δx and Δp_x in formula (1) as the root-mean-square deviations (r.m.s. for short), within large ensembles, of quantities which are otherwise well defined and not as statements 'imposing limitations upon the attainable precision of measurement' ([83], 224). This point of view seems to derive further support from von Neumann's systematic application of the frequency theory to wave mechanics whose statistical ensembles 'make again possible an objective interpretation (which is independent of . . . whether one measures, in a given state, the one or the other of two not simultaneously measurable quantities)' ([80], English version, p. 300). Hence,

> if we start from the assumption that the formulae which are peculiar to the quantum theory are . . . statistical statements, then it is difficult to see how prohibitions of single events could be deduced from a statistical theory of this character. . . The belief that single measurements can contradict the formula of quantum physics seems logically untenable; just as untenable as the belief that a contradiction might one day be deduced between a formally singular probability statement . . . say 'the probability that the throw k will be five equals 1/6' and one of the following two statements . . . 'the throw is in fact a five', or . . . 'the throw is in fact not a five'. [[83], 228f]

This is of course completely correct reasoning provided the elements of the collectives with which we are dealing in the quantum theory are all in a state that is well defined from a classical point of view, i.e. provided we already know what kinds of entities are to be counted as the elements of the collectives. Only if it is assumed that these elements are systems which are in classically well-defined states can we derive Popper's interpretation from the statistical character of the quantum theory, as exhibited in Born's rules.

of the problem of observation, was destined to reveal the subjective character of all the concepts of classical physics' ([8], 97), by which is meant no more than the fact that classical physics describes the world as it appears to an observer (the physical bulk of the observer, his senses included) moving at small velocities while relativity gives an account of 'the objective reality of the phenomena open to observation' ([8], 97).

However, it is also evident that this statistical character, taken by itself, is never sufficient for deriving such an assumption. This is most emphatically stated by Popper himself who accuses Bohr and Heisenberg of precisely such a confusion between statements about ensembles and statements about their elements ([20], 28). For from the fact that a theory is statistical we can infer that it works with collectives of events, processes, transitions or what have you. We cannot draw any inference about the individual properties of these objects, events, processes, etc. nor can we infer the category of the elements (i.e. whether they are for example objects, or transitions). Any such information would have to be given in addition to the laws characterizing the relation between the frequencies in the collectives. Does Born's interpretation, which after all establishes a connection between the formalism and reality, provide such information? It does not – or at least it should not, if proper care is taken. For Born's rules, properly interpreted, need not make any assertion about the character of the elements of the quantum-mechanical collectives. All that is needed are assertions concerning the expectation values which these elements exhibit under certain well-defined conditions (such as occur, for example, in a well-designed measurement).

There are therefore still at least two alternatives open to us: (1) the elements possess their values before the conditions are realized and retain them when they are put into effect; (2) the elements do not possess their values before the realization of the relevant conditions, but are transformed, by the conditions (by the measurement) into a state containing these values in a well-defined manner. However great the empirical success of the statistical interpretation, it does not provide any means for deciding between (1) and (2). And this is a well-known characteristic of all statistical theories, as we have said: even death statistics do not allow us to draw any conclusion concerning the manner in which death has occurred, nor do they allow us to infer that human beings are entities whose traits are independent of observation, i.e. which can be assumed to be either alive, or dead, independently of the occasions on which they were found to be either alive or dead. (Like vampires they might be dwelling in an entirely different state in between observations.) In the case of human beings we possess, of course, evidence about their permanence, in a well-defined state, between the moments of observation. The point is that this information is independent of the fact that in death statistics we are dealing with a statistical theory.[90]

The same is true in the quantum theory: Popper's idea that an elementary particle always possesses a well-defined value of all the magnitudes that can be measured in it does not follow from Born's interpretation.

---

[90] As was shown in [37] this point also invalidates von Neumann's famous 'proof'.

It is an additional idea that should be justified, or at least discussed. It is precisely this idea that has been found to be inconsistent with the laws of interference and the individual validity of the conservation laws and that was therefore replaced by the assumption of the relational character of all dynamical magnitudes. To sum up: Popper's argument is invalid, and its conclusion is false.

This being the case we must now look for a new interpretation of the elements of quantum-mechanical collectives. We must admit that what is being counted is not the number of systems possessing a certain well-defined property; what is counted is rather the number of transitions from certain partly ill-defined states into other partly ill-defined states. And neither the frequency approach which deals with collectives only and is neutral with respect to either the first or the second choice of elements, nor Born's rules which also leave this choice open, can be used as an argument against the customary, i.e. Bohr's, interpretation of formula (1).

According to the second objection, which Popper develops in some detail in [20], the precise notions of position and momentum (and of other dynamical magnitudes) are needed if we want to test the statistical predictions of the quantum theory and especially the uncertainty relations. The choice in the interpretation of Born's rules is therefore not really left open. For tests of statistical relations are possible only if each individual measurement is more precise than the r.m.s. of the magnitude in the ensemble in which it takes place and, more especially, if it is 'more precise than the range or width of the scatter' asserted by the uncertainty relations ([20], 20). Whatever dependence exists between dynamical variables and the experimental arrangement must therefore be negligible in comparison with the uncertainty relations. Or, to express it differently, microscopic systems must possess all the relevant dynamical properties with a precision far greater than the uncertainties in Heisenberg's formula. In practice this means that the elements of our tests have all 'the properties of [classical] particle[s]' ([20], 19).

Now it is quite correct that a r.m.s. of magnitude Δx can be ascertained only if each single position measurement shows an error $\delta x \ll \Delta x$. The same applies to momentum. But the inference that the individual system must have a position and a momentum which are more precisely defined than is stated in the uncertainty relations is permissible only if we assume that the measurement does not introduce new conditions, that is, if we are allowed to use all our dynamical variables in exactly the same manner, before, during, and after the measurement. *If* we make this particular assumption, then $\delta x \ll \Delta x$ and $\delta p \ll \Delta p$ will indeed entail the existence of ensembles and of individuals for which $\delta x \, \delta p \ll h$ ([20], 22f). The question is therefore: should we make such an assumption? Is such an assumption forced upon us by our wish to test the statistical predictions of the quantum theory and, more especially, of the uncertainty relations? It certainly is not.

To show this let us develop Bohr's view a little further. We assume with Bohr that a system possesses certain properties, such as localizability, only while being part of some well-defined experimental arrangement, P; and that it possesses some other property, such as well-defined motion, only while being part of a different experimental arrangement, M. Given an arrangement $A \neq P, M$ the system is neither perfectly localizable, nor can its motion be defined with absolute precision (all this has been said with great clarity by Landé, in his [71]). Now consider a class of systems S, all of them being connected with conditions A. Choose a subclass S' and change the conditions for this subclass from A to P. After the change the positions can be determined with utmost precision. Let their r.m.s. be Δx. Choose another subclass S'', such that S' and $S'' = \Lambda$ and again change the conditions for the elements of this subclass from A to M. Then momenta can be determined with the utmost precision. Let their r.m.s. be Δp. Now it is asserted that whatever the original state of the system and whatever the conditions A under which this state was originally asserted, a Δx and a Δp determined in the manner just explained will be related to each other as in formula (1).

The terminal statement of the last paragraph is a testable statement. The tests can be carried out although it is assumed that the Δ refer to an objective lack of precision in each individual system. The tests are of formula (1). They say that the r.m.s. obtained in the manner described will obey the same relation as the individual uncertainties asserted in (1). It follows that Popper's second argument, too, is a *non sequitur*. Bohr's interpretation of formula (1) is compatible both with Born's rules and with the assertion of the testability of (1).

The last paragraph shows, incidentally, that Popper is quite correct when implying that there may exist two different interpretations of formula (1). However, he is mistaken when inferring that there must therefore be a 'quantum muddle'. On the first interpretation (a), formula (1) expresses how the r.m.s. obtained in complementary experiments (on large collectives of systems which are originally dwelling in the same state and are subjected to identical conditions) are related to each other. On the second interpretation (b), formula (1) expresses some objective indefiniteness of the individual systems, viz. the fact that the usual dynamical magnitudes inhere in them with a certain spread which is a non-statistical feature, though it is capable of statistical test. Interpretation (b) is a consequence of the hypothesis of the relational character of the quantum-mechanical states taken together with (A) and (B) which embody the laws of interference and the conservation laws. *It is an abbreviated statement of all these laws*, utilizing also Einstein's $E = h\nu$ (uncertainty relations for time and energy) and de

Broglie's $p = h/\lambda$. Interpretation (a) is about the statistical properties of experimental results or, if you wish, about the frequency of transitions of a certain kind under measurement. Now the important point is that (b) *may be regarded as a test as well as an explanation* of (a).[91] It is an explanation as it makes it clear why (a) is never contradicted by statistical results. That it is also a test may be seen in the following manner: if the uncertainties obtained with the help of the semi-quantitative arguments of section 7 were considerably larger than the uncertainties derived from the quantum theory together with Born's interpretation, then a serious objection would arise against the latter. It could no longer be said to be compatible with all the laws that are known to be valid on the microdomain and which are contained, in an abbreviated form, in interpretation (b). And this was indeed the way in which Bohr used (b) when proceeding to test the formalism of quantum electrodynamics [14].[92] It is in this connection that the idea of complementarity becomes important. This idea expresses precisely what has been said above, namely, that physical properties inhere in the experimental arrangement and not in the system itself, adding an account, in accordance with (A) and (B), of the laws according to which the applicable properties change when the relevant conditions are realized. It is therefore much richer than the propensity interpretation, as we already had occasion to point out. The propensity interpretation says that probabilities change when the conditions change. Period. Complementarity allows us first to see how propensities can be incorporated into the quantum theory (text to n.36 above) and then informs us what properties are related to what conditions and how they change in the presence of forces or of other processes compatible with the conditions of their application. Thus it is said that once the suitable conditions are satisfied we can use the whole rich apparatus of classical wave theory to analyse, explain, and make predictions.[93] (When Bohr speaks of 'pictures' he simply means visual-

[91] In this way, says Bohr, it has been possible 'to elucidate many paradoxes appearing in the application of the quantum postulate, and to a large extent to demonstrate the consistency of the symbolic method' ([8], 73).

[92] 'An application of the conservation laws to the process [of scattering of radiation by particles] implies that the accuracy of definition of the energy-momentum vector is the same for the radiation and the electron. In consequence, according to relation [(1)], the associated space-time regions can be given the same size for both individuals in interaction' ([8], 61). Cf. also Rosenfeld's report in ([82], 70ff) as well as in ([87], 114ff).

[93] 'This view is already clearly brought out by the much discussed questions of the nature of light and the ultimate constituents of matter. As regards light its propagation in space and time is adequately expressed by the electromagnetic theory. Especially the interference phenomena *in vacuo* and the optical properties of material media are completely governed by the wave theory superposition principle. Nevertheless, the conservation of energy and momentum during the interaction between radiation and matter, as evident in the photoelectric and Compton effect, finds its adequate expression just in the light quantum idea put forward by Einstein. As is well known the doubts regarding the validity of the superposition principle, on the one hand, and of the conservation laws, on the other, which were suggested by this apparent contradiction, have been definitely disproved through direct experiments,

izable theories so that Popper's long objection ([20], 11–15) is beside the point. Besides – why say that '"pictures" are unimportant' ([20], 14)? They may be an aid in calculations and considerably help the researcher to get a quick survey of a puzzling problem. The same is true of the particle 'picture' which is another theory (concerning trajectories), applicable under other conditions. To see how little all this has to do with subjectivism and with an alleged misinterpretation of probabilities let us consider a two-slit experiment with metal slits. In this case the photons will occasionally interact with the slits and eject electrons. We cannot control this process 'since the limitation [due to the uncertainty relations] applies to the [material] agencies of observation as well as to the [material] phenomena under investigation' ([8], 11). We do not know when it will occur and to which photon. But the fact that it does occur shows itself in the interference pattern which will be covered by a faint Gaussian haze consisting of all those photons which have interacted with a particular slit (which one, we do not know) and were thereby subjected to objective conditions favouring their particle aspects.

I come now to the 'great quantum muddle' ([20], 18f, 28), to the 'grave logical blunder[s]' ([20], 29) allegedly committed by Bohr and by his followers. I am afraid this 'muddle', and these 'blunders', are nothing but a piece of fiction. According to Popper, Bohr and Heisenberg project properties of statistical ensembles into the individual particle. This, Popper thinks, is the way in which complementarity and interpretation (b) of the uncertainty relations have arisen. Now I hope it has become clear, from our presentation above, that interpretation (b) has an entirely different origin. One starts with the individual particle (example: the individual conservation laws) as well as with those parts of the statistics which, having the form of selection rules, do indeed allow an inference to the individual case. Moreover, Bohr and Heisenberg, well aware of the existence of careless readers, quite explicitly warn us against the errors ascribed to them by Popper. 'The reader must be warned', writes Heisenberg ([52], 47 and mathematical appendix, paragraph 8) 'against an unwarrantable confusion of classical wave theory with the Schroedinger theory of waves . . .' and he makes it clear that the 'wave picture' refers to the three-dimensional waves of classical optics and not to the ψ-waves. Bohr emphasizes the 'symbolical character of Schroedinger's method' ([8], 79) and opposes to it the 'immediate reality' of the classical waves used in the discussion of interpretation (b) of the uncertainty relations. It is only Popper's neglect of

([8], 55. 'The two views of the nature of lights are . . . to be considered as different attempts at an interpretation of experimental evidence in which the limitation of classical concepts is expressed in complementary ways' ([8], 56). For the translation of this remark from the 'formal mode of speech' into the terminology of existence and experimental conditions (which was introduced about five years later) cf. nn.85 and 87.

the dynamics of the individual particle and his unwarranted assumption that the quantum theory is pure statistics which make him suspect that whatever is said about the individual particle must have been snatched from some collective.

9. THE CASE OF EINSTEIN, PODOLSKY AND ROSEN

Finally, one sees that Bohr's suggestions provide an immediate solution to the famous case of Einstein, Podolsky and Rosen (EPR).[94] This solution which, according to Einstein, comes 'nearest to doing justice to the problem' ([88], 681) may be briefly presented as follows. Einstein, Podolsky and Rosen assume without argument that what we determine when all disturbance has been eliminated is a property of the system examined that belongs to it irrespective of the conditions of the experiment (this is the content of their famous 'criterion of reality'). As opposed to this Bohr maintains, in accordance with his hypothesis, that quantum-mechanical state descriptions are relations between systems and experimental arrangement and are therefore dependent on the latter. It is easily seen how this assumption deals with EPR. For while a property cannot be changed except by direct interference with the system that possesses the property, a relation can be changed without such interference ('the board "knows"', Popper replies, to a similar problem arising in the case of the classical pinboard ([20], 33)). Thus the 'state' of 'being longer than b' of a rubber band may change when we compress it, i.e. when we physically interfere with it. But it may also change when we change b without at all interfering with the rubber band. Hence, lack of physical interference excludes change of state only if it has already been established that positions and momenta and other magnitudes are properties of systems, belonging to them under all circumstances.[95] But just this assumption had to be rejected and had to

[94] 'The Einstein–Podolsky–Rosen experiment has since become a real experiment' says Popper ([20], 28). He fails to mention that this real experiment has corroborated Bohr's interpretation and refuted Einstein (who thought that the quantum-mechanical formulation of the many body problem might break down at large distances). For details cf. ch. 17.

[95] The logic of situations such as those arising in the case of EPR was made clear, long ago, by Galileo in his argument about comets. The distance of a comet, he pointed out, can be determined from its parallax only after its nature has been determined by independent argument, i.e. only after 'it is first proved that comets are not reflexions of light [such as for example a rainbow], but are unique, fixed, real, and permanent objects' ('Discourse on Comets' by Mario Giuducci, quoted from [30], 39). EPR's 'criterion of reality' would infer reality from the possibility of measuring parallax, thus inverting the natural order of the argument. The same remark applies to Popper's attempt to infer definite classical states from the existence of statistical ensembles. To put it in a few words: EPR (and, of course, Popper) put forth their 'principle of reality', but they have not a single argument to offer in its favour. Bohr offers a variety of arguments for his own, contrary, position (relationalism of states). Hence, from the point of view of a critical rationalism, Bohr is ahead of EPR however bad his own arguments and the defendants of EPR are dogmatists, however efficient they may be in pointing out the weakness of Bohr's position.

be replaced by Bohr's hypothesis; and it had to be rejected because of difficulties which are quite independent of the case of EPR. Bohr's solution of this case is therefore not only not *ad hoc*,[96] but it adds further support to the idea of complementarity.

10. CONCLUSION: BACK TO BOHR!

Popper's criticism of the Copenhagen Interpretation, and especially of Bohr's ideas, is irrelevant, and his own interpretation is inadequate. The criticism is irrelevant as it neglects certain important facts, arguments, hypotheses and procedures which are necessary for a proper evaluation of complementarity and because it accuses its defenders of 'mistakes', 'muddles' and 'grave errors', which not only have not been committed, but against which Bohr and Heisenberg have issued quite explicit warnings. His own positive view which is interesting, especially if extrapolated as described in section 2, and which is very relevant in probability theory, is quite inadequate as a remedy for the special problems of the quantum theory and too simplistic to be regarded even as a possible alternative to complementarity (let alone a preferable one). Altogether it is a big and unfortunate step back from what had already been achieved in 1927. His accusation of dogmatism, too, is quite ill founded. There was hardly a physicist who was so intent on seeing all sides of a problem,[97] and so fond of qualifying his remarks, as was Bohr. Besides, Bohr was prepared to consider and even to develop views which could not easily be reconciled with his own philosophy as is shown by the theory of Bohr, Kramer and Slater. *He was prepared to test his philosophy with the help of alternatives*.[98] It is true that Bohr's attitude hardened somewhat after the refutation of Bohr–Kramers–Slater, but this is hardly sufficient reason to accuse him of dogmatism. It is also correct that many contemporary physicists of the younger generation take complementarity for granted without examining it and perhaps even without understanding it. They may well have committed some of the 'mistakes' so dramatically described by Popper. But the remedy for such aberrations surely does not lie in the propagation of even more inadequate views; nor does it lie in the invention of depressing historical myths.[99] Quite

---

[96] As is asserted by Popper ([83], 445f). For a somewhat different and more detailed examination see my 'Problems of Microphysics' ([22], 4–6).

[97] Schiller's 'Nur die Fülle führ zur Klarheit, und im Abgrund wohnt die Wahrheit'; was one of Bohr's favourite verses ([82], 31, footnote).

[98] In this respect Bohr was much less dogmatic that Galileo who proceeded along a single train of thought only. Popper's remark on the 'New Betrayal' of the Science of Galileo should be seen in this light also. Cf. n.78.

[99] What, for example, shall we say, if we compare Bohr's ideas with the orgy of name-calling, dogmatic pronouncements, inane lyricisms, accompanied copiously by arid formalisms which we find in Bunge's contribution to the volume where Popper's paper appears? This is indeed a 'Turn of the Tide' (the title of Bunge's introduction to [20] – from an imaginative

the contrary, it would seem to lie in a clear presentation of all the arguments which have led to the idea of complementarity in the first place and, if one does not like their result, in their replacement by better arguments. It follows that the first step in our attempt to achieve progress in microphysics will have to be a return to Bohr.[100]

REFERENCES

1. Blumenthal, O. ed. *Das Relativitätsprinzip* (Leipzig, 1923).
2. Bohm, D. *Quantum Theory* (Princeton, 1952).
3. Bohm, D. 'A Suggested Interpretation of Quantum Theory in Terms of "Hidden Variables" ', *Phys. Rev.*, 85 (1952), 166ff, 180ff.
4. Bohm, D. and Bub, J. 'A Proposed Solution of the Measurement Problem in Quantum Mechanics in Hidden Variable Theory', *Rev. mod. Phys.*, 38 (1966), 453ff.
5. Bohr, N. *Theory of Spectra and Atomic Constitution* (Cambridge, 1922).
6. Bohr, N. 'Über die Anwendung der Quantentheorie auf den Atombau. I. Die Grundpostulate der Quantentheorie', *Z. Phys.*, 13 (1923), 117ff.
7. Bohr, N. *Ueber die Quantentheorie der Linienspektren* (Braunschweig, 1923).
8. Bohr, N. *Atomic Theory and the Description of Nature* (Cambridge, 1932).

philosophy that has advanced physics and is nourished by an abundance of physical examples to a new dogmatism which, apart from a few philosophical phrases, is almost devoid of intellectual content.

[100] This first step has been made by Bohr, Vigier, and others who are taking into account the features which Popper, Bunge, and others neglect. The lattice-theoretical interpretations of quantum theory, which have led to some very interesting suggestions, can also be traced to Bohr; some authors in this field explicitly refer to Bohr.

Added 1980. Lakatos has criticized this essay but apparently without having read it. He says I 'misrepresent Popper's, Landé's and Margenau's critical attitude to Bohr' ([67], 60, n.1) – but I nowhere deal with the *attitude* of these gentlemen; I deal with their *assertions* (about Bohr and the Copenhagen Interpretation) and their *arguments* and show that the first are almost all false while the second, at least in the case of Popper, are of a childlike naivite. Lakatos says that I 'give insufficient emphasis to Einstein's opposition'. But I do mention Einstein's most powerful argument against Bohr and show why Bohr's reply, which according to Einstein himself comes 'nearest to doing justice to the problem' (see above, text to n.94), must be regarded as a decisive objection against Einstein's interpretation of the case. It is not my fault that the matter can be dealt with in a page. Lakatos says I use 'some inconsistencies and waverings in Bohr's position for a crude apologetic falsification of Bohr's philosophy'. This is the language of the rationalist who looks at research from the safe distance of his office and wants to run science according to one method and base it on one philosophy and who fails to understand that researchers adapt their ideas to the always changing research situation. Bohr does not 'waver' because he does not know what to do, but because new discoveries are constantly being made and because he takes them into account. Finally, Lakatos accuses me of having 'completely forgotten that in some of (my) earlier papers (I was) more Popperian than Popper on the issue'. He seems to assume that there exists a Popperian 'position' in quantum mechanics. The above essay has shown that such is not the case: propensity was introduced by Bohr long before Popper started thinking about the matter, and for the rest we have pompous declarations and sophomoric mistakes which were elevated into a 'position' by Popper's disciples. Unfortunately (for Lakatos) I never participated in these machinations and I criticized them when they came to my notice. Cf. ch. 17, n.83 and text.

9. Bohr, N. 'Can Quantum Mechanical Description of Physical Reality be Considered Complete?', *Phys. Rev.*, 48 (1935), 696ff.

10. Bohr, N. 'On the Notions of Causality and Complementarity', *Dialectica*, 2 (1948), 312ff.

11. Bohr, N. In a review of [71] by L. Rosenfeld. *Nucl. Phys.*, 1 (1956), 133f.

12. Bohr, N. *Atomic Physics and Human Knowledge* (New York, 1958).

13. Bohr, N., Kramers, H. A., and Slater, J. C. 'The Quantum Theory of Radiation', *Phil. Mag.*, 47 (1924), 785ff.

14. Bohr, N. and Rosenfeld, L. 'Zur Frage der Messbarkeit der elektromagnetischen Feldgrössen', *K. danske Vidensk. Selsk. Mat.-fys. Meddr*, 12 (1933), 1ff.

15. Bopp, E. ed. *Werner Heisenberg und die Physik unserer Zeit* (Braunschweig, 1961).

16. Born, M., *Vorlesungen ueber Atommechanik* (Berlin, 1925).

17. Bothe, W. and Geiger, H. 'Ein Weg zur experimentellen Nachprüfung der Theorie von Bohr, Kramers und Slater', *Z. Phys.*, 26 (1924), 44.

18. Bothe, W. and Geiger, H. 'Ueber das Wesen des Comptoneffekts; ein experimenteller Beitrag zur Theorie der Strahlung', *Z. Phys.*, 32 (1925), 639ff.

19. Bunge, M. ed. *The Critical Approach, Essays in Honour of Karl Popper* (New York, 1964).

20. Bunge, M. ed. *Quantum Theory and Reality* (New York, 1967).

21. Caldirola, P. ed. *Ergodic Theories* (New York, 1961).

22. Colodny, R. ed. *Frontiers of Science and Philosophy* (Pittsburgh, 1962).

23. Colodny, R. ed. *Mind and Cosmos* (Pittsburgh, 1966).

24. Compton, A. H. 'The Corpuscular Properties of Light', *Naturwiss.*, 17 (1929), 507ff.

25. Compton, A. H. and Simon, A. W. 'Detected Quanta of Scattered X-Rays', *Phys. Rev.*, 26 (1925), 289ff.

26. Crombie, A. C. ed. *Scientific Change* (London, 1963).

27. Daneri, A., Loinger, A., and Prosperi, G. M. 'Quantum Theory of Measurement and Ergodicity Conditions', *Nucl. Phys.*, 33 (1962), 297ff.

28. Daneri, A., Loinger, A., and Prosperi, G. M. 'Further Remarks on the Relations Between Statistical Mechanics and Quantum Theory of Measurement', *Nuovo Cim.*, 44B (1966), 119ff.

29. Dirac, P. A. M. *The Principles of Quantum Mechanics* (Oxford, 1947).

30. Drake, S. and O'Nally, C. D. eds. *The Controversy of the Comets of 1618* (Philadelphia, 1960).

31. Einstein, A. 'Zum gegenwärtigen Stand des Strahlungsproblems', *Z. Phys.*, 10 (1909), 185ff.

32. Einstein, A. 'Zur Quantentheorie der Strahlung', *Z. Phys.*, 18 (1917), 121ff.

33. Einstein, A. 'Zur Elektrodynamik Bewegter Körper', in [1].

34. Einstein, A. and Ehrenfest, P. 'Quantentheoretische Bemerkungen zum Experiment von Stern und Gerlach', *Z. Phys.*, 11 (1922), 31ff.

35. Einstein, A., Podolsky, B., and Rosen, N. 'Can Quantum Mechanical Description of Physical Reality be Considered Complete?', *Phys. Rev.*, 47 (1935), 777f.

36. Epstein, P. S. 'Zur Theorie des Starkeffektes', *Annln Phys.*, 50 (1916), 489ff.

37. Feyerabend, P. K. 'Eine Bemerkung zum Neumannschen Beweis', *Z. Phys.*, 145 (1956), 421ff.

38. Feyerabend, P. K. 'On the Quantum-Theory of Measurement', in [62], 121ff.
39. Feyerabend, P. K. 'Patterns of Discovery', *Phil. Rev.*, 69 (1960), 247ff.
40. Feyerabend, P. K. 'Problems of Microphysics', in [22], 189ff.
41. Feyerabend, P. K. 'Consolations for the Specialist', ch. 8 in vol. 2.
42. Feyerabend, P. K. and Maxwell, G. eds. *Mind, Matter and Methods* (Minneapolis, 1966).
43. Feynman, R. P., Leighton, R. B., and Sands, M. *The Feynman Lectures in Physics*, iii (Massachusetts, 1965).
44. Fock, W. A. ed. *Philosophische Probleme der Modernen Naturwissenschaft* (Berlin, 1962).
45. Fowler, A. 'The Spectra of Helium and Hydrogen', *Nature, Lond.*, 92 (1913), 95f.
46. Fowler, A. 'The Spectra of Helium and Hydrogen', *Nature, Lond.*, 92 (1913), 231ff.
47. Hanson, N. R. *Patterns of Discovery* (Cambridge, 1961).
48. Havas, P. 'Four-Dimensional Formulations of Newtonian Mechanics and Their Relation to the Special and the General Theory of Relativity', *Rev. mod. Phys.*, 36 (1964), 938ff.
49. Hegel, G. W. F. *Philosophie der Geschichte* ed. Brunstädt, (Leipzig, 1908).
50. Heisenberg, W. 'Über den anschaulichen Inhalt der quantentheoretischer Kinematik und Mechanik', *Z. Phys.*, 43 (1927), 172ff.
51. Heisenberg, W. 'Die Entwicklung der Quantentheorie 1918–1928', *Naturwissenschaften*, 17 (1929), 490ff.
52. Heisenberg, W. *The Physical Principles of the Quantum Theory* (Chicago, 1930).
53. Heitler, W. H. *The Quantum Theory of Radiation* (Oxford, 1954).
54. Holton, G. 'Influences on Einstein's Early Work in Relativity Theory', *Organon*, 3 (1966), 225.
55. Hund, F. 'Göttingen, Kopenhagen, Leipzig im Ruckblick', in [15].
56. Jammer, M. *The Conceptual Development of Quantum Mechanics* (New York, 1966).
57. Jauch, J. M. 'The Problem of Measurement in Quantum Mechanics', *Helv. Phys. Acta*, 37 (1964), 293ff.
58. Jauch, J. M., Wigner, E. P., and Yanase, M. M. 'Some Comments Concerning Measurements in Quantum Mechanics', *Nuovo Cim.*, 48B (1967), 144ff.
59. Kaufmann, W. *Hegel, A Reinterpretation* (New York, 1966).
60. Kelvin, Lord, *Baltimore Lectures* (London, 1904), Appendix B, 491f.
61. Klein, M. J. 'Thermodynamics in Einstein's Thought', *Science*, 157 (1967), 509ff.
62. Körner, S. ed. *Observation and Interpretation* (London, 1957).
63. Kramers, H. A. 'Das Korrespondenzprinzip und der Schalenbau der Atoms', *Naturwissenschaften*, 11 (1923), 550ff.
64. Kuhn, T. S. 'The Function of Dogma in Scientific Research', in [26], 347ff.
65. Kuhn, T. S. *The Structure of Scientific Revolutions* (Chicago, 1962).
66. Kuhn, T. S. Heilbron, J. L. and Forman, P. L. *Sources for History of Quantum Theory* (Philadelphia, 1967).
67. Lakatos, I. *The methodology of scientific research programmes. Philosophical Papers* i, ed. J. Worrall and G. Currie (Cambridge, 1978).

68. Lakatos, I. and Musgrave, A. *Criticism and the Growth of Knowledge*, (Cambridge, 1968).

69. Landau, L. and Lifshitz, S. *Quantum Mechanics* (New York, 1958).

70. Landé, A. *Die Neuere Entwicklung der Quantentheorie* (Leipzig, 1926).

71. Landé, A. *Foundations of Quantum Theory* (New Haven, 1955).

72. Landé, A. 'Comments on the review of "Foundations of Quantum Mechanics, a Study in Continuity and Symmetry"', *Nucl. Phys.*, 3 (1957), 132ff.

73. Ludwig, G. 'Gelöste und ungelöste Probleme des Messprozesses in der Quantenmechanik', in [15].

74. Margenau, H. 'Advantages and Disadvantages of Various Interpretations of the Quantum Theory', *Physics Today*, 7 (1954), 6ff.

75. Margenau, H. 'Measurements and Quantum States', I and II, *Phil. Sci.*, 30 (1963), 1ff, 138ff.

76. Meyer-Abich, K. M. *Korrespondenz, Individualitaet, und Komplementaritaet* (Wiesbaden, 1965).

77. McKnight, J. L. 'The Quantum Theoretical Concept of Measurement', *Phil. Sci.*, 24 (1957), 321ff.

78. Møller, P. M. *En Dansk Students Eventyr* (Copenhagen, 1893).

79. Moore, Ruth, *Niels Bohr* (New York, 1966).

80. Neumann, J. von. *Mathematische Grundlagen der Quantenmechanik* (Berlin, 1932).

81. Pauli, W. 'Quantum Theory of Magnetism: The Magnetic Electron', in [97], 194ff.

82. Pauli, W. ed. *Niels Bohr and the Development of Physics* (London, 1955).

83. Popper, K. R. *The Logic of Scientific Discovery* (New York, 1959).

84. Popper, K. R. *Conjectures and Refutations* (New York, 1962).

85. Rosen, N. 'The Relation Between Classical and Quantum Mechanics', *Am. J. Phys.*, 32 (1964), 597ff.

86. Rosenfeld, L. 'An Anti-Review', *Nucl. Phys.*, 3 (1957), 132.

87. Rozenthal, S. ed. *Niels Bohr, His Life and Work as Seen by his Friends* (New York, 1967).

88. Schlipp, P. A. ed. *Albert Einstein, Philosopher-Scientist* (Evanston, 1948).

89. Schrödinger, E. 'Die gegenwärtige Situation in der Quantenmechanik', *Naturwissenschaften*, 23 (1935), 807ff, 823ff, 844ff.

90. Schrödinger, E. 'Measurement of Length and Angle in Quantum Mechanics', *Nature, Lond.*, 173 (1954), 442.

91. Schrödinger, E. 'The Philosophy of Experiment', *Nuovo Cim.*, 1 (1955), 1ff.

92. *Scientific Papers Presented to Max Born* (Edinburgh, 1953).

93. Seelig, C. *Albert Einstein* (Zürich, 1960).

94. Shankland, R. S. 'Conversations with Albert Einstein', *Am. J. Phys.*, 31 (1963), 47ff.

95. Slater, J. C. 'Radiation and Atoms', *Nature, Lond.*, 113 (1924), 307–8.

96. *Solvay Conference, Proceedings of Vth* (Paris, 1928).

97. *Solvay Conference, Proceedings of VIth* (Paris, 1930).

98. *Solvay Conference, Proceedings of XIIth* (New York, 1962).

99. Sommerfeld, A. *Atombau Und Spektrallinien* (Leipzig, 1922).

100. Waerden, van der, B. L. *Sources of Quantum Mechanics* (Amsterdam, 1967).

101. Wigner, E. P. 'The Problem of Measurement', *Am. J. Phys.*, 31 (1963), 6ff.

I7

Hidden variables and the argument of Einstein, Podolsky and Rosen

1. THE ARGUMENT

Opponents of Bohr's interpretation often refer to an argument by Einstein, Podolsky and Rosen, (EPR) according to which, the formalism of wave mechanics is such that it demands the existence of exact simultaneous values of non-commuting variables. Clearly, if this should be the case, then Bohr's interpretation of the uncertainty relations would have to be dropped and it would have to be replaced by the interpretation of Einstein and Popper. At the same time all those difficulties which prompted Bohr to invent the hypothesis of the indefiniteness of state descriptions would reappear. Even worse, it would seem that an inconsistency has been discovered in the very foundations of the quantum theory. For if it should indeed be the case that the only way of combining duality, the quantum postulate and the conservation laws consists in assuming indefiniteness of state descriptions, then the case of Einstein, Podolsky and Rosen would show that wave mechanics is intrinsically unable to allow for a coherent account embracing these three experimental facts. We now turn to a closer analysis of the argument.

Assume that a system S (coordinates $q, q', q'' \ldots, q^n; r, r', r'', \ldots, r^m$; or (qr) for short) which is in the state $\Phi (qr)$ has been (either mentally, or by physical separation) divided into two subsystems, $S' (q)$ and $S'' (r)$. It can be shown[1] that it is always possible to select a pair of observables $\alpha (q)$, and $\beta (r)$ (corresponding to sets of mutually orthogonal situations in S' and S'' respectively) with sets of eigenstates $\{|\alpha_i (q) >\}$; $\{|\beta_i (r) >\}$ such that

$$|\Phi (qr) > = \Sigma_i c_i \, | \, \alpha_i > | \, \beta_i > \qquad (1)$$

A pair of observables with the property just mentioned will be called *correlative* with respect to Φ, S' and S''.

The special case discussed by Einstein, Podolsky and Rosen is characterized by the following three conditions.[2]

---

[1] J. von Neumann, *Mathematical Foundations of Quantum Mechanics* (Princeton, 1955), 429ff. *Added 1980.* For an up-to-date account and literature concerning the problems that are dealt with in this chapter, the reader is advised to consult M. Jammer, *The Philosophy of Quantum Mechanics* (New York, 1974).

[2] For proof see E. Schrödinger, 'Probability Relations between Separated Systems', *Proc.*

(i) There exists more than one pair of observables which are correlative with respect to Φ, S' and S''.

(ii) Assume that $(\alpha\beta)$; $(\gamma\delta)$; $(\varepsilon\xi)$; ... are pairs of observables which are correlative with respect to Φ, S' and S''; then there is at least one pair of pairs, say $(\alpha\beta)$, and $(\gamma\delta)$, such that

$$\alpha\gamma \neq \gamma\alpha, \quad \beta\delta \neq \delta\beta \tag{2}$$

(i) and (ii) are satisfied if and only if the constants c_i in (1) satisfy the conditions

$$|c_i|^2 = \text{a constant} \tag{3}$$

(iii) The systems S' (q) and S'' (r), i.e. the spatial regions defined by the (q) and (r), are separated such that no physical interaction is possible between them.[3]

From (i) it follows that, if $|\gamma_i>$ and $|\delta_i>$ are the eigenfunctions of γ and δ in S' and S'' respectively, then Φ may also be represented in the form

$$|\Phi \ (qr) > \ = \ \Sigma_i c'_i \ |\gamma_i > \ |\delta_i > \tag{4}$$

Now assume that the magnitude corresponding to α is measured in S' with the result α_k. From (1), as well as from the assumption that

any measurement leaves the system in an eigenstate of the variable measured

$$\tag{5}$$

it follows that the state of S'' after the measurement will be the state $|\beta_k>$, i.e. it will be an eigenstate of β.

Assume, on the other hand, that γ is measured in S' with the result $|\gamma_i>$. It then follows from (4) and (5) that the state of S'' after the measurement will be $|\delta_i>$, i.e. it will be an eigenstate of δ.

Camb. Phil. Soc., 31 (1935), 555ff; 32 (1936), 446ff as well as H. J. Groenewold, 'On the Principles of Quantum Mechanics', *Physica*, 12 (1948), 405ff.

[3] This condition suggests choosing the α, β, γ, δ in such a manner that they all commute both with the (q) and with the (r). An example with this property has been described by D. Bohm in his *Quantum Theory* (Princeton, 1951), ch. 22. If such a choice is not made then it is always possible to assail the argument because of the specific form in which it is presented. For such an unintentionally irrelevant attack see de Broglie, *Une Tentative d'Interprétation Causale et non Linéaire de la Mécanique Ondulatoire* (Paris, 1956), 76ff. De Broglie discusses Bohr's example ('Can the Quantum Mechanical Description of Physical Reality be considered Complete?', *Phys. Rev.*, 48 (1935), 696ff) of a pair of particles with known $q_1 - q_2$ and $p_1 + p_2$ respectively. Now the state just described (which satisfies (i) and (ii) above, q_1 and p_1 being the relevant variables) is realized only as long as the pair dwells near the two-slit screen whose slit-distance defines $q_1 - q_2$. However, in this case, (iii) will be violated. The argument is of course irrelevant as Einstein's point is independent of the way in which the state satisfying (i) and (ii) has been created. It can also be shown (cf. Schrödinger, 'Die gegenwaertige Lage in der Quantenmechanik', *Naturwissenschaften*, 23 (1935) 807ff, 824ff, 844ff) that t seconds after the pair has left the slit Einstein's argument can be raised with respect to the variables p_1 and $q_1 - (P_1/m) \, t$. Still, it is an advantage to possess an example that cannot be criticized for the reasons just mentioned. Also Bohm's example is not beset by the difficulties of the continuous case (for these difficulties see D. Bohm and Y. Aharonov, 'Discussion of Experimental Proof of the Paradox of Einstein, Rosen and Podolsky', *Phys. Rev.*, 108 (1957), 1070ff, appendix).

Adding (iii), we may now derive the following result. First step: if S' and S''' are separated in space such that no physical interaction is possible, then it is impossible that a physical change of S' will influence the state of S'''. More especially, it is impossible that a measurement in S' (which leads to a physical change of S') will influence the state of S'''. More especially still, if there is a measurement in S' whose result allows us to infer that S''' is in a particular physical state, then we have to assume that S''' was in the inferred state already *before* the performance of the measurement, and would have been in this state even if the measurement had never been carried out. Second step, applying the above result to the measurement of α and γ we reason as follows: immediately after the measurement of α which yields the result α_k, S''' is in the state $|\beta_k\rangle$, and this is, according to what has been said above, also the state of S''' immediately before the measurement. Hence, S''' must be in some eigenstate of β whether or not a measurement of α has been performed (we assume, of course, that all observables are complete commuting sets). By the same argument it must also be in some eigenstate of δ, whether or not a measurement of γ has been performed. In short, the system under consideration – *and every object that is capable of interacting with some system* – must always be in a classically well-defined state. Bohr's hypothesis of intrinsic indeterminacy, and (2), entail that this is impossible. This contradiction between Bohr's assumption of the indefiniteness of state descriptions and the argument just presented has been called the paradox of Einstein, Podolsky and Rosen. It is preferable, however, not to talk of a paradox, but rather of an argument.[4]

For Einstein (and Popper) the argument refutes Bohr's hypothesis. Having presented it, Einstein therefore feels justified in regarding the quantum theory 'as an incomplete and indirect description of reality' such that 'the ψ-function . . . relates . . . to an "ensemble of systems" in the sense of statistical mechanics'.[5] I shall soon deal with the merits of the argument, but before doing so I wish to discuss some of the objections

---

[4] The paradoxical aspect arises as soon as the argument is combined with the completeness assumption. In this case we obtain the result that changes of state may occur which are (1) well predictable (although their exact outcome is not predictable); and which (2) occur in places which are very far from the reach of physical forces. It is this feature of the argument that among other things has led to the assumption of a sub-quantum-mechanical level involving laws different from the laws of the quantum theory which allow for the occurrence of coordinated fluctuations. Such an assumption would of course lead to predictions that in some respect are different from the predictions of the present theory. For example, it would imply a disturbance of the correlations if the measuring apparatus interacting with the first system is turned around very rapidly. Such a difference of predicted results is far from undesirable, however. It is a guide to new experiments which then will be able to decide as to which point of view should be adopted in the end. Cf. D. Bohm, *Causality and Chance in Modern Physics* (London, 1957), chs. 3, 4, as well as 'A Proposed Explanation of Quantum Theory in Terms of Hidden Variables at a Sub-Quantum-Mechanical Level' in *Observation and Interpretation*, ed. Körner (London, 1957), 33ff, 86f.

[5] See *Albert Einstein, Philosopher-Scientist*, ed. P. A. Schilpp (Evanston, Ill. 1948), 666ff.

which have been raised against it and all of which seem to be unsatisfactory.

A. D. Alexandrow and J. L. R. Cooper have asserted that the paradox cannot arise, for as long as $[S'S'']$ is described by a wave-function which cannot be broken up into a product, (iii) does not hold.[6] The reply is that within (iii) 'separated' and 'interaction' refer to *classical fields*, or at least to fields which contribute to the energy present in a certain space–time domain. Alexandrow and Cooper seem to assume that the ψ-field can be interpreted as such a field which is not borne out by the facts.

A much more typical objection is contained in a paper by Furry.[7] This paper has the advantage of being clear and straightforward and it also seems to reflect the attitude of a good many physicists. The physical process (interaction between S' and S'') which terminates in the paradoxical state is construed by Furry in two ways, A and B. According to assumption A the process leads to transitions in the quantum-mechanical states of S' and S''. The transitions occur according to the laws of probability, but they terminate in well-defined states for both S' and S'', which are related to each other in such a way that a measurement in S' will indeed lead to unambiguous information about the state of S'' (and vice versa). The correlation is between *states which are already there* or, to put it formally, it is due to the fact that the state of $S' + S''$ after the interaction is given by the mixture

$$P_{\Phi'} = \Sigma_i \mid c_i \mid^2 \mid \alpha_i ><\alpha_i \mid \beta_i ><\beta_i \mid \tag{6}$$

We see that in this mixture the pair $(\alpha\beta)$ is such that if α is measured in S' with the result α_k, then β will be certain to exhibit, on measurement, the value β_k.

Proof: $\mathrm{Tr}\{P_{\mid\alpha_k>} P_{\Phi'} P_{\mid\alpha_k>} P_{\mid\beta_l>}\}/\mathrm{Tr}\{P_{\mid\alpha_k>} P_{\Phi'}\} = \delta_{kl}$

(Another feature of the case is that it leads to the violation of some conservation laws.)[8] Furry alleges that A is the assumption made by Einstein.

Method B which, he says, is the method adopted by the quantum theory and which preserves the conservation laws, implies that an interaction between two systems, S' and S'', will in general lead to a pure state

$$\mid\Phi> = \Sigma_{ik}c_{ik} \mid\alpha_i> \mid\beta_k> \tag{7}$$

and that correlations between the values of α and β will occur if and only if $c_{ik} = c_i \delta_{ik}$, or if Φ has form (1). The difference between method A and

[6] A. D. Alexandrow, *Proc. Acad, USSR*, 84 (1952), 253ff. J. L. R. Cooper, 'The Paradox of Separated Systems in Quantum Theory', *Proc. Camb. Phil. Soc.*, 46 (1951), 620ff.

[7] 'A Note on the Quantum Mechanical Theory of Measurement', *Phys. Rev.*, 49 (1936), 397ff; cf. also Groenewold, 'On the Principles of Quantum Mechanics'. M. H. L. Pryce, too, has used the argument in private discussions.

[8] Thus in Bohm's example in n.3 angular momentum will not be conserved.

method B, therefore, consists in the fact that method A represents a state with the above properties by (6) while method B represents it by (2). It is easily shown that the two methods do not lead to the same observational results.

Proof: for $[\alpha\gamma]$; $[\beta\delta] \neq 0$, in general

$$\text{Tr}\{P_{|\gamma\rangle} P_\Phi P_{|\gamma\rangle} \delta\} \neq \text{Tr}\{P_{|\gamma\rangle} P_\Phi{}'_{|\gamma\rangle} \delta\}$$

Method A is therefore inconsistent with the method adopted by the quantum theory, which latter is a direct consequence of the superposition principle. This many physicists regard as a refutation of Einstein's idea (which was silently applied in the first step of the derivation of the paradox) that 'if without in any way disturbing a system we can predict with certainty the value of a physical quantity, then there exists an *element of physical reality* corresponding to this physical quantity',[9] i.e. then this value belongs to the system whether or not a measurement has been, or can be, carried out.

Concerning this refutation we must consider two points. To start with, the fact that A differs from the method which is adopted by the quantum theory can be regarded as detrimental only if it has first been shown that in the special case dealt with by E P R the predictions of the quantum theory are empirically successful whereas the predictions implied by method A are not. The fact that the quantum theory has been confirmed by a good many experimental results is of no avail here, for we are now asking for the behaviour of systems under conditions which have not yet been tested. And it was indeed Einstein's guess that the current formulation of the many-body problem in quantum mechanics might break down when particles are far enough apart.[10] The second point is that Einstein's argument would stand unassailed even if B were to give the correct statistical predictions. For it is not an argument for a particular description, *in terms of statistical operators,* of the state of systems which are far apart. It is rather an argument against the assumption that any description in terms of statistical operators can be regarded as complete, or against the assumption that 'the ψ-function is . . . unambiguously coordinated to the physical state'.[11]

In the literature this last assumption has become known as the *completeness assumption.* As has been indicated in the preceding paragraph the completeness assumption implies Bohr's hypothesis. One may therefore interpret E P R both as an argument against the completeness assumption and as an argument against Bohr's hypothesis. In what follows both these interpretations will be used.

[9] A. Einstein, B. Podolsky and N. Rosen, 'Can the Quantum Mechanical Description of Physical Reality be considered Complete', *Phys. Rev.,* 47 (1935), 777ff.

[10] See Bohm and Aharonov, 'Discussion of Experimental Proof', 1071, as well as Bohm in *Observation and Interpretation,* 86ff.

[11] Einstein, 'Physics and Reality' in *Ideas and Opinions* (New York, 1954), 317.

As regards the first point, we may additionally point that it has been attempted to decide experimentally between assumption A and assumption B. The experiment was carried out by C. S. Wu,[12] and it consisted in studying the polarization properties of correlated photons. Such photons are produced in the annihilation radiation of positron–electron pairs. Each photon is emitted in a state of polarization orthogonal to that of the other which is similar to Bohm's example of EPR. The experiment refuted method A and confirmed B, the method that is in accordance with the general validity of the superposition principle. However, as we have shown above, this result cannot be used for refuting the contention, which is the core of Einstein's argument, that what is realized in *both* cases is an ensemble of classically well-defined systems rather than a single case.[13]

Similar remarks apply to the analysis given by Blochinzev.[14] It is true Blochinzev emphasizes that 'within quantum theory we do not describe the "state in itself" of the particle, but rather its relation to the one, or the other (mixed or pure) collective'.[15] He points out that this relation is of a completely objective character and that any measurement is to be regarded as a process which separates certain subcollectives from the collective in which they were originally embedded. Now, the assertion that the quantum theory describes the elementary particles only insofar as they are elements of a collective would seem to make Blochinzev's interpretation coincide with Einstein's interpretation, which also asserts that the present quantum theory is a theory of collectives. However, the important difference is that for Blochinzev the individual particle does not possess any state-property over and above its membership in a particular collective. 'Our experiments', he writes,[16] 'are precise enough to show us that the pair (p, q) of a single particle does not exist in nature.' This means, of course, that Blochinzev, too, accepts the completeness assumption, which is just the point at issue. Again, he has not shown that Einstein's argument contradicts the quantum theory, or that it contradicts 'experience'; he has only shown that it contradicts the completeness assumption, i.e. he has shown that it serves the purpose for which it was constructed.

We see that all the arguments against EPR which we have discussed so

[12] C. S. Wu and I. Shaknov, 'The Angular Correlation of Scattered Annihilation Radiation', *Phys. Rev.*, (77) (1950), 136ff. That the case is equivalent to the one discussed in EPR is shown in W. Heitler, *Quantum Theory of Radiation* (Oxford 1957), 269. Cf. also the analysis by Bohm and Aharonov, 'Discussion of Experimental Proof'.

[13] A similar position is held by D. R. Inglis, 'Completeness of Quantum Mechanics and Charge Conjugation Correlations of Theta Particles', *Rev. mod. Phys.*, 33 (1961), 1ff, especially the last section.

[14] 'Kritik der philosophischen Anschauugen der sogenannten "Kopenhagener Schule" in der Physik', *Sowjetwissenschaft*, 4 (1954), 545ff. *Grundlagen der Quantenmechanik* (Berlin, 1953), 497ff.

[15] 'Kritik der philosophischen Anschauungen', 564.

[16] *Grundlagen*, 50.

far fail to refute it.[17] Apart from those suggestions which consist in introducing a sub-quantum-mechanical level, and which deny the *correctness*, and not only the completeness, of the present theory, the only argument left seems to be the one that led to Bohr's hypothesis in the first place and which does not depend on the more detailed formal features of either the elementary quantum theory, or of any one of its improved alternatives (Dirac's theory of the electron; field theories). However, with respect to these qualitative considerations many writers seem to be of the opinion that they count little when compared with the impressive utilization, by EPR, of the powerful formal apparatus of wave mechanics. In order to dispel this impression (which seems to reflect an overconfident and uncritical attitude to mathematical formalisms) I shall now discuss the difficulties which arise when the *conclusion* of EPR is assumed to be correct, i.e. when it is assumed that the elementary theory is correct and that a more detailed description of state can be given that is admitted by the completeness assumption. A state referred to by such a more detailed description will be called a *superstate*.[18]

2. SUPERSTATES

Given the laws of quantum theory, are superstates possible? In the present section I shall develop some arguments (i.e. additional to the qualitative arguments of ch. 16) against the possibility of superstates. The *first argument* will show that, dependent on the way in which they have been introduced, superstates either contain redundant elements, or are empirically inaccessible. To show that they contain redundant parts a few explanations must be given concerning the role of a state in physical theory.

What is the role of a dynamical state in a physical theory? The reply is that it contains part of the initial conditions which, taken together with other initial conditions such as mass and charge, with boundary conditions (properties of the acting fields), as well as with some theories, help us to explain and to predict the behaviour of the physical system to which it applies. According to this definition an element of state is superfluous if it does not play a role in any prediction and explanation. It is even more superfluous if this applies not only to the *future* properties and behaviour of the system, but also to the properties and the behaviour it possesses at the very moment at which the occurrence of the element is being asserted. In this case there exists no possibility whatever of testing the assertion that the element has occurred and it may properly be called *descriptively redundant*. To show that a superstate contains descriptively redundant elements let us consider the case of a particle with total spin σ. Assume that σ and σ_x have

---

[17] This is admitted, implicitly, by D. R. Inglis, 'Completeness of Quantum Mechanics'.
[18] The first to use this expression seems to have been Groenewold, 'On the Principles of Quantum Mechanics'.

been measured and the values of σ' and $\sigma_x{}'$ obtained. Then an immediate repetition of the measurement will again give these values. On the other hand, assume that σ_y is measured when the measurement of σ and σ_x has been completed. Then the formalism of the theory tells us that *any* value of σ_y may be obtained which shows quite clearly that adding *a specific value* of σ_y (and, for that matter of σ_z) to the set $[\sigma' \ \sigma'_x]$ does not in the least change the information content of our assertion. By generalizing, we obtain the following result: a set of magnitudes specifying the outcome of the measurement of a complete set of commuting observables has maximum informative content. Any addition to this set is descriptively redundant, *whatever the method* (E P R or other) used for obtaining it, provided, of course, the method did not involve a disturbance of the state already realized.

The superstates we have been discussing so far had the property that only part of them could be used for deriving information about the actual state of the physical system, i.e. assuming **P** and **Q** to be two different complete commuting sets pertaining to the same system, the superstate **[PQ]** was chosen in such a manner that Prob **(P/[PQ])** = 1 and Prob **(Q[PQ])** = a constant $\neq$ 1, or the other way round. We may now want to define a pair $[PQ]$ of new variables in the following manner.

$$\text{Prob } (P/[PQ]) = \text{Prob } (Q/[PQ]) = \text{Prob } (P/[QP]) = \text{Prob } (Q/[QP]) = 1$$
(8)

What are the consequences of such a definition (which has been adopted, implicitly, by Bohm)?[19]

To start with it should be noted that the first and the last equations in (8) are part of the *definition* of a superstate of this new kind, whereas in the case of the **P, Q** these equations follow the quantum theory. Note further that we are here dealing with a minimum condition which is trivially satisfied in the classical case; the condition is that a series of statements describing a superstate should be sure that it allows for the derivation of the value of any one of the elements of the superstate. We have not yet considered any dynamical law. But it is clear that if (8) is not satisfied, then deterministic laws will not be possible. Thus the conditions (8) are a necessary presupposition of determinism in the quantum theory (and in any other theory). Let us now see where these conditions lead.

(8) entails that Prob $([QP]/P)/\text{Prob }([PQ]/P) = \text{Prob }([QP])/\text{Prob }([PQ])$. If we now postulate that the absolute probabilities of the superstates be independent of the order of their elements (and we indicate adherence to this postulate by writing 'PQ' instead of '$[PQ]$'), then we obtain Prob $(P/[PQ]) = \text{Prob }(P/[QP]) = \text{Prob }(P/(P/PQ) = 1$ (from (8)) $= \text{Prob }(P/P)$ for any pair PQ satisfying (8) above.

[19] 'A Suggested Interpretation of the Quantum Theory in Terms of "Hidden Variables"', Part 1', *Phys. Rev.*, 85 (1951), 166ff.

In other words, the elements of the newly introduced superstates are statistically independent. (9)

Now let us assume, in order to fill this abstract scheme with some empirical content, that

$$P \longleftrightarrow p \qquad (10)$$

where p is a complete set of commuting observables in the sense of the quantum theory. We shall also assume that the systems discussed are fully described by the complementary sets p and q.

On the basis of Bayes' theorem we obtain,

$$\text{Prob } (q/PQ) = \text{Prob } (P/qQ). \text{ Prob } (q/Q)/\text{Prob } (P/Q)$$

which leads to

$$\text{Prob } (q/Q) = \text{Prob } (P). \text{ Prob } (q/pQ)/\text{Prob } (p/q)$$

(Here we have used (9) as well as (10)).

Now we have from the quantum theory that

Prob $(p/q) = \text{Prob } (p/q \ldots) = | \langle p|q \rangle |^2$, therefore
Prob $(q/Q) = \text{Prob } (P)$ (11)

In a completely analogous manner,

Prob $(p/PQ) = \text{Prob } (P/pQ). \text{ Prob } (p/Q)/\text{Prob } (P/Q)$ leads to
Prob $(p/Q) = \text{Prob } (P)$ (12)

Now,

Prob $(qQ) = \text{Prob } (P). \text{ Prob } (Q) = \text{Prob } (PQ)$
Prob $(pQ) = \text{Prob } (P). \text{ Prob } (Q) = \text{Prob } (PQ)$, hence
Prob $(p) = \text{Prob } (q)$ (13)

Finally,

Prob $(p) = \text{Prob } (P)$ (by virtue of (11)) $= \text{Prob } (q/Q) = \text{Prob } (q)$ (using (13)).

That is to say q *and* Q (and, as can be easily shown, also p and Q) *are statistically independent.* Result: if we assume that there exist superstates which satisfy conditions (8); and if we also assume that one of the elements of these superstates is accessible to experimental investigation as it is provided by quantum-mechanical measurements, then the rest of the superstate will be statistically independent of any physical magnitude that can be measured in the system under consideration and cannot therefore be said to possess any empirical content. Again it emerges that the maximum of information producible about a quantum-mechanical system is given by the assertion that one of its complete sets of commuting observables possesses a certain value. Any additional assertion is arbitrary and not accessible to independent experimental test. Adopting Bohr's hypothesis of indefinite state descriptions, we can easily explain this fact by pointing out that this inaccessibility is not due to the intricacies of the measuring process which forbid us obtain more detailed information about nature, but rather to the *absence of more detailed features of nature itself*. This is explained in detail in ch. 16.6.

At this stage one might still be inclined to say, in opposition to Bohr's hypothesis, that the use of superfluous information, although not very elegant, and certainly metaphysical, can at most be rejected on the basis of considerations of 'taste'.[20] That this is not so is shown by the second argument against the admissibility of superstates. According to this second argument (cf. ch. 11.7), superstates are incompatible with the conservation laws and with the dynamical laws in general. In the case of the energy principle this becomes evident from the fact that for the single electron $E = p^2/2m$, so that after a measurement of position any value of the energy may emerge and, after a repetition of the measurement, any different value. One may try to escape this conclusion, as Popper apparently has been inclined to,[21] by declaring that the energy principle is only statistically valid. However, it is very difficult to reconcile this hypothesis with the many independent experiments (spectral lines; experiment of Franck and Hertz; experiment of Bothe and Geiger) which show that energy is conserved also in the quantum theory and this not only on the average, *but for each single process of interaction*. The difficulty of Einstein's point of view, which works with superstates of the first kind, i.e. with superstates that do not satisfy (8) and whose elements are determined by successive observations, becomes very obvious if we apply it to the case which is known as the *penetration of the potential barrier*. We obtain rather drastic violations of the principle of the conservation of energy here,[22] if we assert it in the form that for any superstate $[Eqp]$, as determined by three successive measurements, E will satisfy the equation $E = p^2/2m + V(q)$.[23] Adding these difficulties to the arguments leading up to Bohr's hypothesis of indefinite state description we obtain very powerful reasons indeed against the conclusion of EPR.[24] This

---

[20] This is W. Heisenberg's attitude. See *The Physical Principles of Quantum Theory* (Chicago, 1930), 15.

[21] I am here referring to discussions I had with Karl Popper.

[22] In conversation, Landé has expressed the hope that further development of this point will lead to a satisfactory account of the case of the penetration of a potential barrier. This is quite possible. However, what I am concerned with here is to show the strength of the Copenhagen Interpretation to those who are of the opinion that the transition to a different interpretation is more or less a matter of philosophical taste rather than of physical inquiry.

[23] For a numerical evaluation cf. Blochinzev, *Grundlagen*, 505. Cf. also Heisenberg, *Physical Principles*, 30ff.

[24] It ought to be mentioned that Bohm (*Causality and Chance*, n.15) has shown how superstates which obey conditions (8) can be made compatible with the dynamical laws. However, the unsatisfactory feature remains that these superstates violate the principle of independent testability and that their introduction must therefore be regarded as a purely verbal manoeuvre. Yet it is important to repeat that von Neumann (*Mathematical Foundations*, 326) thought that his 'proof' would be strong enough to exclude even such verbal manoeuvres.

makes it imperative to show how the argument can be adapted to that hypothesis. An attempt in this direction and, to my mind, a quite satisfactory attempt, has been made by Bohr.[25]

3. THE RELATIONAL CHARACTER OF THE QUANTUM-MECHANICAL STATES

If I understand Bohr correctly, he asserts that the logic of a quantum-mechanical state is not as is supposed by EPR.[26] EPR seem to assume that what we determine when all interference has been eliminated is a *property* of the system investigated. As opposed to this, Bohr maintains that state descriptions of quantum-mechanical systems are *relations* between the systems and measuring devices in action, and are therefore dependent upon the existence of other systems suitable for carrying out the measurement. It is easily seen how this second basic postulate of Bohr's point of views makes indefiniteness of state description compatible with EPR. For while a property cannot be changed except by a physical *interference* with the system that possessed that property, a relation can be changed without such interference. Thus the state 'being longer than *b*' of a rubber band may change when we compress the rubber band, i.e. when we physically interfere with it. But it may also change when we change *b* without at all interfering with the rubber band. Hence, lack of physical interference excludes changes of state only if it has already been established that positions and momenta and other magnitudes are properties of systems, rather than relations between them and suitable measuring devices. 'Of course', writes Bohr, referring to Einstein's example,[27] 'there is in a case like the one . . . considered no question of a mechanical disturbance of the system under investigation . . . But even at this stage there is essentially the question of *an influence on the very conditions which define the possible types of prediction regarding the future behaviour* of the system', and he compares this influence with 'the dependence on the reference system, in relativity theory, of all readings of scales and clocks'.[28]

I would like to repeat, at this stage, that Bohr's argument is not supposed to *prove* that quantum-mechanical states are relational and indeterminate; it is only supposed to show under what conditions the indefiniteness assumption, *which is assumed to have been established by independent arguments*, can be made compatible with the case of Einstein, Podolsky and Rosen. If this is overlooked one may easily get the impression that the argument is either

[25] Einstein, too, regards Bohr's attempt as coming 'nearest to doing justice to the problem'. *Albert Einstein, Philosopher-Scientist*, 681.

[26] 'Can the Quantum Mechanical Description of Physical Reality be considered Complete?' Cf. also D. R. Inglis, 'Completeness of Quantum Mechanics'.

[27] 'Can the Quantum Mechanical Description of Physical Reality be considered Complete?'

[28] *Ibid.*, 704.

circular, or *ad hoc*. That the argument is circular has been asserted by Hilary Putnam.[29] In the case of relativity, says Putnam, we may set up two different reference systems and obtain *simultaneously* two different readings for the *same* physical system. This is not possible in the quantum theory, for it would presuppose, what is denied by Bohr, that we can make simultaneous measurements of position and momentum in S', or that we can even *imagine* that position and momentum both possess definite values in S'. But the appearance of circularity disappears when we realize that the hypothesis of indefiniteness of state descriptions is *presupposed* and that a way is sought to make it compatible with E P R. A similar remark applies to Popper's criticism that the argument is *ad hoc*.[30] One must as it were approach the argument from the realization that superstates cannot be incorporated into wave mechanics without leading to problems. Once this is admitted there arises the need for a proper interpretation of the very surprising case discussed by E P R. It is in this connection that Bohr's suggestion proves so extremely helpful.[31] Finally, we ought to discuss briefly the assumption which is silently made by almost all opponents of the Copenhagen point of view, that E P R creates trouble for this point of view but not for the quantum theory (the elementary theory, that is) itself. This overlooks the fact that there is no interpretation available that gives as satisfactory an account of all the facts united by the theory as does the idea of the indefiniteness of state descriptions. If we therefore interpret E P R as fatal for this idea, then we are forced to the conclusion *that the theory itself is in trouble*. (This conclusion has been drawn by Bohm and by Schrödinger.)[32]

It is very important to realize the far-reaching consequences of Bohr's

---

[29] Private communication.

[30] *The Logic of Scientific Discovery* (New York, 1959), 445ff.

[31] It ought to be pointed out, however, that there is one assumption in the earlier speculations about the nature of microscopic objects which has been definitely refuted by E P R. It is the assumption that 'the most important difference between quantum theory and the classical theories consists in the fact that in the case of an observation we must carefully consider the disturbance, due to experiment, of the system investigated' (Heisenberg, 'Die Entwicklung der Quantentheorie 1918–28', *Naturwissenschaften*, 17 (1929), 495; cf. also Bohr, *Atomic Theory and the Description of Nature* (Cambridge, 1952), 5, 11, 15, 54, 68, 93, 115; also 'Causality and Complementarity', *Dialectica*, (1948), (315). And it is the corresponding assumption that the indeterminacy of the state of quantum-mechanical systems is essentially due to *this* disturbance (cf. Bohm and Aharonov, 'Discussion of Experimental Proof', 1070ff as well as Popper, *The Logic of Scientific Discovery*, 445ff). What is shown by E P R is that physical operations, such as measurements, may lead to sudden changes in the state of systems which are *in no physical connection whatever* with the domain in which the measurement is being performed. Unfortunately the attitude of the adherents of the Copenhagen point of view with respect to this argument has very often been that the reply which was given by Bohr (and which cost him, as is reported, some headaches), was already implicit in the earlier ideas, which would mean that these ideas were much more vague than one would at first have been inclined to believe.

[32] Bohm, *Causality and Chance*. According to Schrödinger the paradox is an indication of the fact that the elementary quantum theory is a non-relativistic theory. See 'Die gegenwaertige Lage in der Quantenmechanik', especially the last section.

hypothesis. Within classical physics the interaction between a measuring instrument and an investigated system can be described in terms of the appropriate theory. Such a description allows for an evaluation of the effect, upon the system investigated, of the measurement, and it thereby allows us to select the best possible instrument for the purpose at hand. Within classical physics, the classification of the measuring instruments is therefore achieved, at least partly, by the theory that is being investigated. Now according to Bohr, a quantum-mechanical state is a relation between (microscopic) systems and (macroscopic) devices. Also a system does not possess any properties over and above those contained in its state description (this is the completeness assumption). This being the case it is not possible, even conceptually, to speak of an *interaction* between the measuring instrument and the system investigated. The logical error committed by such a manner of speaking would be similar to the error committed by a person who wanted to explain changes of velocity of an object created by the transition to a different reference system as the result of an interaction between the object and the reference system. This has been made very clear by Bohr ever since the publication of EPR which refutes the earlier picture,[33] where a measurement glues together, with the help of an indivisible quantum of action, *two different* entities, viz. the apparatus on the one hand and the investigated object on the other.[34] But if we cannot separate the microsystem from its relation to a classical apparatus, then the evaluation of a measuring instrument can no longer be based on the type of *interaction* that occurs.[35] This has led to the assertion that the classification of measuring instruments that is used by the quantum theory can at most consist in giving a list without being able to justify the presence of any member in the list.[36] Such an assertion does not seem to be correct. First of

---

[33] Cf. n.31.

[34] This is sometimes obscured by the fact that Bohr's account of measurement is not the only one. Physicists often rely on a simplified version of von Neumann's theory where the relation between the measuring instrument and the system under investigation is indeed treated as an interaction (this theory will be discussed later in the present chapter, especially in sections 6 and 7), or else they use a theory of measurement similar to the one explained by Bohm (*Quantum Theory*, ch. 22) which is also a theory of interaction. Heisenberg had treated measurements as interactions from the very beginning and he had also pointed to the fact, which is proved in von Neumann's theory, that the 'cut' ('Schnitt') between the object and the measuring device can be shifted in an arbitrary manner. Such more formal accounts have often been regarded as elaborations of Bohr's own point of view. This is not the case. Bohr's theory of measurement and von Neumann's theory (or any other theory that treats measurement as an interaction) are *two entirely different theories*. As will be shown later von Neumann's theory encounters difficulties which do not appear in Bohr's account. Bohr himself does not accept von Neumann's account (private communication, Ascov, 1949). A *formal* theory which is very close to Bohr's own point of view has been developed by Groenewold. Cf. his essay in S. Körner, *Observation and Interpretation*, 196ff.

[35] Cf. n.4 as well as section 4 of my paper 'Complementarity', *Proc. Arist. Soc. Suppl.* vol. 32 (1958), 75ff.

[36] This assertion was made by Hilary Putnam.

all a proper application of the correspondence principle will at once provide means of measurement for position and momentum. Speaking more abstractly, we may also say that now a measurement in a system whose ψ-function is an element of a Hilbert space H leads to a destruction of coherence between certain subspaces H', H'', H''' of H and can be characterized by operators P', P'', P''', effecting projection into exactly these subspaces. It is, of course, required to give an interpretation of the Ps – but this problem is identical with the corresponding problem in classical physics which is the interpretation of the primitive descriptive terms of the theory.[37]

We may sum up the results of the foregoing investigation (and those of ch. 16) in the following manner. We first presented a physical hypothesis which was introduced by Bohr in order to explain certain features of microscopic systems (e.g. their wave properties). It was pointed out that this physical hypothesis is of a purely objective character and that it is also needed, *in addition to Born's rules*, for a satisfactory interpretation of the formalism of wave mechanics.[38] The argument of EPR then showed

[37] For the specific difficulties of the quantum-mechanical case, see section 6 of the present paper.

[38] This means, of course, that the uncertainty relations can be derived in two entirely different manners. The first derivation is of a fairly qualitative character. It makes use of the considerations described in ch. 16.2, 16.3 and 16.7 and introduces the quantum of action with the help of de Broglie's formula $p = h/\lambda$. This derivation makes it very clear to what extent the existence of duality and the quantum of action forces us to restrict the application of such classical terms as position, momentum, time, energy, and so on, and it thereby transfers some intuitive content to the uncertainty relations. The second derivation makes use of the commutation relations of the elementary theory (cf. for example H. Weyl, *Gruppentheorie und Quantermechanik* (Berlin, 1931), 68 and 345). Now it is very important to realize that the result of the first and intuitive derivation *may be regarded as a test of the adequacy of the wave mechanics* and indeed of any future quantum theory. For assume the wave mechanics produces an uncertainty that is much smaller that the one derived with the help of duality (which is a highly confirmed empirical fact), de Broglie's relation (which is also a highly confirmed empirical fact) and Bohr's assumption of the indefiniteness of state descriptions (which is the only reasonable hypothesis that allows for the incorporation of the quantum postulate and the conservation laws). This would amount *to a refutation of wave mechanics*, i.e. it would amount to the proof that the wave mechanics is not capable of giving an adequate account of duality, the quantum postulate and the conservation laws. On the other hand, the agreement between the qualitative result and the quantitative result now transfers an intuitive content to the formalism.

The fact that the uncertainty relations can be derived in two different ways and that the quantum theory combines both derivations has been realized by various thinkers. Thus Popper (*The Logic of Scientific Discovery*, 224) points out 'that Heisenberg's formulae . . . result as logical conclusions from the theory; but the *interpretation* of these formulae as rules limiting attainable precision of measurement, in Heisenberg's sense, does not follow from the theory'. And E. Kaila, *Zur Metatheorie der Quantenmechanik* (Helsinki, 1950) has made the existence of various interpretations of (1) the basis of an attack against the quantum theory. Now as against Popper it must be pointed out that the interpretation in question *does* follow from the theory provided the theory has been interpreted in accordance with the intentions of Bohr and Heisenberg. For in this case the interpretation uses, in addition to Born's rules, also the hypothesis of the indefiniteness of state descriptions. Popper regards such an addition to Born's rules as illegitimate and as a result of positivistic inclinations. This is

that a further assumption must be introduced in order to make Bohr's hypothesis compatible with this formalism. According to this further assumption, the state of a physical system is a relation rather than a property and it asserts that an adequate measurement must be performed to make statements about it meaningful. By a 'measurement' is meant, in this connection, a certain type of macroscopic process – a terminological peculiarity which is rather unfortunate and which must be blamed for the many subjectivistic conclusions that have been drawn from Bohr's ideas.

It would be incorrect to say that the presentation of the point of view of Bohr and of his followers is completed with the presentation of the two ideas we have just explained . For as is well known it has been attempted, both by Bohr, and by some other members of the Copenhagen circle, to give greater credibility to these ideas by incorporating them into a whole philosophical (ontological) system that comprises physics, biology, psychology, sociology and perhaps even ethics. Now the attempt to relate physical ideas to a more general background and the correlated attempt to make them intuitively plausible is by no means to be underestimated. Quite the contrary, it is to be welcomed that these physicists undertook the arduous task to adapt also more general philosophical notions to two physical ideas which have some very radical implications. However, the philosophical backing of physical ideas that emerged from these more general investigations has led to a situation that is by no means desirable. It has led to the belief in the uniqueness and the absolute validity of both of Bohr's assumptions. It is, of course, admitted that the quantum theory may have to undergo some very decisive changes in order to cope with new phenomena (the first step here is the transition from the elementary theory to Dirac's theory of the electron; the second step is the transition to the field theories). But it is also pointed out that, however large these changes may be, they will always leave untouched the two elements mentioned, viz. the indefiniteness of state descriptions and the relational character of the quantum-mechanical states, which, so it is added, cannot be replaced by different ideas without creating formal inconsistencies, or inconsistencies with experiment.[39] 'The

incorrect and there are physical reasons which demand indefiniteness. Unfortunately these physical reasons are almost always presented in positivistic language which creates the impression that the peculiarity of the quantum theory, i.e. the features which are ascribed to it over and above the Born interpretation, are indeed due to an epistemological manoeuvre. Cf. ch. 11.7 and ch. 16.7f.

[39] For this sentiment see W. Pauli, *Dialectica*, 8 (1954), 124; L. Rosenfeld, *Louis de Broglie, Physicien et Penseur* (Paris, 1953), 41, 57; P. Jordan, *Anschauliche Quantentheorie* (Berlin, 1936), 1, 114f, 276; G. Ludwig, *Die Grundlagen der Quantenmechanik* (Berlin, 1954), 165ff.

In the last footnote I showed how the adequacy of the formal uncertainties, i.e. of the uncertainties that follow from the commutation relation, can be tested by qualitative considerations concerning the dual character of *elementary particles*. As has been shown by Bohr and Rosenfeld ('Zur Frage der Messbarkeit der elektromagnetischen Feldgrössen', *K. danske Vidensk. Selsk. Mat.-fys. Meddr*, 12, (1933) no. 8, as well as *Phys. Rev.*, 78 (1950), 794ff), the adequacy of the *field theories* and their consitency with the required restriction of the new conceptions', asserts L. Rosenfeld, 'which we need [in order to cope with new phenomena] will be obtained . . . by a rational extension of the quantum theory',[40] which *preserves* the indeterminacies; and the new theories of the microcosm will therefore be increasingly indeterministic. Today this dogmatic *philosophical* attitude with respect to fundamentals seems to be fairly widespread,[41] In the remainder of this chapter I shall try to give my reasons why I believe it to be completely unfounded and why I moreover regard it as a very unfortunate feature of part of contemporary science.[42]

However, before going into details, the following remarks seem to be in order. The particular interpretation of the microscopic theories (and especially of the quantum theory of Schrödinger and Heisenberg), which results from the combination of these theories with Bohr's two hypotheses and with the more general philosophical background referred to above, has been called the *Copenhagen Interpretation*. A close look at this interpretation shows that it is not *one* interpretation, but a variety of them. True, the indefiniteness assumption and, to a lesser extent, the assumption of the relational character of the quantum-mechanical states always play an important role, and so do the uncertainty relations. Yet the exact interpretation of these assumptions and of Heisenberg's formulae is neither *clear*, nor is there a *single* such interpretation. Quite the contrary – what we find is that all philosophical creeds, from extreme idealism (positivism, subjectivism) to dialectical materialism, have been imposed upon the physical elements. Heisenberg and von Weizsaecker present a more Kantian version;[43] Rosenfeld has injected dialectics into his account of the

applicability of the classical terms can be shown in a similar manner. Cf. also L. Rosenfeld, 'On the Quantum Electrodynamics' in *Niels Bohr and the Development of Physics*, ed. W. Pauli (London, 1955), 70ff, as well as Heitler, *Quantum Theory of Radiation* (Oxford, 1957), 79ff.

[40] *Observation and Interpretation*, 45. For the idea of a 'rational extension', or a 'rational generalization' cf. ch. 2.4, and n.4. A 'rational extension' of the quantum theory would be any formalism that is consistent with the qualitively derived uncertainties. Cf. also my paper 'Complementarity'.

[41] It is interesting to note the we are here presented with a dogmatic *empiricism*. Which shows that empiricism is no better antidote against dogmatism than is, say, Platonism. It is easily seen why this must be so: both empiricism and Platonism (to mention only one philosophical alternative) make use of the idea of *sources* of knowledge; and sources, be they now intuitive ideas, or experiences, are often assumed to be infallible, or at least very nearly so. Only a little consideration will show, however, that neither can give us an undistorted picture of reality as neither our brains, nor our senses, can be regarded as faithful mirrors.

[42] For a more detailed account see my papers 'Complementarity'; ch. 16 above; and 'Niels Bohr's Interpretation of the Quantum Theory', in *Current Issues in the Philosophy of Science* ed. G. Maxwell and M. Scriven (New York, 1960).

[43] See Heisenberg, *Physics and Philosophy* (New York, 1958). In their physics, too, Heisenberg and Bohr went different ways. 'Bohr tried to make the dualism between the wave picture and the particle picture the starting point of a physical interpretation', writes Heisenberg (*Theoretical Physics in the Twentieth Century, A Memorial Volume to Wolfgang Pauli*, ed. Fierz and Weisskopf (New York, 1960), 45), 'whereas I attempted to continue on the way of the quantum theory and Dirac's transformation theory without trying to get any help from the

matter;[44] whereas Bohr himself is reported to have criticized all these versions as not being in agreement with his own point of view.[45] Quite obviously the fictitious unity conveyed by the term 'Copenhagen Interpretation' must be given up. Instead we shall try to discuss only those philosophical ideas which Bohr himself has provided, and we shall refer to other authors only if their contributions can be regarded as an elaboration of such ideas. The outline of the general background will be started with a discussion of the idea of complementarity.

4. COMPLEMENTARITY

Bohr's hypothesis of indefinite state descriptions referred to description in terms of *classical concepts*, i.e. it referred to description in terms of either Newtonian mechanics (including the different formulations which were provided later by Lagrange and Hamilton), or of theories which employ contact action, or field theories. The hypothesis amounted to the assertion that description in terms of *these* concepts must be made 'more liberal' if agreement with experiment is to be obtained.[46] The principle of complementarity expresses in more general terms this peculiar restriction, forced upon us by experiment, in the handling of the classical concepts. In the form in which this principle is applied it is based mainly upon two empirical premises as well as upon some further premises which are neither empirical, nor mathematical, and which may therefore be properly called 'metaphysical'.[47] The *empirical premises* are (apart from the conservation

wave mechanics.' 'Bohr', writes Heisenberg at a different place (*Niels Bohr and the Development of Physics*, 15), 'intended to work the new simple pictures, obtained by wave mechanics, into the interpretation of the theory, while I for my part attempted to extend the physical significance of the transformation matrices in such a way that a complete interpretation was obtained which would take account of all possible experiments.' On the whole Bohr's approach was more intuitive, whereas Heisenberg's approach was more formalistic, indeed so much so that Pauli felt called upon to demand that 'it must be attempted to free . . . Heisenberg's mechanics a little more from the flood of formalism characteristic for the Göttinger savants [vom Göttinger formalen Gelehrsamkeitsschwall]'; (letter from Pauli to Kronig of October 9, 1925; quoted from *Theoretical Physics in the Twentieth Century*, 26). Von Weizsaecker's point of view is most clearly explained in his book *Zum Weltbild der Physik* (Leipzig, 1954).

[44] See his article in *Louis de Broglie*. As opposed to Rosenfeld, P. Jordan (*Anschauliche Quantentheorie*) and Pauli seems to represent a purely positivistic position.

[45] D. Bohm and H. J. Groenewold, private communication.

[46] N. Bohr, *Atomic Theory*, 3.

[47] I use here the word 'metaphysical' in the same sense in which it is used by the adherents of the Copenhagen point of view, viz. in the sense of 'neither mathematical nor empirical'. That the Copenhagen Interpretation contains elements which are metaphysical in this sense has been asserted, in slightly different words, by Heisenberg who declared in 1930 (*Die Physikalischen Grundlagen der Quantentheorie*, 15) that its adoption was 'a question of taste'. This he repeated in 1958 in the now more fashionable linguistic terminology (cf. *Physics and Philosophy*, 29f).

laws) (1) the dual character of light and matter; and (2) the existence of the quantum of action as expressed in the laws

$$p = h/\lambda, \text{ and } E = h\nu^{48}$$

I do not intend in this paper to discuss all the difficult considerations which finally led to the announcement of the dual character of light and matter. Although these considerations have sometimes been criticized as being inconclusive, they seem to me to be essentially sound. It is also beyond the scope of the present paper to explain how duality can be used for providing a coherent account of the numerous experimental results which form the confirmation basis of the contemporary quantum theory.[49] I shall merely state the principle of duality and make a few comments upon it. Duality means that all the experimental results about light and matter divide into two classes. Facts of the first class, such as the Compton-effect and the photoelectric effect, while contradicting any wave theory, can be completely and exhaustively explained in terms of the assumption that light (or matter) consists of particles. Facts of the second class, such as interference phenomena, while contradicting any particle theory, can be completely and exhaustively explained in terms of the assumption that light (or matter) consists of waves. There exists, at least at the present moment, no system of physical concepts which can provide us with an explanation that covers and is compatible with *all* the facts about light and matter.

The following comments should be made. First, by a particle theory we understand, in the present context, any theory that works with entities of the following kind: they exert influence upon and are influenced by small regions of space only;[50] and they obey the principle of momentum conservation. No further assumptions are made about the nature of these particles and about the laws they obey. By a wave theory we understand, on the other hand, a theory that works with entities of the following kind: they are extended, their states at different places are correlated by a phase, and the phase obeys a (linear) superposition principle. It is the superposition

---

[48] The assumption of the existence of the quantum of action is very often given an interpretation that goes beyond these two equations; however, I agree with Landé and Kaila (*Zur Metatheorie der Quantenmechanik* (Helsinki, 1950), 48) who have both pointed out, though with somewhat different reasons, that a more 'substantial' interpretation of the quantum of action than is contained in these two equations is neither justified, nor tenable. The original view according to which the quantum of action is an indivisible 'link' between interacting systems which is responsible for their mutual changes was refuted by Einstein, Podolsky and Rosen. For this see n.31.

[49] For an account of these results cf. the literature in ch. 11.7 and ch. 16. It is worth pointing out, by the way, that duality is only one of various ordering principles that are needed to give a rational account of the facts upon the atomic level, and especially of the properties of atomic spectra. It took some time to separate the facts relevant for the enunciation of the principle of duality from other facts which had to be explained in a different manner, e.g. by Pauli's exclusion principle and the assumption of an electronic spin.

[50] This explanation is given by Heisenberg, *Physics and Philosophy*, 7.

principle that forms the core of all wave theories. What is refuted by either the Compton-effect or by the photoelectric effect is not a *particular* wave theory (which may be characterized by a particular equation of motion for the waves), but the much more general assumption that light consists of extended and superimposable entities. Secondly, it should be pointed out that the cross-relation between experimental evidence and theories is essential for duality as well as for the idea of complementarity that is based upon it. I doubt whether anything like this exists in those domains in which complementarity has now become a kind of saviour from trouble, such as in biology, psychology, sociology and theology. Thirdly, it must be emphasized – and this remark will prove to be of great importance later on – that the wave theories (in the general sense explained above) and the particle theories do not only serve as devices which allow us to *summarize*, and to *unify*, a host of experimental results in an economical way. Without the key terms of either theory these results *could neither be obtained, nor could they be stated*. To take an example, interference experiments work with coherent or partly coherent light only. Hence, in preparing them proper attention must be paid to the relative phases of the incoming wave-train which means that we have to apply the wave theory already in the preparation of the experiment. On the other hand such facts as the localizability of interaction between light and matter and the conservation of momentum in these interactions cannot be properly described without the use of concepts which belong to some particle theory. Using the term 'classical' for concepts of either a wave theory or a particle theory, we may therefore say that 'only with the help of classical ideas is it possible to ascribe an unambiguous meaning to the results of observation'.[51]

Duality is regarded by Bohr and by his followers as an experimental fact which must not be tampered with and upon which all future reasoning about microphysical events is to be based. As a physical theory is acceptable only if it is compatible with the relevant facts, and as 'to object to a lesson of experience by appealing to metaphysical preconceptions is unscientific',[52] it follows that a microphysical theory will be adequate and acceptable only if it is compatible with the fact of duality, and that it must be discarded if it is not so compatible. This demand leads to a set of very general conditions to be satisfied by any microscopic theory. We are now going to state these conditions.

First of all, the wave concepts and the particle concepts are the only concepts available for the description of the character of light and matter. Duality shows that these concepts cannot any more be applied generally, but can serve only for the description of what happens under certain experimental conditions. Using familiar terms of epistemology this means

[51] Niels Bohr, *Atomic Theory*, 16.
[52] L. Rosenfeld in *Observation and Interpretation*, 42.

that the description of the *nature* of light and matter has now to be replaced by a description of the way in which light and matter *appear* under certain experimental conditions. Secondly, a change from conditions allowing for the application of, say, the wave picture to conditions allowing for the application of the particle picture will, in the absence of more general and more abstract concepts which apply under all conditions, have to be regarded as an *unpredictable jump*. The statistical laws connecting events in the first picture with events in the second picture will therefore not allow for a deterministic substratum, they will be *irreducible*. Thirdly, the combination of duality with the second set of empirical premises introduced above (the Einstein–de Broglie relations) shows that the duality between the wave properties and the particle properties of matter may also be interpreted as a duality between two sets of variables (e.g. position and momentum), which in the classical theory are both necessary for the complete description of the state of a physical system. We are forced to say that a system can never be in a state in which all its classical variables possess sharp values. If we have determined with precision the position of a particle, then its momentum is not only undetermined, it is even meaningless to say that the particle possesses a well-defined momentum. Clearly, the *uncertainty relations* now indicate the domain of meaningful applicability of classical functors (such as the functor 'position'), rather than the mean deviations of their otherwise well-defined values in large ensembles. This is Bohr's hypothesis of the indefiniteness of state descriptions. The relational character of state descriptions results from the need to restrict the application of any set of concepts to a certain experimental domain. This is how the more general point of view explained here is related to the two specific hypotheses which we discussed in the preceding sections (cf. also ch. 16.7).

The argument is generally valid. It follows, and this is Bohr's contention, that it will hold for any theory into which Planck's h enters in an essential way. Hence, any future microscopic theory will have to be descriptive of appearances only, it will contain irreducible probabilites, and it will have to work with commutation relations between variables which are only partly well defined and meaningful. The development of microphysics can only lead to greater indeterminacy. It will never again return to a state of affairs where we are able to give a complete, objective and deterministic description of the nature of physical systems and physical events. In the interest of economy of thought and effort, theories of this kind should therefore be forever excluded from consideration.

I must repeat that in the above two paragraphs only a very sketchy outline has been given of the argument of Bohr and his followers and that it has not at all been shown what great variety of experimental facts is covered, and explained, by the two hypotheses which follow from this argument. This bare outline is not at all sufficient for making understandable the influence Bohr's ideas have had upon physicists and philosophers. But I think that it contains all the essential elements of the Copenhagen point of view, and that it will serve well as a starting point for criticism.

The argument proceeds from what seems to be a mere truism; it proceeds from the assertion that, duality being an experimental fact, it must not be tampered with, but must be regarded as an unalterable basis for any further theoretical consideration. After all, facts are the building stones out of which a theory may be constructed and therefore they themselves neither can nor should be modified. To proceed in this way seems to be the truly scientific attitude, whereas any interference with the facts shows what can only be called the first step towards wild and unwarranted speculation. It is not surprising that this starting point of the argument is frequently taken for granted, as it seems to be the natural procedure to adopt for a scientist. Did not Galileo start modern science by eliminating speculation and by directly putting questions to nature? And do we not owe the existence of modern science to the fact that problems were finally dealt with in an empirical manner rather than on the basis of groundless speculation?

It is here, at the very beginning, that the position of the orthodox must be criticized. For what is regarded by them as a truism is neither correct nor reasonable; and their account of history, too, is at variance with the actual development. Things were just the other way round. It was the Aristotelian theory of motion which was defended by reference to experimental results, while Galileo who was not prepared to take these results at their face value, but insisted that they be analysed and be shown to be due to the interplay of various and as yet unknown factors.

> If we are seeking to understand [the] birth of modern science we must not imagine that everything is explained by the resort to an experimental mode of procedure, or even that experiments were any great novelty. It was commonly argued, even by the enemies of the Aristotelian system, that that system itself could never have been founded except on the footing of observation and experiment. . . We may [also] be surprised to note that in one of the dialogues of Galileo it is Simplicius, the spokesman of the Aristotelians – the butt of the whole piece – who defends the experimental method of Aristotle against what is described as the mathematical method of Galileo.[53]

Indeed the whole tradition of science, from Galileo (or even from Thales) up to Einstein and Bohm[54] is incompatible with the principle that 'facts' should be regarded as the unalterable basis of any theorizing. In this tradition the results of experiment are not regarded as unalterable and unanalysable building stones of knowledge. They are regarded as capable

---

[53] H. Butterfield, *The Origins of Modern Science* (London, 1957), 80. Butterfields book contains a very valuable account of the role of the experimental method in the seventeenth century.
[54] Cf. the latter's *Causality and Chance*.

of analysis, of improvement (after all, no observer, and no theoretician collecting observations is ever perfect), and it is assumed that such analysis and improvement is absolutely necessary. What would be a more obvious observational fact that the difference between celestial motions (regularity) and terrestrial motions (irregularity)? Yet from the earliest times the attempt was made to explain both on the basis of the same laws. Again, what would be a more obvious observational fact than the great variety of substances and phenomena found on the surface of the earth? Yet from the very beginning of rational thinking it was attempted to explain this variety on the basis of the assumption that it was due to the working of a few simple laws and a few simple substances, or perhaps even a single substance. Also the new theory of motion which was developed by Galileo and Newton cannot possibly be understood as a device for establishing relations between our experiences, or between laws which are directly founded upon our experiences, and this for the simple reason that the laws expressing these observable motions (such as the law of free fall, or Kepler's laws) were asserted to be incorrect by this theory. This is quite in order. Our senses are no less fallible than our thoughts and no less open to deception. The Galilean tradition, as we may call it, therefore proceeds from the very reasonable point of view that our ideas *as well as* our experiences (complicated experimental results included) may be erroneous, and that the latter give us at most an *approximate* account of what is going on in reality.[55] Hence, within this tradition the condition to be satisfied by a future theory of the microcosm is not that it be simply *compatible* with duality and the other laws used in the above argument, but that it be compatible with duality *to a certain degree of approximation* which will have to depend on the precision of the experiments used for establishing the 'fact' of duality.

A completely analogous remark holds for the assertion that Planck's constant will have to enter *every* microscopic theory in an essential way. After all, it is quite possible that this constant has meaning only under certain well-defined conditions (just as the density of a fluid, or its viscosity, or its diffusion constant can be meaningfully defined only for not too small a volume), and that the experiments we have made so far explore only part of these conditions.[56] Obviously the invariance of h in all *these* experiments cannot be used as an argument against such a possibility. But if neither the constancy of h nor duality can be guaranteed to hold in new domains of research, then the whole argument is bound to break down: it does not guarantee the persistence of the familiar features of complementarity, of probabilistic laws, of quantum-jumps, of the commutation relations in future investigations.

It ought to be pointed out, by the way, that the above two paragraphs

[55] See vol. 2, ch. 2 for details.
[56] Cf. Bohm, *Causality and Chance*, ch. 4.

cannot be regarded as a *refutation* of the principle that our theories must never contradict what at a certain time counts as an experimental fact. After all, it may well be possible (and it has been possible) to construct theories which satisfy this demand of maximal empirical adequacy with respect to a given set of observations. Part of Aristotle's theory of motion was of this kind. However, it is very doubtful whether this restriction of research would ever allow for theories of the universality, the precision, and the formal accomplishment of Newton's celestial mechanics, or of Einstein's general theory of relativity, both of which lead to a correction of previously existing experimental laws.[57]

Let us now assume, for the sake of argument, that a radically empiricistic point of view has been adopted, i.e. let us regard duality and the constancy of *h* as holding with absolute precision. Is the argument now valid? This at once introduces the second 'metaphysical' assumption that is used by Bohr and his followers. According to this second assumption the classical concepts are the only concepts we have. As we cannot construct a theory or a description of fact out of concepts which we do not have, and as the classical concepts cannot any more be applied in an unrestricted way, we are stuck with the complementary mode of description. Against this argument which has been elaborated in some detail by Heisenberg,[58] and by von Weizsaecker,[59] it is sufficient to point out that introducing a set of concepts is not something that occurs independently of and prior to the construction of theories. Concepts are introduced as part of a theoretical framework, they are not introduced by themselves. However, with respect to theories it must be asserted that man is not only capable of *using* them and the concepts which they embody for the construction of descriptions, ex-

---

[57] See vol. 2, ch. 1 for more general arguments.

[58] *Physics and Philosophy*, 56: 'It has sometimes been suggested that one should depart from the classical concepts altogether and that a radical change of the concepts used for describing the experiments might possibly lead back to a ... completely objective description of nature. This suggestion, however, rests upon a misunderstanding ... Our actual situation in science is such that we *do* use the classical concepts for the description of the experiments. There is no use discussing what could be done if we were other beings than we are.'
This is an astounding argument indeed! It asserts, in fact, that a language that is used for describing observational results and that is fairly general cannot possibly be replaced by a different language. How, then did it happen that the Aristotelian physics (which was much closer to the everyday idiom and to observation than the physics of Galileo and Newton) was replaced by the point of view of the classical science? And how was the theory of witchcraft replaced by a reasonable psychology, based as it was upon innumerable *direct* observations of demons and demonic influence? (Think of the phenomenon of split personality which lends very direct support to the idea of demonic influence.) On the other hand why should we not try to improve our situation and thereby indeed become 'other beings than we are'? Is it assumed that the physicist has to remain content with the state of human thought and perception as it is given at a certain time and that he cannot (or should not) attempt to change, and to improve upon that state?

[59] *Zum Weltbild*, 110: 'Every actual experiment we know *is* described with the help of classical terms and we do not know how to do it differently.' The obvious reply is, of course: 'Too bad; try again!'

ɔerimental and otherwise, but that he is also capable of *inventing* them. How ɛlse could it have been possible, to mention only one example, to replace the Aristotelian physics and the Aristotelian cosmology by the new physics of Galileo and Newton? The only conceptual apparatus then available waş the Aristotelian theory of change with its opposition of actual and potential properties, form and matter, the four causes and the like. This conceptual apparatus was much more general and universal than the physical theories of today for it contained a general theory of change, spatiotemporal and otherwise. It also seems to be closer to everyday thinking and was therefore more firmly entrenched than any succeeding physical theory, classical physics included. Within this tremendously involved conceptual scheme Galileo's (or rather Descartes') law of inertia does not make sense. Should, then, Galileo have tried to get on with the Aristotelian concepts as well as possible because these concepts were the only ones in actual use and as 'there is no use discussing what could be done if we were other [i.e. more ingenious] beings that we are?'[60] By no means! What was needed was not improvement, or delimitation of the Aristotelian concepts in order to 'make room for new physical laws';[61] what was needed was an *entirely new theory*.

[60] Cf. the quotation in n.58.
[61] For a very clear presentation of this idea behind the correspondence principle see G. Ludwig, *Die Grundlagen der Quantenmechanik* ch. 1. As Aage Petersen has pointed out to me, Bohr's ideas may be compared with Hankel's principle of the permanence of rules of calculation in new domains (for this principle cf. ch. 4 of F. Waismann's *Einfuehrung in das Mathematische Denken* (Vienna, 1947)). According to Hankel's principle the transition from a domain of mathematical entities to a more embracing domain should be carried out in such a manner that as many rules of calculation as possible are taken over from the old domain to the new one. For example, the transition from natural numbers to rational numbers should leave unchanged as many rules of calculation as possible. In the case of mathematics, this principle has very fruitful applications. Its application to microphysics is suggested by the fact that some important classical laws remain *strictly valid* in the quantum domain. A *complete* replacement of the classical formalism seems therefore unnecessary. All that is needed is a modification of that formalism which retains the laws that are found to be valid and makes room for those new laws which express the specific behaviour of the quantum-mechanical entities. According to Bohr the modification must be based upon a more 'liberal attitude towards' the classical concepts (*Atomic Description*, 3). We must realize that these concepts are '*idealizations*' (5; original italics), or 'abstractions' (63) whose suitability for description or explanation depends upon the relative smallness of the quantum of action and which must therefore be 'handled with caution' (66) in new experimental domains. 'Analysis of the elementary concepts' (66) has to reveal their limitations in these new fields (4, 5, 8, 13, 15, 53, 108) and new rules for their use have to be devised 'in order to evade the quantum of action' (18). These rules must satisfy the following demands: (a) they must allow for the description of any conceivable experiment in classical terms – for it is in classical terms that results of measurement and experimentation are expressed; (b) they must 'provide room for new laws' ('Can Quantum Mechanical Description of Physical Reality Be Considered Complete?', 701ff; *Atomic Theory*, 3, 8, 19, 53), and especially for the quantum of action (18); (c) they must always lead to correct predictions. (a) is needed if we want to retain the idea that experience must be described in classical terms; (b) is needed if we want to avoid any clash with the quantum of action; (c) is needed if this set of rules is to be as powerful as a physical theory in the usual sense. Any set of rules satisfying (a), (b) and (c) is called by Bohr a 'natural generalization of the classical mode of description' (pp. 4, 56, 70, 92, 110;

Now at the time of Galileo human beings were apparently able to do this extraordinary thing and become 'beings different from what they were before' (and one should again realize that the conceptual change that was implied was much more radical than the conceptual change necessitated by the appearance of the quantum of action). Are there (apart from pessimism with respect to the abilities of contemporary physicists) any reasons to assume that what was possible in the sixteenth and seventeenth centuries will be impossible in the twentieth century? As far as I can understand, it is Bohr's contention that such reasons do indeed exist, that they are of a logical rather than of a sociological character,[62] and that they are connected with the peculiar nature of classical physics.

'Causality and Complementarity', 316; 'Discussions with Einstein' in *Albert Einstein, Philosopher-Scientist*, 210, 239), or a 'reinterpretation . . . of the classical electron theory' (*Atomic Theory*, 14). 'The aim of regarding the quantum theory as a rational generalization of the classical theories', writes Bohr (*Atomic Theory*, 70, 37, 110), 'has led to the formulation of the . . . correspondence principle.' The correspondence principle is the tool by means of which the generalizations may be and have been obtained.

Now it is very important to realize that a 'rational generalization' in the sense just explained does not admit of a realistic interpretation of any one of its terms. The classical terms cannot be interpreted in a realistic manner for their application is restricted to a description of experimental results. The remaining terms cannot be interpreted realistically either, for they have been introduced for the explicit purpose of enabling the physicist to handle the classical terms properly. The instrumentalism of the quantum theory is therefore not a philosophical manoeuvre that has been wilfully superimposed upon a theory which would have looked much better when interpreted in a realistic fashion. It is a demand for theory construction that was imposed from the very beginning and, in accordance with which, part of the quantum theory was actually obtained. Now at this point the historical situation becomes complicated for the following reason: the *full* quantum theory (and by this we mean the full elementary theory) was created by Schrödinger who was a realist and who claimed to have found a theory that was more than a 'rational generalization of classical mechanics' in the sense just explained. That is, the full quantum theory we owe historically to a metaphysics that was diametrically opposed to the philosophical point of view of Niels Bohr and his disciples. This is quite an important historical fact for the adherents of the Copenhagen picture very often criticize the metaphysics of Bohm and Vigier by pointing out that no *physical theory* has as yet been developed on that basis. (For such a criticism see N. R. Hanson, 'Five Cautions for the Copenhagen Critics', *Phil. Sci.*, 26 (1959), 325ff, esp. 334ff.) They forget that the Copenhagen way of thinking *has not produced a theory either*. What it *has* produced is the proper interpretation of Schrödinger's wave mechanics *after* this theory had been invented. For it turned out that Schrödinger's wave mechanics was just that complete rational generalization of the classical theory that Bohr, Heisenberg and their collaborators had been looking for and parts of which they had already succeeded in developing.

[62] As will be evident from the quotation in n.58, Heisenberg and von Weizsaecker seem to base their argument upon the *sociological fact* that the majority of contemporary physicists use the language of classical physics as their observation language. Bohr seems to go further. He seems to assume that the attempt to use a different observation language *can never succeed*. His arguments are similar to the arguments of transcendental deduction used by Kant. The fact that Heisenberg and von Weizsaecker seem to represent a less dogmatic and more practical position has prompted Hanson to distinguish between two different wings, as it were, in the Copenhagen school; the extreme Right, represented by Bohr, which regards the attempt to introduce a new observation language as *logically* impossible; ans the Centre, represented by von Weizsaecker and Heisenberg, where such an attempt is only regarded as being *practically* impossible. I deny that this distinction exists. First of all, the difference between *logical* impossibility and *sociological* impossibility (or *practical* impossibility), although regarded

Bohr's first argument in favour of this contention proceeds from the situation, outlined above, that we need the classical concepts not only for giving a *summary* of facts, but for stating them at all. Like Kant before him, he observes that even our experimental statements are always formulated with the help of theoretical terms and that the elimination of these terms must lead, not to the 'foundations of knowledge' as the positivists would have it, but to complete chaos. 'Any experience', he asserts, 'makes its appearance within the frame of our customary points of view and forms of perception',[63] and at the present moment the forms of perception are those of classical physics.

But does it follow, as is asserted by Bohr, that we can never go beyond the classical framework and that all our future microscopic theories must have duality built into them?

It is easily seen that the use of classical concepts for the description of experiments in contemporary physics can never justify such an assumption. For a theory may be found whose conceptual apparatus, when applied to the domain of validity of classical physics, would be just as comprehensive and useful as the classical apparatus without yet coinciding with it. Such a situation is by no means uncommon. The behaviour of the planets, of the sun and of the satellites was described by the Babylonians, by Plato, and by Ptolemy; it can be described by Newtonian concepts and by the concepts of general relativity. The order introduced into our experiences by Newton's theory and its exotic predecessors is retained *and improved upon* by relativity. This means that the concepts of relativity are sufficiently rich for the formulation even of all the *facts* which were stated before with the help of Newtonian physics. Yet the two sets of concepts are completely different and bear no logical relation to one another.

An even more striking example is provided by the phenomena known as the 'appearance of the devil'. These phenomena are accounted for both by the assumption that the devil exists, and by some more recent psychological (and psycho-sociological) theories.[64] The concepts used by these two

with awe by a good many philosphers, is too subtle to impress any physicist. Neither will the assertion of logical impossibility deter him from trying to achieve the impossible (for example, to achieve a relative theory of space and time); nor will he feel relieved when he is being offered practical impossibility instead of logical impossibility. But we also find that the distinction which Hanson wants to draw between Heisenberg and Bohr is not really one which Heisenberg himself would recognize, or at least so it appears from his writings. For on p. 132 of *Physics and Philosophy* the possibility of an alternative to the Copenhagen point of view is equated with the possibility that 2 times 2 may equal 5, that is, the issue is now made a matter of logic. For this point see also the discussion between Hanson and myself in *Current Issues in the Philosophy of Science*, 390ff. There *do*, of course, exist some very decisive differences between Bohr's approach and Heisenberg's approach, but these differences lie in an entirely different field. Cf. ch. 16.6.

[63] *Atomic Theory*, 1.

[64] Cf. Huxley's highly interesting discussion of the merits of the Cartesian psychology as a means for the explanation of demonic appearances as well as his account of what is and what

schemes of explanation are in no way related to each other. Nevertheless the abandonment of the idea that the devil exists does not lead to experiential chaos, as the psychological scheme can account for at least part of the order already introduced.

To sum up, although in reporting our experiences we make use, and must make use, of certain theoretical terms, it does not follow that different terms will not do the job equally well, or perhaps even better, because more coherently. And as our argument was quite general, it seems to apply to the classical concepts as well.

This is where Bohr's second argument comes in. According to this second argument, which is quite ingenious, we shall have to stay with the classical concepts, as the human mind will never be able to invent a new and different conceptual scheme. As far as I can make out, the argument for this peculiar inability of the human mind rests upon the following *premises*. (a) We invent (or should use) only such ideas, concepts, theories, as are suggested by observation; 'only by observation itself', writes Bohr, 'do we come to recognize those laws which grant us a comprehensive view of the diversity of phenomena'.[65] (b) Because of the formation of appropriate habits, any conceptual scheme employed for the explanation and prediction of facts will imprint itself upon our language, our experimental procedures, our expectations, as well as our experiences. (c) Classical physics is a universal conceptual scheme, i.e. it is so general that no conceivable fact falls outside the domain of its application. (d) Classical physics has been used long enough for the formation of habits, referred to under (b), to become operative. The *argument* itself runs as follows: if classical physics is a universal theory (premise c) and has been used long enough (premise d), then all our experiences will be classical (premise b) and we shall therefore be unable to conceive any concepts which fall outside the classical scheme (premise a). Hence the invention of a new conceptual scheme which might enable us to circumvent duality is impossible.

That there must be something amiss with the argument is seen from the fact that all the premises, except perhaps the first one, apply also to the Aristotelian theory of motion. As a matter of fact the very generality of this theory would seem to make it a much stronger candidate for the argument than classical physics could ever be. However, the Aristotelian theory *has been* superseded by a very different conceptual apparatus. Clearly, this new conceptual apparatus was then not suggested by experience *as interpreted in the Aristotelian manner* and it was therefore a 'free mental creation'.[66] This

is not unthinkable at a certain time and within a certain point of view in ch. 7 of his *Devils of Loudun* (New York, 1952).

[65] *Atomic Theory*, 1.

[66] Albert Einstein in *Ideas and Opinions* (London, 1954), 291 (reprint of an article that was first published in 1936); cf. also H. Butterfield, *The Origins of Modern Science*.

refutes (a). That (b) needs modifying becomes clear when we consider that a scientist should always keep an open mind and that he should therefore always consider possible alternatives along with the theory that he is favouring at a certain moment.[67] If this demand is satisfied, then the habits cannot form, or at least they will not any longer completely determine the actions of the scientist. Furthermore, it cannot be admitted that the classical scheme is universally valid. It is not applicable to such phenomena as the behaviour of living organisms (which the Aristotelian scheme did cover), personal consciousness, the formation and the behaviour of social groups, and many other phenomena. We have to conclude, then, that Bohr's arguments against the possibility of alternatives to the point of view of complementarity are all inconclusive.

This result is exactly as it should be. Any restrictive demand with respect to the form and the properties of future theories can be justified only if an assertion is made to the effect that certain parts of the knowledge we possess are absolute and irrevocable. Dogmatism, however, should be alien to the spirit of scientific research, and this quite irrespective of whether it is now grounded upon 'experience' or upon a different and more 'aprioristic' kind of argument.

What has been refuted so far is the contention that complementarity is the *only possible point of view* in matters microphysical and that the only successful theories will be those which work with inbuilt uncertainties that are interpreted in accordance with Bohr's two hypotheses. Still, it has not been shown that complementarity is not *a possible point of view*. Quite on the

---

[67] J. Agassi has pointed out to me that this principle was consciously used by Faraday in his research work. Against the use of such a procedure it has been argued, by T. S. Kuhn (private communication), that the close fitting between the facts and the theory that is a necessary presupposition of the proper organization of the observational material can be achieved only by people who devote themselves to the investigation of one single theory to the exclusion of all alternatives. For this *psychological* reason he is prepared to defend the (dogmatic) rejection of novel ideas at a period when the theory which stands in the centre of discussion is being built up. I cannot accept this argument. My first reason is that many great scientists seemed to be able to do better than just devote themselves to the development of one single theory. Einstein is the outstanding recent example. Faraday and Newton are notable examples in history. Kuhn seems to be thinking mainly of average scientists who may well have difficulties when asked not only to work out the details of some fashionable theory but also to consider alternatives. However, even in this case I am not sure whether this inability is 'innate', as it were, and incurable, or simply due to the fact that the *education* of the 'average scientist' is in the hands of people who subscribe, implicitly, to Kuhn's doctrine of the necessity of concentration. My second reason for not being able to accept this argument is as follows: assume that it is indeed correct that human beings are not able at the same time to work out the details of one theory *and* to consider alternatives. Who says, then, that details are more important than alternatives which, after all, keep us from dogmatism and are a very concrete and lively warning of the limitations of all our knowledge? If I had to choose between a very detailed account of the fabric of the universe at the expense of not being able to see its limitations and between a less detailed account whose limitations, however, were very obvious, then I would at once choose the latter. The details I could gladly leave to those who are interested in practical application. Cf. also vol. 2, ch. 8.

contrary, we have tried to exhibit the advantages of Bohr's point of view and we have also defended this point of view against irrelevant attacks. I shall now turn to an examination of the *adequacy* of Bohr's point of view. However, first a few remarks should be made on attempts to present it in a more formal manner.

5. VON NEUMANN'S INVESTIGATION

There are many physicists who would readily admit that Bohr's *reasoning* is not very convincing and that it may even be invalid. But they will point out that there exists a much better way of arriving at its *result,* viz. von Neumann's proof to the effect that the elementary quantum theory is incompatible with hidden variables. This proof has not only been utilized by those who found the metaphysical elements in Bohr's philosophy not to their taste, it has also been used by members of the Copenhagen school in order to show that what Bohr had derived on the basis of qualitative arguments could be proved in a rigorous way. However, one ought to keep in mind that the relation between the point of view of Bohr and the point of view of von Neumann is by no means very close. For example, Bohr has repeatedly emphasized that the measuring device must be described in *classical* terms,[68] whereas it is essential for von Neumann's theory of measurement that both the system investigated and the measuring device be described with the help of a ψ-function. The latter procedure leads to difficulties which do not arise in Bohr's treatment. Hence, when dealing with von Neumann's investigations, we are not dealing with a refinement, as it were, of the arguments of Bohr – we are dealing with a new approach.

In ch. 16.8 I have described the proof itself and then had occasion to point out that it involves an illegitimate transition from the properties of ensembles to the properties of the elements of these ensembles. In the present section, I shall assume the proof to be correct and show that even then it cannot be used as an argument to the effect that the atomic theory will forever have to work with inbuilt uncertainties.

The proof consists in the derivation of a certain result from the quantum theory (the elementary theory) in its present form and interpretation. It follows that even if the result were the one claimed by von Neumann it could not be used for excluding a theory according to which the present theory is only approximately correct, i.e. agrees with experiments in some respects but not in others. However simple this argument, the fact that the present theory is confirmed at all has created such a bias in its favour that a little more explanation seems to be required. Assume for that purpose that somebody tries to utilize von Neumann's proof in order to show that *any*

---

[68] Cf. n.34. Cf. also Bohr's definition of a 'phenomenon' in 'Causality and Complementarity', 317, as well as in *Albert Einstein, Philosopher-Scientist*, 237f. See also ch. 16, sections 2–4 above.

future theory of the microcosm will have to work with irreducible probabilities. If he wants to do this then he must assume that the principles upon which von Neumann bases his result are valid under *all* circumstances that future research might uncover. Now the assertion of the absolute validity of a physical principle implies the denial of any theory that contains its negation. For example, the assertion of the absolute validity of von Neumann's premises implies the denial of any theory that ascribes to these premises a limited validity in a restricted domain only. But how could such a denial be justified by *experience* if the denied theory is constructed in such a way that it gives the same predictions as the defended one wherever the latter has been found to be in agreement with experiment? And that theories of the kind described can indeed be constructed has been shown by David Bohm.

Apart from this error with respect to the result of his proof, von Neumann himself was completely aware of the limitations of this alleged result. 'It would be an exaggeration', he writes, 'to maintain that causality has thereby [i.e. by the proof of the two theorems referred to in ch. 13.2] been done away with: quantum mechanics has, in its present form, several serious lacunae and it may even be that it is false.'[69] Not all physicists have shared this detached attitude. Thus, having outlined the proof, Max Born makes the following comment: 'Hence, if any future theory should be deterministic, it cannot be a modification of the present one, but must be essentially different. How this should be possible without sacrificing a whole treasure of well-established results I leave the determinist to worry about.'[70] Does he not realize that precisely that same argument could be used for the retention of absolute space in mechanics, or against the introduction of the statistical version of the second law? And has not the fact that very different theories, such as Newtonian mechanics on the one hand and general realitivity on the other, can be used for describing the same facts (for example, the path of Jupiter), already made it clear that theories can be 'essentially different' without a 'sacrifice of a whole treasure of well-established results' being involved? There is therefore no reason why a future atomic theory should not return to a more classical outlook without contradicting actual experiment, or without leaving out facts already known and accounted for by wave mechanics. Von Neumann's imaginary results cannot in any way be used as an argument against the application at the microlevel of theories of a certain type (for example, deterministic theories).

Hanson's attitude is still less comprehensible. He, too, tries to defend indeterminism and the absence of hidden parameters by a combined reference to von Neumann's proof and 'nature'.[71] But he also realizes, as did

[69] *Mathematical Foundations*, 327.
[70] *Natural Philosophy of Cause and Chance* (Oxford, 1948), 109.   [71] 'Five Cautions', 332.

von Neumann, that the elementary theory on which the proof is based is 'but a programmatic sketch of something more comprehensive',[72] and that it is empirically unsatisfactory. He even admits that a more comprehensive and really satisfactory theory 'simply does not exist'.[73] Now if all that is granted, how can he still try to make use of von Neumann's argument, whose result will be correct and satisfactory only if the premises are correct, satisfactory and complete, i.e. only if the elementary theory is correct, satisfactory and complete? After all, who can now say that the observational and other difficulties of the elementary theory are *not* due to the fact that hidden parameters *do* exist and have been omitted from consideration?

What we have shown so far is that all the arguments which have been used in the literature against alternatives to complementarity are invalid. There does not exist the slightest reason why we should assume that the proper road to future progress consists in devising theories which are even more indeterministic than wave mechanics, and that the appropriate formalism will forever contain inbuilt commutation relations. All the way through, the question has been left undecided as to whether the more general ideas of the point of view of complementarity give an adequate account of the *existing theories*, i.e. whether these ideas give an adequate account of the elementary theory and of the field theories. The answer to this question, which exhibits various difficulties, will be given in the remaining sections of the paper, where we shall also have an opportunity to consider some of the more formal alternatives to the ideas of Bohr.

6. OBSERVATIONAL COMPLETENESS

It was the intention of Bohr and Heisenberg, but notably of the latter, to develop a theory which was thoroughly observational in the sense that a sentence expressing an unobservable state of affairs could not be formulated in it. According to the point of view of complementarity, the mathematics of the theory is to be regarded only as a means for transforming statements about observable events into statements about other observable events, and it has no meaning over and above that function. Everyday language and classical physics do not have this property. Both allow for the existence of physical situations which cannot be discovered by any observation whatever. As an example,[74] consider the case of two banknotes, both printed with the help of the same printing press, the one under legal circumstances, the other by a gang of counterfeiters who used the same press at night illegally. If we assume that the banknotes were printed within a very short interval of time and that they show the same numbers, and if we further assume that they somehow got mixed up, then we shall have to say

---

[72] *Ibid.*, 329.   [73] *Ibid.*
[74] The example is due to K. R. Popper.

that by virtue of their different history they possess certain properties, different for both, which we shall never be able to distinguish. Another example frequently referred to is the intensity of an electromagnetic field at a certain point.[75] The usual methods of measuring this field use bodies of finite extension and finite charge and they can therefore inform us only about average values, not about the exact values of the field components. And as there exist laws of nature according to which there is a lower limit to the size of test bodies, it is even physically impossible to perform a measurement which would result in such information. A third example which should be even more instructive is the disappearance of historical evidence in the course of time. That Caesar sneezed twice on the morning of April 5, 67 b.c. should either be true or false. However, as it is very unlikely that this event was recorded by any contemporary writer, and as the physical traces it left in the surroundings as well as the memory traces it left in the brains of the bystanders have long since disappeared (in accordance, among other things, with the second law of thermodynamics), we now possess no evidence whatsoever. Again we are presented with a physical situation which exists (existed), and yet cannot be discovered by any observational means.

A physicist or a philosopher who is biased in favour of a radical empiricism will quite naturally regard such a situation as unsatisfactory. He will be inclined to reject statements such as those contained in our examples by pointing out that they are observationally insignificant, and in doing so he will be guided by the demand that one should not allow talk about situations which can be shown to be inaccessible to observation. Classical physics does not satisfy this demand automatically. It allows for the consistent formulation of sentences with no observational consequences, *together with* the assertion that such consequences do not exist. An attempt to enforce the radical empiricist's demand will therefore have to consist in an *interpretation* of classical physics according to which some of its statements are cognitively meaningful, whereas others are not. This means that the exclusion of the undesirable sentences will have to be achieved by a philosophical manoeuvre which is superimposed upon physics. Classical physics itself does not provide means for excluding them.[76]

There exist, however, *philosophical* theories which possess exactly this character. An example is Berkeley's theory of matter (if we omit the *ad hoc* hypothesis that objects unperceived by human beings are still perceived by God). According to this theory material objects are bundles of sensations and their existence consists in their being perceived or observed. If this theory is developed in a formally satisfactory manner then it does not allow

[75] Cf. E. Kaila, *Zur Metatheorie der Quantenmechanik*, 34.
[76] It is worth while pointing out that the Aristotelian physics was much closer to the crude experiences of everyday life than is classical physics.

for the consistent formulation of any statement about material objects in which it is asserted that there is a situation which is not accessible to perception. One may call such a theory *observationally complete*. When formulating matrix mechanics, Heisenberg had the intention of constructing a *physical* theory that was observationally complete in exactly this sense, observation with the help of classically well-defined apparatus replacing the more direct form of observation with the help of one's senses. It is assumed in the more general ideas held by the members of the Copenhagen school, and notably by Bohr, that the elementary quantum theory in its present form and interpretation corresponds with this intention. The assumption is not justified or, at least, nobody has as yet shown it to be correct. However, let us first examine an apparently very strong argument in its favour.

Consider a state $|\Phi\rangle$ which is such that it cannot be characterized by the values of any complete set of commuting observables. Such a state would be truly unobservable. For first of all, there is no measurement (in the sophisticated sense of the quantum theory) which can bring it about; and secondly, there is no measurement which on immediate repetition would lead to the result characteristic for this state, as we have assumed that there is no such result. If we still want to assert the existence of the state then we must regard it as an element of Hilbert space (we adopt von Neumann's formalism) and it must be possible to represent it in the form

$$|\Phi\rangle = \Sigma_i \, c_i \, |a_i\rangle$$

where the $|a_i\rangle$ form a complete orthonormal set which is connected with the complete commuting set of observables a. Now incorporate $|\Phi\rangle$ into an orthonormal set $\{|\Phi_i\rangle\}$ in such a manner that the set $\{|\Phi\rangle\} + \Sigma_i\{|\Phi_i\rangle\}$ is complete. Then for any $|a_k\rangle$

$$|a_k\rangle = \Sigma_i \, |\Phi_i\rangle \, \langle\Phi_i|a_k\rangle + |\Phi\rangle \, \langle\Phi|a_k\rangle$$

which on measurement of the observable corresponding to the set $\{|\Phi\rangle\} + \Sigma_i\{|\Phi_i\rangle\}$ would yield $|\Phi\rangle$ unless $\langle\Phi|a_k\rangle = \langle a_k|\Phi\rangle^* = c_k^* = 0$. As $\langle a_k$ may be any eigenstate of a it follows that $|\Phi\rangle = 0$. *States which are not accessible to observation do not exist.*

Now if this argument is supposed to prove observational completeness with respect to *classical states of affairs* then the formal scheme of it must be filled with empirical content. More especially, we must make the following assumptions. First, it must be assumed that there exist changes of states which can transform any state into a mixture of the eigenstates of an observable a, or into an a-mixture as we shall call it. This demand is a purely theoretical demand which must be satisfied by the *formalism* of the theory and which is independent of the interpretation of this formalism. Secondly, it must be assumed that states or observables can be characterized in a purely classical manner. Thirdly, we must demand that for any observable thus interpreted here there exists a classical device capable of transforming *any* state into an α-mixture (again classically interpreted). Finally, the methods of measurement in actual use today must produce α-mixtures with respect to the observables they are supposed to measure, or else the numbers obtained are of no relevance whatever.

As regards the first assumption, it must be pointed out that it can be discussed only if a definite meaning has been given to the phrase 'any state', i.e. if the class of all states has been well defined. As is well known, there is no unanimity on this point. The usual attitude is to neglect the question altogether and to decide it differently in different cases. The trouble with such a procedure is, of course, that it leads to a breakdown of the universal applicability of the Born-interpretation in the sense that no theoretical justification will be available for the comparison of probabilities that have been obtained in different cases, or even in different treatments of one and the same case.[77] On the other hand, the only presentation of the theory which gives a definite account of the manifold of states to be used, von Neumann's presentation, has sometimes been regarded as being too narrow, for it excludes as illegitimate procedures for which it provides no equivalent whatever, and which yet seem to be necessary for the calculation of some of the most important experimental applications of the theory (problem of scattering).[78] We see that one of the basic presuppositions of the first assumption is still far from clear. However, suppose that a satisfactory way has been found of delimiting the totality of allowed states. Is it then possible to justify the assumption that there exist changes from pure states into mixtures?

There exist two attempts at such a justification. The first attempt is based upon the Born interpretation. This interpretation associates a certain probability with each transition from originally given states into one of the eigenstates of the observable measured (let us assume that α is this observable). The probabilities will have to be obtained on the basis of counting all those systems which after measurement possess identical eigenvalues α', α'', α''' etc. From these two assumptions it is inferred that immediately after a measurement of a state that is not an eigenstate of α, the state of the system has turned into an α-mixture (this consequence has sometimes been

---

[77] Cf. E. L. Hill, 'Function Spaces in Quantum Mechanical Theory', *Phys. Rev.*, 104 (1956), 1173ff.

[78] J. M. Cook, *J. Math. Phys.*, 36 (1957), 82ff shows that within Hilbert space the scattering problem can be solved only if $\int\int\int |V(xyz)|^2 \, dxdydz < \infty$ which excludes the Coulomb case. I owe this reference to Professor E. L. Hill. In a discussion of the above paper, Bolsterli has pointed out that this does not invalidate the applicability of von Neumann's approach to the problem of scattering. The solution lies in not working with the complete Coulomb field, but in using a suitable cutoff. This is, of course, a possible procedure; but it is not at all satisfactory as long as a general procedure for determining the size of a cutoff is not available.

called the *projection postulate*). But a 'transcendental deduction' of this kind works only on condition that Born's statistical interpretation is universally applicable, and this is exactly what we want to find out. For it is the universal applicability of Born's rules which guarantees the observational completeness of the theory. What we have shown is that this universal applicability can be guaranteed only if we add the projection postulate to the theory: the projection postulate is a necessary condition of the observational completeness of the theory. However, is it possible to justify this postulate in an independent way?

An attempt at an empirical justification of the postulate which also brings in the empirical considerations demanded by the third assumption is due to von Neumann.[79] Von Neumann interprets the Compton-effect as a quantum-mechanical measurement. The quantity to be measured is any coordinate of the place of collision P. One way of measuring P (M_1) consists in determing the path of the light quantum after the collision. Another way of measuring P (M_2) consists in determining the path of the electron after the collision. Now assume that M_1 has been performed. Before, we could only make statistical assumptions about its outcome (e.g. about ϑ, the angle of deflection for the light quantum, or about P). But as $tg\alpha = (\lambda/(\lambda + \lambda_C))$ tg $(\vartheta/2)$ (λ the wavelength of the incident photon which is assumed to be known, λ_C the Compton wavelength, α the angle of deflection for the electron, ϑ the angle of deflection for the photon) the outcome of M_2 is certain once M_1 has been performed and the result used: M_2 leads to exactly the same result as M_1. It follows that a state in which the value of P was not well defined is transformed by M_1, i.e. by a measurement of this very quantity, into a state in which its value is well defined. Or, by generalization: a general state is transformed into an eigenstate of the quantity measured.

The simplicity of this argument and the force derived from it is only apparent. It is due to the fact that a rather simple way has been used of describing what happens before and after the interaction of the electron and the photon. The application of wave mechanics to the problem shows that the interpretation of the result in terms of quantum-jumps is only a first approximation,[80] unless one has already introduced this hypothesis and used it during the calculation. Also the detailed account is much too complicated to allow for a simple argument such as the one just presented. We know too little about processes of interaction to be able to make any experimental result the basis for an argument with regard to certain features of the theory. Another, completely different argument against the validity of von Neumann's empirical derivation of quantum jumps is the following: the state of the system [electron + photon] after the interaction is

[79] *Mathematical Foundations*, 212.
[80] Cf. for example, Sommerfeld, *Atombau and Spektrallinien* (Braunschweig, 1939), ch. 8; W. Heitler, *Quantum Theory of Radiation*, 3rd edition (Oxford, 1957), sec. 22ff.

a state of type (1) (section 1) with α and ϑ being correlated in a manner similar to the manner in which α and β are correlated in (1). Now an observation of a specific value of α can lead to the prediction of the correlated value of ϑ (or of β in the example of section 1) only if immediately after the observation the state of the system [electron + photon] has been reduced to a state in which both α and ϑ have sharp values, i.e. only if the projection postulate has already been applied. Which shows that even if von Neumann's rather simple assumptions about the interaction process were acceptable, even then the argument would have to be rejected as being circular.

However, we have not yet discussed all the difficulties. Even if the argument were admissible, and even if it represented the situation in a sufficiently detailed way, and without circularity, it would still only show that the projection postulate is *compatible with experience;* it would not remove the *theoretical difficulties* which are connected with it. These theoretical difficulties which I shall call the *fundamental problem of the quantum theory of measurement* consist in the following: (1) the Schrödinger equation transforms pure states into pure states; (2) in general the situation described by a mixture cannot be described by a single wave function; hence (3) the projection postulate cannot be explained on the basis of Schrödinger's equation alone, and it is even inconsistent with it if we assume that this equation is a process equation which governs all physical processes at the microlevel. The discussion of this fundamental problem leads straight into the quantum theory of measurement.

7. MEASUREMENT

A measurement is a physical process which has been arranged either for the purpose of testing a theory or for the purpose of determining some as yet unknown constant of a theory. A complete account of measurement will give rise to at least three sets of problems. First of all, there is the problem of whether the statements obtained with the help of the experiment are relevant, i.e. whether they indeed concern the theory or the constants in question, and under what conditions. Problems of this kind may be called *problems of confirmation.* They will not be dealt with in a systematic manner. Another set of problems concerns the question whether the observable elements of the process (or of the equilibrium state in which it usually terminates) stand in a one to one relationship to the (not necessarily directly observable) elements whose properties we are investigating when performing the measurement. This set of problems may be split into two parts, viz. (1) the question under what circumstances a situation may be called observable; and (2) the question whether it is possible *by physical means* to bring about a situation in which observable states are correlated with non-observable states in such a manner that an inference is possible from the structure of the first to the structure of the second. Question (1) is a question of psychology. It is the question whether and how human beings, assisted or not by instruments, react towards situations of a certain kind. Problems connected with question (1) will be called *observer problems*. Question (2) is a question of physics. It is the question whether the physical conditions which must be satisfied by a well-designed measurement are compatible with the laws of physics. Problems connected with question (2) will be called *physical problems of measurement*. Obviously the physical problems admit of a solution only if a physical characterization, in terms of some theory, is available of those states of affairs which are observable.

In the present section we shall be mainly concerned with the physical problems of measurement in the quantum theory, although we may occasionally also deal with observer problems and problems of confirmation. More especially, we shall be dealing with the question whether and how the physical conditions which must be satisfied by any well-designed measurement of a quantum-mechanical observable can be made compatible with the dynamical laws, and especially with Schrödinger's equation. This essentially is our problem: the theory is observationally complete only if the projection postulate is added to its basic postulates. The projection postulate is incompatible with the unrestricted validity of Schrödinger's equation. How can this apparent inconsistency be resolved?

According to Popper this problem is only an apparent one and it 'arises in all probability contexts'.[81] For example,

> assume that we have tossed [a] penny and that we are shortsighted and have to bend down before we can observe which side is upmost. The probability formalism tells us then that each of the possible states has a probability of 1/2. So we can say that the penny is half in one state and half in the other. And when we bend down to observe it the Copenhagen spirit will inspire the penny to make a quantum jump into one of its two eigenstates. For nowadays a quantum jump is said . . . to be the same as a reduction of the wave packet. And by 'observing' the penny we induce exactly what in Copenhagen is called a 'reduction of the wave packet'.[82]

This is a very seductive proposal indeed, for there is no uneasiness combined with the classical case. But this is due to the fact that *in the classical case we are not dealing with wave-packets*. What we are dealing with are mutually exclusive alternatives only one of which, as we know, will be realized in the end. The classical description allows for this possibility, for it is constructed in such a manner that the statement 'the probability that the outcome is a head is 1/2' is compatible with 'the outcome is actually a head'. We may therefore interpret the transition from the first statement to the second statement as the transition from a less definite description to a more definite one, as a transition that is due to a change in our knowledge but has no implications whatever as regards the actual physical state of the system which we are describing. The second statement does not assert the occurrence of a process that is denied by the first statement. Hence, the 'jump' that occurs is a purely subjective phenomenon rather a harmless phenomenon at that.

This is not so in the quantum-mechanical case. Consider for this purpose a measurement whose possible outcomes are represented by the states Φ' and Φ'', and which occurs when the system is in a state $\Phi = \Phi' + \Phi''$. In this case Φ cannot be regarded as an assertion to the effect that one of two mutually exclusive alternatives, Φ' or Φ'', occurs for when Φ is realized physical processes may occur which do not occur when Φ' is realized, nor when Φ'' is realized. This, after all, is what interference amounts to. The transition, on measurement, from Φ to, say, Φ'', is therefore accompanied by a change in physical conditions which does not take place in the classical case. Even worse: what the Schrödinger equation yields when applied to the system [Φ + a suitable measuring apparatus] is, strictly speaking, another pure state $\Psi = A\Phi' + B\Phi''$, so that now *a mere look* at the apparatus (after which we assert that we have found Φ'') seems to lead to a physical change, namely to the destruction of interference between $A\Phi'$ and $B\Phi''$. It is this characteristic of the quantum-mechanical case that is completely overlooked in Popper's analysis.[83]

Another fairly simple suggestion is due to Landé. Landé tries to solve what we have called the fundamental problem of measurement by denying that Schrödinger's equation is a process equation. In order to show this, he proceeds from the usual interpretation of $\langle \Phi | a_i \rangle$ and $\langle \Phi | \beta_k \rangle$ (α and β being two complete sets of commuting observables) as different expectation functions belonging to *one and the same state,* and he suggests that this interpretation also be used in the case of the temporal development of states, i.e. he suggests that also Φ (t') and Φ (t'') be interpreted as two different expectation functions belonging to the same state rather than as two different states.[84]

If this interpretation is adopted,[85] then Schrödinger's equation is no longer a process equation which transforms states into other ('later') states, but becomes an equation which transforms expectation functions of one state into other expectation functions *of the very same state*. In this interpretation, states never change unless we perform a measurement in which latter

[83] I discussed this difficulty with Popper on a beautiful summer morning driving from London to Glyndebourne and he seemed then to agree with my arguments (1958). Wigner has shown that the problem remains even if the initial state should happen to be a mixture. 'The Problem of Measurement', *Am. J. Phys.*, 31 (1963), 6ff.

[84] 'Heisenberg's Contracting Wave Packets', *Am. J. Phys.*, vol. 27 (1959), as well as 'Zur Quantentheorie der Messung', *Z. Phys.*, 153 (1959), 389ff.

[85] Cf. my critical note in *Am. J. Phys.*, 28 (1960), 28.

case there occurs a sudden transition from a state into one of the eigenstates of the observable measured. This procedure only apparently removes the temporal changes. Strictly speaking it is nothing but a verbal manoeuvre. By pushing the temporal changes into the representatives it now makes the dynamical variables time-dependent, whereas in the usual presentation, which is criticized by Landé, the variables do not change in time. However, a representation of the quantum theory in terms of stable states and moving variables is well known: it is the Heisenberg representation.[86] And as this latter representation can be shown to be equivalent to the one Landé wants to abandon (the Schrödinger representation) it follows that his criticism, and his alternative suggestion completely lose their point.

We now turn to a very brief examination of von Neumann's theory of measurement.[87] In von Neumann's investigations the projection postulate and Schrödinger's equation are given equal importance. The process of measurement itself is regarded as an *interaction* between a macrosystem (represented by some wave-function) and a microsystem (represented by another wave-function). The main result is that the projection postulate is compatible with the formalism of wave mechanics and Born's interpretation. This theory has been attacked for epistemological, physical, and mathematical reasons, and it seems now clear that it cannot be regarded as a satisfactory account of the process of measurement in the quantum theory. The *epistemological* difficulty of von Neumann's theory consists in this:[88] the theory allows for the application of the projection postulate even on the macrolevel, and it then leads to the paradoxical result that by simply taking notice of a macroscopic trace the observer may destroy interferences and thus influence the physical course of events.[89] The source of this difficulty is easily seen: it lies in the fact that a *micro-account* is given both of the measuring apparatus and of the system investigated although only the latter can be said to have microscopic dimensions and although only the latter is investigated in such a detailed manner that its microscopic features

---

[86] Landé's valuable investigations (cf. especially *Foundations of Quantum Theory, A Study in Continuity and Symmetry* (New Haven, 1955)) are based on the Heisenberg representation. They may be regarded as an attempt to derive the Born-interpretation and the completeness assumption (which Landé uses in the above-mentioned book, pp. 24f, but which he unfortunately drops later on) as well as some other characteristic features of the quantum-level (quantum-jumps, superposition) from thermodynamical considerations and plausible philosophical assumptions. At this stage the result of Landé's investigations is much closer to the Copenhagen point of view than both Landé and the orthodox are prepared to admit (cf. a similar judgement by H. Mehlberg, *Current Issues in the Philosophy of Science*, 368). This is only one of the many instances in the history of the quantum theory where people passionately attack each other when they are in fact doing the same thing.

[87] *Mathematical Foundations*, ch. 6.   [88] For a more detailed discussion see ch. 13.

[89] It was Schrödinger who first drew attention to this paradoxical consequence of the theory. See his article 'Die gegenwaertige Lage in der Quantenmechanik', 812. See also the discussion by P. Jordan , 'On the Process of Measurement in Quantum Mechanics', *Phil. Sci.*, 16 (1949), 269ff. For details see ch. 16.3 and ch. 13.

and their dual nature become apparent. Clearly such a procedure will not reflect properly the behaviour of macro-objects *as seen by a macro-observer*, as it does not contain the approximations which are necessary for a return to the classical level. One of the most obvious *physical* consequences is that the entropy of the total system [micro + macrosystem] remains unchanged as long as the projection postulate is not applied. This is very different from the corresponding result in the classical case, a difference which cannot be ascribed to the appearance on the microlevel of the quantum of action.[90] It seems therefore advisable to employ a greater latency of description when discussing the process of measurement. A very similar suggestion emerges from considerations due to Elsasser:[91] The observation of the macroscopic movement which terminates the process of measurement will usually take some time interval Δt of macroscopic dimensions. During this time the measuring apparatus is supposed to retain its main *classical* properties, or, to express it differently, it is supposed to remain an element of a statistical ensemble which is defined in a way which depends on the imprecision of macroscopic operations. Now if we assume that the measuring apparatus is in a pure state in which variables, complementary to the variables on which the main apparatus variables depend in a decisive manner, possess sharp values, then such constancy of classical properties over a classically reasonable time interval cannot be guaranteed. Suppose, for example, that the pure state in which the apparatus allegedly dwells is one in which all the elementary constituents of the apparatus possess a well-defined position. Then the corresponding momenta will range over all possible values, i.e. the system will disintegrate in a very short time.[92] Hence, 'if systems of many degrees of freedom are involved the possibility of giving a unique quantum mechanical representation of a system by a pure state, and the possibility of leaving it in approximately the conditions under which it appears as sample of a given collective, will in general exclude each other'.[93] Finally, I would like to draw attention to the *mathematical* fact, first pointed out by Wigner,[94] that only an approximate measurement is possible of operators which do not commute with a conserved quantity.[95] All these

---

[90] As a matter of fact, we at once obtain an *H*-theorem when we introduce the usual subdivision of phase space expressing the limitations of measurement on the macrolevel. For a first application of a procedure of this kind see J. von Neumann, 'Beweis des Ergodensatzes und des H-Theorems in der neuen Mechanik', *Z. Phys.*, 57 (1929), 80ff.

[91] 'On Quantum Measurement and the Role of Uncertainty Relations in Statistical Mechanics', *Phys. Rev.*, 52 (1937), 987ff.

[92] Cf. also Ludwig's considerations concerning the possibilities of measurement and of actually *creating* such a state in *Grundlagen der Quantenmechanik*, 171f.

[93] Elsasser, 'On Quantum Measurement and the Role of Uncertainty Relations', 989.

[94] 'Die Messung Quantenmechanischer Operatoren', *Z. Phys.*, 131 (1952), 101ff.

[95] For a more detailed account see the paper by H. Araki and M. M. Yanase 'On the Measurement of Quantum Mechanical Operators', *Phys. Rev.*, 120 (1960), 622ff, which the latter author was kind enough to let me have prior to its publication.

results taken together make it very clear that the problem of measurement demands application of the methods of statistical mechanics *in addition to* the laws of the elementary theory. A similar suggestion seems to emerge from the analyses of P. Jordan,[96] and of H. Margenau.[97] In a very suggestive paper, Jordan has pointed out that the application of statistical considerations may lead to the elimination, on the macroscopic level, of the very troublesome interference terms which in von Neumann's account were removed with the help of the projection postulate. Margenau, on the other hand, has drawn attention to the fact that no real stage of a real measurement is correctly described by the projection postulate. The postulate does not correctly describe the state of the system investigated *after* the result of the measurement has been recorded (in the form of, e.g. a macroscopic trace on a photographic plate). The reason is that the process of recording often destroys the system.[98] Nor does the postulate describe the state of the system, *before* the recording. The reason is that at this moment the beams corresponding to the various eigenfunctions of the observable measured are still capable of interfering so that the system cannot be said to dwell in any one of them to the complete exclusion of dwelling in a different one.[99] Taking all this into account, Margenau drops the projection postulate and assumes that state-functions are objectively real probabilities which are *tested* by a measurement without being *transformed* by it into a different state-function. What is important in such a test is not the fact that the state

---

[96] *Anschauliche Quantentheorie.*

[97] 'Advantages and Disadvantages of Various Interpretations of Quantum Theory', *Physics Today*, 7 (1957); 'Philosophical Problems Concerning the Meaning of Measurement in Physics', *Phil. Sci.*, 25 (1958), 23ff. Cf. also John McKnight 'The Quantum Theoretical Concept of Measurement', *Phil. Sci.*, 24 (1957), 321ff, and Loyal Durand III, *On the Theory and Interpretation of Measurement in Quantum Mechanical Systems* (Princeton, 1958).

[98] I have not been convinced that this difficulty which plays a central role in Margenau's considerations is a difficulty of principle rather than a technical difficulty that can be superseded by the construction of a more efficient measuring apparatus. Margenau himself has indicated that in the case of the position measurement of either a photon or an electron the Compton-effect may be used in such a manner that the system investigated is not destroyed. We may also use, at least theoretically, the mutual scattering of photons or the correlations between polarization which exist in the radiation created by the annihilation of a positron–electron pair. Concerning the use of a polarizer the following procedure suggests itself: take the example of a photon and examine whether it left the polarizer P (which we shall assume to be capable of transmitting photons of mutually perpendicular polarization in two different directions, I and II) in direction I. Assume furthermore that only such photons are counted whose passage through a filter F (and through P) can be ascertained by catching Compton recoil electrons that have been emitted by F. Of those photons we know with certainty that they have passed P and will therefore travel along I and II. Now if immediately after the capture of a recoil electron no reaction is observed in a photomultiplier Z that is located in II then we may be certain that the photon left in direction I with the opposite polarization – and this without any interference with the photon itself *after* its passage through P. (This example is an elaboration of a suggestion made by Schrödinger, 'Die gegenwärtige Situation in der Quantenmechanik', *Naturwissenschaften*, 22 (1935), 341).

[99] I would like to point out that this is an attempt, on my part, to reconstruct Margenau's argument which is not too clear on this second point.

of a system has become an eigenstate of the observable measured; what is important is the emergence of a set of numbers which is the one and only result of the measurement and its only point of interest.

It seems to me that none of the objections I have reported in the last paragraph can be raised against the theory that has been developed by G. Ludwig.[100] Ludwig's account is based upon the Schrödinger equation and certain assumptions concerning the formal features (in terms of the elementary theory) of the macroscopic level. The following result is obtained: measurements are complicated thermodynamic processes which terminate in a state in which the macro-observables of the measuring instrument M which are correlated with the microproperties of the system S under investigation possess fixed values, i.e. values that are independent of the nature of the macroprocedure which has led to their determination. The projection postulate is not used anywhere in the calculation of the expected changes of either S or of M. Yet the theory results in something very close to this postulate; for in the equilibrium state that terminates the measurement the expectation values for macroscopic results are identical with the expected values for the correlated properties. It is for this reason that Ludwig regards the projection postulate as an 'abbreviated account of a very complicated process'.[101]

It would seem that a very careful interpretation is needed of this assertion. It cannot mean that the measurement transforms the state of the system S into a state that is very close to, though not identical with, the state predicted by the projection postulate. This interpretation is excluded by virtue of the fact that the measurement leads in many cases to a destruction of the system investigated. All we can say in *these* cases is that the final macroscopic situation adequately mirrors the number of the eigenstates of the observable measured and their relative weights in the state of S *before* the measurement commenced. Such an interpretation, which is also defended by Margenau, would be very close to Bohr's where 'the "properties" of a microscopic object are nothing but possible changes in various macroscopic systems'.[102] Indeed, I must confess that I cannot see a very great difference between Margenau's suggestions and the theory of Bohr where the micro-system and the macrosystem are supposed to form an indivisible block,

[100] *Die Grundlagen der Quantenmechanik*, ch. 5.
[101] 'Die Stellung des Subjekts in der Quantentheorie', in *Veritas, Iustitia, Libertas,* Freie Universität Berlin und Deutsche Hochschule für Politik (Berlin, 1952), 266. This paper also contains a popular account of Ludwig's theory and a comparison of this theory with the customary point of view as expressed by von Weizsaecker.
[102] *Die Grundlagen der Quantenmechanik,* 170. Other accounts which regard the projection postulate as an approximate description of a more complicated process have been given by Bohm, *Quantum Theory,* ch. 22; Green, 'Observation in Quantum Mechanics', *Nuovo Cimento,* 9 (1958), 880ff; A. Daneri and A. Loinger, *Quantum Theory of Measurement* (Milan, 1959), esp. section 9. All these alternative accounts, however, use examples and are by no means as general and detailed as Ludwig's account.

the only changes of which are those than can be described in classical terms. As the above quotations show Ludwig himself regards his own theory as an attempt at a more formal presentation of Bohr's point of view. And this it is, but only to a certain extent.[103]

For whereas in Ludwig's account the properties of the macroscopic level agree with the properties required by the classical physics only to a certain degree of approximation, such a theoretical (and practically negligible) difference is not admitted by Bohr: according to Bohr the measuring instrument is *fully classical* and restrictions occur only if we try to understand the microsystems, by analogy, in classical terms. Also Bohr's account is not beset by the mathematical and philosophical difficulties that are still present in Ludwig's theory.[104] Altogether his semi-qualitative ideas still seem to be preferable to all those very sophisticated mathematical accounts (including Ludwig's) where measurement is treated as an *interaction* between systems that can be described, either exactly, or to a certain degree of approximation, by the formalism of the elementary theory. Let us briefly recall the main features of Bohr's theory: we are concerned with macrosystems which are described in classical terms, and with the calculation of expectation values *in these systems only*.[105] The properties of a micro-object are nothing but possible changes in the macroscopic systems.

This interpretation of measurement removes most of the unsatisfactory features of theories of interaction. However, for this interpretation of the process of measurement – and with this remark we resume our discussion of the observational completeness of the quantum theory – the truth of what we called above the second and the third assumptions now becomes of paramount importance: if all statements of the theory are to be about macroscopic situations then it is decisive indeed to show (a) that for every observable a, there exists a classical device capable of transforming any

---

[103] In their most recent communication (*Quantum Theory of Measurement and Ergodicity Conditions*, *Nuclear Physics*, 33 (1962), 297ff) A. Daneri, A. Loinger and G. M. Prosperi quote a remark by L. Rosenfeld to the effect that 'the conception of Jordan and Ludwig is in harmony with the ideas of Bohr'. As regards Jordan's intuitive approach this may well be the case. But Ludwig's ideas on measurement most definitely deviate from those of Bohr who would never dream of representing the state of a macrosystem by a statistical operator. Considering the difficulties of Ludwig's approach, such as identification would also seem to be unfair to Bohr.

[104] One difficulty which has been pointed out by Ludwig himself (*Z. Naturf.*, 12a (1957), 662ff) and which is also discussed in Daneri and Loinger, *Quantum Theory of Measurement*, is that no satisfactory definition has been given of macro-observables.

According to Hilary Putnam, Ludwig's theory also leads to the following difficulty: the final stage, in Ludwig's theory, is interpreted as a classical mixture only one of whose elements may be assumed really to exist. The derivation, from the formalism of the quantum theory, does *not* give us such a mixture. It rather provides us with a mixture *all* of whose elements have an equal claim to existence. Thus the transition to the usual classical interpretation of the resulting mixture is unaccounted for.

[105] For a detailed account along these lines cf. H. J. Groenewold's essay on measurement in *Observation and Interpretation*.

state into an α-mixture; and (b) that the elements of the resulting mixture can again be observed as macroscopic modifications of measuring instruments. It is equally important to show that all the magnitudes that are customarily used for the description of quantum-mechanical systems are observables in the sense of the theory (or hypermaximal operators, if von Neumann's approach is adopted).

Needless to say, this problem is far from solved. But the situation is even worse. The difficulties of the problem of observation in the quantum theory seems to be much greater than the difficulties of the analogous problem within say, classical point mechanics, despite the fact that a great deal of the former theory was constructed with the explicit purpose of not admitting anything unobservable. The reason can easily be seen. Fundamentally any property of a classical system of point particles can be calculated from the positions and momenta of the elements. In the quantum theory, the existence of the commutation relations necessitates the use of a new instrument for any function of non-commuting variables. This greatly increased the number of measuring instruments required for giving meaning to the main terms of the theory. It can be shown that this number must be Aleph One. If we now realize that so far measuring instruments have been found only for the simplest magnitudes, and that there does not seem to exist any way of finding instruments for the measurement of more complicated magnitudes such as, for example, the angle between two mutually inclined surfaces of a crystal,[106] then we must admit that the idea of the observational completeness of the quantum theory is not far from being a myth. Also the empirical content of a theory contains the preceding theories as approximations. Despite the many assertions to the contrary and despite the fact that the idea of a 'rational generalization' is built in such a manner that a transition to the classical level seems to be an almost trivial affair, no proof is as yet avilable to the effect that the *existing* theories contain the classical point mechanics as a special case. This further reduces their empirical content.

Taking all this into account we seem to arrive at the following paradoxical result: more than any other theory in the history of physics (Aristotelian physics, perhaps, excluded), the quantum theory has been connected with a radically empiricistic outlook. It has been asserted that we have here finally arrived at a theory which directly deals with observations (observations of the classical kind, that is). It now turns out that this theory is much further removed from what it regards as its own empirical basis, viz. classical observation results, than were any of the theories which preceded it. 'Quantum mechanics', writes Schrödinger,

---

[106] Cf. E. Schrödinger, in 'Measurement of Length and Angle in Quantum Mechanics', *Nature, Lond.*, 173 (1954), 442.

claims that it deals ultimately, and directly, with nothing but actual observations since they are the only real things, the only source of information, which is only about *them*. The theory of measurement is carefully phrased so as to make it epistemologically unassailable. But what is all this epistemological fuss about if we have not to do with actual, real findings 'in the flesh', but only with imagined findings?[107]

A similar sentiment is expressed by Bridgman.[108] According to him, a first glance at the quantum theory seems to show that it is a 'thoroughly operational theory' which impression 'is achieved by labelling some of the mathematical symbols "operators", "observables", etc. But in spite of the existence of a mathematical symbolism of that sort, the exact corresponding physical manipulations are . . . obscure, at least in the sense that it is not obvious how one would construct an idealized laboratory apparatus for making any desired sort of measurement.'

It is therefore not only incorrect, and dogmatic, to say that complementarity is the *only possible* point of view in matters microphysical, there also exist grave doubts as to whether it is even a *possible* point of view, i.e. there exists grave doubts as to whether it accurately represents the one fully developed quantum theory of today, viz. the elementary quantum theory of Schrödinger and Heisenberg. We may also say that the empiricistic and positivistic objections which some of the followers of the Copenhagen point of view have raised against alternative interpretations apply with full force to the elementary theory which, they claim, is correctly represented by complementarity. This does not diminish the great merits of this interpretation as regards our understanding of the microscopic level. It only shows that like so many other things it has its faults and should therefore not be regarded as the last, the final, and the only possible word in matters microphysical.

[107] *Nuovo Cimento* (1955), 3.
[108] *The Nature of Physical Theory* (Dover, 1936), 188f.

Sources

The publishers thank the editors and copyright holders who have given permission for these essays to be reprinted here either entirely, or in part: chapter 2, 'An Attempt at a Realistic Interpretation of Experience', from *Proceedings of the Aristotelian Society*, n.s., volume 58, 1958, pp. 143ff; chapter 3, 'On the Interpretation of Scientific Theories', *Proceedings of the 12th International Congress in Philosophy*, volume 5, 1960, pp. 151ff; chapter 4, 'Explanation, Reduction and Empiricism', *Minnesota Studies in the Philosophy of Science*, volume 3, 1962, pp. 28ff; chapter 5, 'On the "Meaning" of Scientific Terms', *Journal of Philosophy*, volume 12, 1965, pp. 266ff; chapter 6, 'Reply to Criticism: Comments on Smart, Sellars and Putnam', *Boston Studies in the Philosophy of Science*, volume 2, 1965, pp. 223ff; chapter 7, 'Science without Experience', *Journal of Philosophy*, volume 66, 1969, pp. 791ff; chapter 9, 'Linguistic Arguments and Scientific Method', *Telos*, volume 2, 1969, pp. 43ff; chapter 10, 'Materialism and the Mind–Body Problem', *The Review of Metaphysics*, volume 17, 1963, pp. 49ff; chapter 11, 'Realism and Instrumentalism: Comments on the Logic of Factual Support', *The Critical Approach to Science and Philosophy*, ed. M. Bunge, Free Press, 1964, pp. 280ff; chapter 12, 'A Note on the Problem of Induction', *Journal of Philosophy*, volume 61, 1964, pp. 349ff; chapter 13, 'On the Quantum Theory of Measurement', *Observation and Interpretation*, ed. S. Körner, Butterworth, 1957, pp. 121ff; chapter 14, 'Professor Bohm's Philosophy of Nature', *British Journal for the Philosophy of Science*, volume 10, 1960, pp. 321ff; chapter 15, 'Reichenbach's Interpretation of Quantum Mechanics', *Philosophical Studies*, volume 9, 1958, pp. 49ff; chapter 16, 'Bohr's World View', appeared in two parts under the title 'On a Recent Critique of Complementarity', *Philosophy of Science*, volume 35, 1968, pp. 309ff and volume 36, 1969, pp. 82ff; chapter 17, 'Hidden Variables and the Argument of Einstein, Podolsky and Rosen', as part of 'Problems of Microphysics', *Frontiers of Science and Philosophy*, ed. R. G. Colodny, Englewood Cliffs, New Jersey, 1962, pp. 208ff.

Name index

Subject index